Operational Research for Renewable Energy and Sustainable Environments

Joshua Thomas
University of Wollongong Malaysia, Malaysia

Gerhard Wilhelm Weber
Poznan University of Technology, Poland

Roman Rodríguez Aguilar
Universidad Panamericana, Mexico

Elias Munapo
North West University, South Africa

Pandian Vasant
Ton Duc Thang University, Vietnam

IGI Global
PUBLISHER of TIMELY KNOWLEDGE

A volume in the Practice, Progress, and Proficiency in Sustainability (PPPS) Book Series

Published in the United States of America by
IGI Global
Engineering Science Reference (an imprint of IGI Global)
701 E. Chocolate Avenue
Hershey PA, USA 17033
Tel: 717-533-8845
Fax: 717-533-8661
E-mail: cust@igi-global.com
Web site: http://www.igi-global.com

Library of Congress Cataloging-in-Publication Data

Names: Thomas, J. Joshua, 1973- editor.
Title: Operational research for renewable energy and sustainable environments / edited by Joshua Thomas, Gerhard Wilhelm Weber, Roman Rodriguez Aguilar, Elias Munapo, and Pandian Vasant.
Description: Hershey, PA : Engineering Science Reference, [2024] | Includes bibliographical references and index. | Summary: "Operational Research for Renewable Energy and Sustainable Environments is a critical scholarly resource that examines efficient use of modern electrical technology and renewable energy sources which have a positive impact on sustainable development. Highlighting topics such as cogeneration thermal modules, photovoltaic (PV) solar, and renewable energy systems (RES) application practices, this publication is geared towards academics, advocates, government officials, policy makers, humanized managers, practitioners, professionals, and students interested in the latest research on renewable energy and clean technology for sustainable rural development"-- Provided by publisher.
Identifiers: LCCN 2023012734 (print) | LCCN 2023012735 (ebook) | ISBN 9781668491300 (h/c) | ISBN 9781668491317 (s/c) | ISBN 9781668491324 (eISBN)
Subjects: LCSH: Renewable energy sources. | Sustainable development. | Operations research.
Classification: LCC TJ808 .O64 2023 (print) | LCC TJ808 (ebook) | DDC 658.4/0340286--dc23/eng/20230727
LC record available at https://lccn.loc.gov/2023012734
LC ebook record available at https://lccn.loc.gov/2023012735

This book is published in the IGI Global book series Practice, Progress, and Proficiency in Sustainability (PPPS) (ISSN: 2330-3271; eISSN: 2330-328X)

British Cataloguing in Publication Data
A Cataloguing in Publication record for this book is available from the British Library.

Practice, Progress, and Proficiency in Sustainability (PPPS) Book Series

Ayman Batisha
International Sustainability Institute, Egypt

ISSN:2330-3271
EISSN:2330-328X

Mission

In a world where traditional business practices are reconsidered and economic activity is performed in a global context, new areas of economic developments are recognized as the key enablers of wealth and income production. This knowledge of information technologies provides infrastructures, systems, and services towards sustainable development.

The **Practices, Progress, and Proficiency in Sustainability (PPPS) Book Series** focuses on the local and global challenges, business opportunities, and societal needs surrounding international collaboration and sustainable development of technology. This series brings together academics, researchers, entrepreneurs, policy makers and government officers aiming to contribute to the progress and proficiency in sustainability.

Coverage

- Strategic Management of IT
- Socio-Economic
- E-Development
- Sustainable Development
- Outsourcing
- Eco-Innovation
- Innovation Networks
- Global Content and Knowledge Repositories
- Global Business
- Knowledge clusters

IGI Global is currently accepting manuscripts for publication within this series. To submit a proposal for a volume in this series, please contact our Acquisition Editors at Acquisitions@igi-global.com or visit: http://www.igi-global.com/publish/.

The Practice, Progress, and Proficiency in Sustainability (PPPS) Book Series (ISSN 2330-3271) is published by IGI Global, 701 E. Chocolate Avenue, Hershey, PA 17033-1240, USA, www.igi-global.com. This series is composed of titles available for purchase individually; each title is edited to be contextually exclusive from any other title within the series. For pricing and ordering information please visit http://www.igi-global.com/book-series/practice-progress-proficiency-sustainability/73810. Postmaster: Send all address changes to above address.

Titles in this Series

For a list of additional titles in this series, please visit: http://www.igi-global.com/book-series/practice-progress-proficiency-sustainability/73810

Teaching and Learning for a Sustainable Future Innovative Strategies and Best Practices
Chai Lee Goi (Curtin University, Malaysia)
Information Science Reference • © 2024 • 341pp • H/C (ISBN: 9781668498590) • US $230.00

Fostering Cross-Industry Sustainability With Intelligent Technologies
Brojo Kishore Mishra (NIST Institute of Science and Technology (Autonomous), India & Saveetha College of Liberal Arts and Sciences, SIMATS, Chennai, India)
Engineering Science Reference • © 2024 • 603pp • H/C (ISBN: 9798369316382) • US $350.00

Global Perspectives on Climate Change, Social Resilience, and Social Inclusion
Aly Abdel Razek Galaby (Alexandria University, Egypt) Mohammed Abo ElEnein (American University, UAE) and Hassan Mohamed (Alexandria University, Egypt)
Engineering Science Reference • © 2024 • 279pp • H/C (ISBN: 9781668489635) • US $275.00

Natural Resource Management Issues in Human-Influenced Landscapes
Dilini Hemachandra (Faculty of Agriculture, University of Peradeniya, Sri Lanka) and Jeevika Weerahewa (Faculty of Agriculture, University of Peradeniya, Sri Lanka)
Engineering Science Reference • © 2024 • 300pp • H/C (ISBN: 9781668470510) • US $225.00

Water-Soil-Plant-Animal Nexus in the Era of Climate Change
Ahmed Karmaoui (Moulay Ismail University, Meknès, Morocco & Moroccan Center for Culture and Science, Morocco)
Engineering Science Reference • © 2024 • 478pp • H/C (ISBN: 9781668498385) • US $240.00

The Role of Women in Cultivating Sustainable Societies Through Millets
Raghvendra Kumar (GIET University, India) and Ishaani Priyadarshini (University of California, Berkeley, USA)
Engineering Science Reference • © 2024 • 293pp • H/C (ISBN: 9781668498194) • US $265.00

Opportunities and Challenges for Women Leaders in Environmental Management
Perfecto Gatbonton Aquino, Jr. (The University of Cambodia, Cambodia) Revenio Cabanilla Jalagat, Jr. (Al-Zahra College for Women, Oman) and Mercia Selva Malar Justin (Xavier Institute of Management and Entrepreneurship, India)
Engineering Science Reference • © 2024 • 259pp • H/C (ISBN: 9781668459867) • US $250.00

IGI Global
PUBLISHER of TIMELY KNOWLEDGE

701 East Chocolate Avenue, Hershey, PA 17033, USA
Tel: 717-533-8845 x100 • Fax: 717-533-8661
E-Mail: cust@igi-global.com • www.igi-global.com

Table of Contents

Preface ... xiv

Chapter 1
Algorithms and Software Packages for Solving Transportation Problems With Intuitionistic Fuzzy Numbers ... 1
P. Senthil Kumar, Amity School of Engineering and Technology, Amity University, Bengaluru, India

Chapter 2
Artificial Intelligence Technologies for Sustainability in Business ... 56
Yesim Can Saglam, Gebze Technical University, Turkey

Chapter 3
Can Fossil Fuels Be Replaced by Entirely Renewable Energy ... 71
Ravi Gorli, Gitam University, India

Chapter 4
Data Envelopment Analysis and Its Application in Energy and Environment ... 83
Saransh Tiwari, Decision Sciences Area, Indian Institute of Management, Lucknow, India
Sanjeet Singh, Decision Sciences Area, Indian Institute of Management, Lucknow, India

Chapter 5
Generation Expansion Planning Using Renewable Energy Resources ... 117
Suriya Ponnambalam, Bharath Institute of Higher Education and Research, India
M. K. Ilampoornan, Bharath Institute of Higher Education and Research, India

Chapter 6
Green Building Certificates in Building Projects ... 129
Vasco Barata, Universidade Lusófona, Portugal
Jorge Vareda Gomes, Universidade Lusófona, Portugal
Mario Romão, ISEG, Universidade de Lisboa, Portugal

Chapter 7
Machine Learning-Based Load Forecast for Energy Markets 147
Mauparna Nandan, Techno Main Saltlake, Kolkata, India
K. S. Sastry Musti, Namibia University of Science and Technology, Nambia

Chapter 8
Modeling of Anti-Islanding Protection for Solar Energy-Based Microgrid 170
Dharmbir Prasad, Asansol Engineering College, India
Rudra Pratap Singh, Asansol Engineering College, India
Anis Hakim, Asansol Engineering College, India
Ranadip Roy, Sanaka Educational Trust's Group of Institutions, Durgapur, India
Md. Irfan Khan, IAC Electricals Pvt. Ltd., Kolkata, India

Chapter 9
Need of 21st Century: A Sustainable Environment Together With a Smart Environment................ 199
Shrikant Tiwari, Galgotias University, Greater Noida, India
Chidambaram N., CHRIST University, Bangalore, India
Amit Kumar Tyagi, National Institute of Fashion Technology, New Delhi, India

Chapter 10
Revolution Towards DC Microgrids: A History 225
Rinki Maurya, Indian Institute of Information Technology, Allahabad, India
Priya Sharma, Indian Institute of Information Technology, Allahabad, India
Ashutosh Kumar Singh, Indian Institute of Information Technology, Allahabad, India
Surya Prakash, Indian Institute of Information Technology, Allahabad, India

Chapter 11
Sustainable Energy Planning and Management Using Metaheuristic Algorithms and Computational Modeling 252
Latifa Dekhici, University of Sciences and Technology of Oran Mohamed Boudiaf (USTO-MB), Algeria
Khaled Guerraiche, LDREI laboratory, Higher School of Electronic and Electrical Engineering of Oran (ESGEEO), Algeria
Karima Belmabrouk, University of Sciences and Technology of Oran Mohamed Boudiaf (USTO-MB), Algeria

Chapter 12
Sustainable Technology Accessing the Software Product Line (SPL) via Model-Based Mutation ... 267
Muskan Singh, University of Petroleum and Energy Studies, India
Sudhanshu Srivastava, University of Petroleum and Energy Studies, India
Shelly Garg, University of Petroleum and Energy Studies, India

Chapter 13
System Modeling and Functional Analysis of Renewable Energies for Management of Smart Grids.. 279
Hamza Wertani, National Engineering School of Carthage, Carthage University, Tunisia

Compilation of References 305

About the Contributors 343

Index 348

Detailed Table of Contents

Preface xiv

Chapter 1
Algorithms and Software Packages for Solving Transportation Problems With Intuitionistic Fuzzy Numbers 1
P. Senthil Kumar, Amity School of Engineering and Technology, Amity University, Bengaluru, India

To determine the optimal value and solution for balanced and unbalanced intuitionistic fuzzy transportation problems (UBIFTPs), two different approaches are proposed in this chapter. Additionally, the parameters of the proposed problems are considered to be triangular intuitionistic fuzzy numbers (TIFNs). Two efficient methods, namely method-I (named linear programming method) and method-II (named PSK method), are presented. Both of them are used to solve the proposed problems. The ideas of these two methods are illustrated with the help of simple examples, and their relevant computer programming is presented. From software such as RStudio, LINGO, RGui, and MATLAB, the obtained solutions for the proposed problems are compared and analyzed with the solutions obtained by the proposed methodologies and already existing methodologies. The unique or superior results, discussions, advantages, and disadvantages of the proposed methods are all presented. Finally, the author's future work is mentioned.

Chapter 2
Artificial Intelligence Technologies for Sustainability in Business 56
Yesim Can Saglam, Gebze Technical University, Turkey

In a fiercely competitive environment, modern technologies have attracted the tremendous attention of businesses for existence. As a consequence of this development, research areas including supply chain management, industrial engineering, operation management, and so on have been influenced by the effect of these cutting-edge technologies. Leading-edge technologies such as artificial intelligence are assisting to remodel the operational management and production systems of businesses. In the literature, increasing consensus and investigations exist that new technologies can provide a variety of advantages for performing and developing sustainability issues. However, empirical insights on how the advantages of artificial intelligence can enhance business sustainability (in terms of economic, environmental, and social sustainability) are relatively scarce. Therefore, this chapter investigates the advantages of artificial intelligence on business sustainability from the angle of three aspects, namely environmental, economic, and social sustainability.

Chapter 3
Can Fossil Fuels Be Replaced by Entirely Renewable Energy 71
Ravi Gorli, Gitam University, India

A renewable energy supply system must be established. This chapter outlines a complete decarbonization strategy. It explains how renewable energy is improving the long-term sustainability of energy sources. A new IEA analysis shows that progress has been made and that additional improvements in global energy efficiency are possible. By 2050, the estimated annual cost of energy shift will be $1.7 trillion. In terms of output, renewable energy sources have surpassed fossil fuels. Never before has cutting-edge tech been so accessible. In 2017, solar thermal electricity with thermal storage hit a record low price of under 10 cents per kWh. Renewable energy utilization is rising. Transportation, heating, and other energy needs account for 20% of overall consumption. Companies can benefit greatly from using green energies like solar thermal, geothermal, and biofuels.

Chapter 4
Data Envelopment Analysis and Its Application in Energy and Environment 83
Saransh Tiwari, Decision Sciences Area, Indian Institute of Management, Lucknow, India
Sanjeet Singh, Decision Sciences Area, Indian Institute of Management, Lucknow, India

The expansion of the industrial and manufacturing sectors is crucial to the economic well-being of all nations. Problems arise because this development is inevitably accompanied by air, water, and other types of pollution that are commonly linked to human health issues and climatic shifts. Thus, we need to think about how to strike a balance between economic growth and pollution reduction in order to keep global society as stable as possible. It is generally agreed that data envelopment analysis (DEA) is one approach of assessing sustainability. Therefore, this chapter is aimed at providing a comprehensive knowledge to the researchers and practitioners on the use of DEA in the field of energy and environment, including a discussion of the key principles and methodology in DEA used in environmental evaluation. This chapter, together with the explanation of conceptual and methodological advances, will serve as a road map for researchers' future investigations into the effects of DEA on energy and environmental problems.

Chapter 5
Generation Expansion Planning Using Renewable Energy Resources 117
Suriya Ponnambalam, Bharath Institute of Higher Education and Research, India
M. K. Ilampoornan, Bharath Institute of Higher Education and Research, India

Generation expansion planning using renewable resources to greenhouse gas emission has been analyzed in this chapter. Hydro, biomass, solar, and wind power plants have been used as renewable energy resources for expansion planning. In the developed model, two scenarios have been generated. GEP scenario represents power system expansion planning without any renewable power plant. On the other hand, the generation expansion with renewable energy resources scenario represents the role of renewable power plant in the system. The role of renewable power plant has been analyzed in two aspects, which are contribution to the production of electricity and to the reduction of GHG emission.

Chapter 6
Green Building Certificates in Building Projects 129
Vasco Barata, Universidade Lusófona, Portugal
Jorge Vareda Gomes, Universidade Lusófona, Portugal
Mario Romão, ISEG, Universidade de Lisboa, Portugal

The importance of the building sector as a key player in the path of innovation associated with sustainable transition has been gaining preponderance at different levels of society. This radical change has been increasingly driven by governmental and non-governmental pressures, encouraging changes not only at the organizational level, but also at the macroeconomic level. Based on this premise, the present work aimed to study the influence of the use of green building certificates (GBC), not only in the innovation process, but also on what it promotes in collaborative processes in building projects. A multiple case study was used, and a comparison was made with academic references. It confirmed the importance of the role of the green project champion who, through the leadership process, has a direct influence on the development of the project in its different phases.

Chapter 7
Machine Learning-Based Load Forecast for Energy Markets 147
Mauparna Nandan, Techno Main Saltlake, Kolkata, India
K. S. Sastry Musti, Namibia University of Science and Technology, Nambia

Time series-based forecasting is one of the most popular machine learning methods due to its effectiveness in estimation of future values based on observations, interpolations, and interpretations of past data. Forecasting the load in power distribution networks is an essential step in energy trading and system operation. Machine learning, specifically time series-based forecasting, can be used in the precise prediction of energy consumption in power networks. Precise predictions have the potential to reduce operating and maintenance expenses, enhance the dependability of power supply and delivery systems, and enable informed decisions for future development endeavors. This chapter employs time series analysis to forecast energy usage in 10-minute intervals specifically for the city of Tétouan in Morocco by applying gradient boosting algorithm. Past and present data trends have been presented along with various accuracy parameters such as mean absolute error (MAE), mean squared error (MSE), and root mean squared error (RMSE).

Chapter 8
Modeling of Anti-Islanding Protection for Solar Energy-Based Microgrid 170
Dharmbir Prasad, Asansol Engineering College, India
Rudra Pratap Singh, Asansol Engineering College, India
Anis Hakim, Asansol Engineering College, India
Ranadip Roy, Sanaka Educational Trust's Group of Institutions, Durgapur, India
Md. Irfan Khan, IAC Electricals Pvt. Ltd., Kolkata, India

Effective management of renewable energy resources becomes a key task with an increasing penetration at the distribution level. Renewable plants may be small or medium low voltage networks but consist of several sources and loads having a combined effect. Increasing interest and investment in renewable energy give rise to rapid development of high penetration solar energy. There are multiple ways to interface solar panel arrays with the power grid. A system can switch from a grid tie mode to an island mode of operation in the event of a utility grid fault. The active local passive approach is a point of common coupling taken into consideration for anti-islanding protection in integrated systems. This work analyzes the effect of grid tied PV systems, which can enhance power system reliability. A grid-connected PV system is modeled in PSCAD software to examine its operation and controlling aspects; however, system capacity sizing for the 5.03 kWp microgrid has been performed using PVsyst tools.

Chapter 9
Need of 21st Century: A Sustainable Environment Together With a Smart Environment................ 199
Shrikant Tiwari, Galgotias University, Greater Noida, India
Chidambaram N., CHRIST University, Bangalore, India
Amit Kumar Tyagi, National Institute of Fashion Technology, New Delhi, India

In this 21st century, the need for a sustainable environment is becoming increasingly essential. This work discusses the importance of adopting sustainable practices and implementing smart technologies to create a useful and eco-friendly living environment. This chapter explains the important need for sustainable practices to mitigate these challenges and ensure a livable future. Smart environment concepts and technologies are then discussed as a means to achieve sustainability. This work discusses the integration of advanced technologies, such as the internet of things (IoT), artificial intelligence (AI), and data analytics, into various aspects of environmental management. Examples include smart energy grids, intelligent waste management systems, and eco-friendly transportation solutions. These technologies enable efficient resource allocation, real-time monitoring, and data-driven decision-making, fostering a sustainable and resilient environment. This work discusses the benefits of adopting a smart environment approach.

Chapter 10
Revolution Towards DC Microgrids: A History.. 225
Rinki Maurya, Indian Institute of Information Technology, Allahabad, India
Priya Sharma, Indian Institute of Information Technology, Allahabad, India
Ashutosh Kumar Singh, Indian Institute of Information Technology, Allahabad, India
Surya Prakash, Indian Institute of Information Technology, Allahabad, India

A DC microgrid is a decentralized electricity system that uses direct current (DC) power for distribution and consumption. It is an alternative to the traditional alternating current (AC) microgrid, which is the dominant electrical distribution system used in most parts of the world. The adoption of DC microgrids has been gaining momentum in recent years due to their numerous advantages over AC microgrids, including improved energy efficiency, reduced power loss during transmission, and the ability to integrate renewable energy sources more easily. In this chapter, the authors review the technical and economic benefits of all types of possible grids and microgrids and discuss the challenges and opportunities associated with their deployment. Overall, the analysis suggests that the transition to DC microgrids is a promising approach for addressing the energy challenges of the 21st century and could play a significant role in the ongoing energy transition.

Chapter 11
Sustainable Energy Planning and Management Using Metaheuristic Algorithms and Computational Modeling 252
Latifa Dekhici, University of Sciences and Technology of Oran Mohamed Boudiaf (USTO-MB), Algeria
Khaled Guerraiche, LDREI laboratory, Higher School of Electronic and Electrical Engineering of Oran (ESGEEO), Algeria
Karima Belmabrouk, University of Sciences and Technology of Oran Mohamed Boudiaf (USTO-MB), Algeria

Sustainable energy planning and management is crucial for achieving a reliable, affordable, and sustainable energy supply, while promoting environmental protection, social equity, and economic efficiency. Optimizing renewable energy systems for sustainability involves the development, deployment, and operation of renewable energy systems that maximize their environmental, social, and economic benefits while minimizing negative impacts. Metaheuristic algorithms and computational modeling techniques are increasingly being used to optimize renewable energy systems for sustainability. This chapter highlights the use of computational modeling in the field of suitable energy planning and management, with a focus on some fields as solar power forecasting and wind turbine design. Two case studies are presented, namely the environmental economic power dispatch and the environmental friendly microgrids redundancy problem.

Chapter 12
Sustainable Technology Accessing the Software Product Line (SPL) via Model-Based Mutation ... 267
Muskan Singh, University of Petroleum and Energy Studies, India
Sudhanshu Srivastava, University of Petroleum and Energy Studies, India
Shelly Garg, University of Petroleum and Energy Studies, India

Due to frequently changing demands and for the sake of customer's satisfaction, engineers were forced to develop SPL also known as "software product line." SPL is used for the product of similar families or having something in common. FMs also known as "feature models" are used for generating the altered versions of the products through linking features with the products. Generation of SPL sometimes leads to the production of billions of products. So, it becomes very challenging to test the quality of test suites. The techniques that exist focus more towards selecting only a limited set of products. If we talk about FMs then they impact test suites in some way. In this work, the authors have paid more attention to the capabilities of the test suites to detect the existing errors. So FMs are prone to errors. Two experiments that they have conducted for calculating mutation score clearly demonstrate that dissimilar test suites are able to detect errors present in FMs very easily as compared to the similar test suites.

Chapter 13
System Modeling and Functional Analysis of Renewable Energies for Management of Smart Grids.. 279
Hamza Wertani, National Engineering School of Carthage, Carthage University, Tunisia

Smart grids studies focus on improving their reliability and reducing the cost of energy, particularly when they are operated onshore. A smart grid can integrate electric vehicles that will greatly modify the urban infrastructure as well as the electrical network. This chapter presents an overview of system modeling for management of smart grids based on renewable energies and functional analysis techniques. In fact, this work allows a discussion of a hybrid system on the one hand and the description of the functional and operational control-command applications method SADT (structured analysis, design technique) and SA-RT (structured analysis real time), on the other hand.

Compilation of References 305

About the Contributors 343

Index 348

Preface

Welcome to the *Operational Research for Renewable Energy and Sustainable Environments*. As editors, we are thrilled to present this comprehensive reference book, a beacon guiding readers through the intricate landscape of modern green technology, artificial intelligence, and innovative science within the realm of renewable energy and sustainable environments.

In collaboration with esteemed contributors, Joshua Thomas, Gerhard Wilhelm Weber, Roman Rodríguez Aguilar, Elias Munapo, and Pandian Vasant, we have curated a collection that delves into the forefront of advancements in electric vehicles, renewable energy optimization, photovoltaic (PV) systems, thermal modules, wind turbines, hydro-power, green technology, geothermal energy, bio-fuel, hydrogen, and related cutting-edge topics. We believe this title will stand as one of the best-selling reference books of 2023, resonating across diverse audiences due to its international perspectives and comprehensive coverage on the global stage.

This scholarly resource serves as a critical tool for those interested in the efficient use of modern electrical technology and renewable energy sources, emphasizing their positive impact on sustainable development. Tailored for academics, advocates, government officials, policy makers, humanized managers, practitioners, professionals, and graduates, our handbook explores the latest research on renewable energy and clean technology for sustainable rural development.

The main body of the text is organized into parts that focus on contemporary and emerging applications of operational research optimization methods in renewable energy and sustainable environments worldwide. Topics span from energy-related modern and innovative technical solutions to addressing global energy problems in United Nations world heritage sites, to the utilization of energy search and metasearch engines for online learning, among other crucial subjects.

This handbook is intended to be a valuable resource for a wide audience, including academics, undergraduates, post-graduates, professionals, executives, ministers, policymakers, researchers, managers, software developers, energy technologists, engineers, chief executives officers, resource staff, and environmental experts. The comprehensive coverage includes key areas such as Renewable Energy, Sustainability, Clean Environment, Climate Change, Optimization, Operational Research, Management Science, Energy Economics, Green Technology, Global Warming, Electrical Vehicles, Green Economy, Sustainable Batteries, Care Economy, Green Supply Chain, Green Manufacturing, Green Factories, and Clean Hydrogen.

We invite you to embark on a journey through the wealth of knowledge contained within these pages, exploring the latest insights and innovations that contribute to a more sustainable and environmentally conscious future.

ORGANIZATION OF THE BOOK

Chapter 1: Algorithms and Software Packages for Solving Transportation Problems With Intuitionistic Fuzzy Numbers

In this chapter, P. Senthil Kumar introduces two approaches for determining optimal values and solutions to balanced and unbalanced intuitionistic fuzzy transportation problems (UBIFTPs). The proposed methods, named linear programming method and PSK method, utilize triangular intuitionistic fuzzy numbers (TIFNs). The author illustrates these approaches with simple examples and presents their computer programming. The chapter compares solutions obtained from software such as RStudio, LINGO, RGui, and MATLAB, providing a comprehensive analysis of the proposed methodologies and existing ones. The discussion covers unique results, advantages, disadvantages, and outlines the author's future work.

Chapter 2: Artificial Intelligence Technologies for Sustainability in Business

Authored by Yesim Can Saglam, this chapter explores how modern technologies, particularly artificial intelligence, influence supply chain management, industrial engineering, and operation management in businesses. The focus is on understanding how AI contributes to the economic, environmental, and social sustainability of businesses. The chapter emphasizes the scarcity of empirical insights on this subject and investigates the advantages of AI in enhancing sustainability across these three dimensions.

Chapter 3: Can Fossil Fuels Replace by Entirely Renewable Energy

Ravi Gorli's chapter outlines a complete decarbonization strategy, emphasizing the improvement of long-term sustainability through renewable energy sources. The author presents an analysis showcasing the progress made in global energy efficiency, surpassing fossil fuels in terms of output. The chapter discusses the accessibility of cutting-edge technologies, the declining cost of solar thermal electricity, and the rising utilization of renewable energy in various sectors.

Chapter 4: Data Envelopment Analysis and Its Application in Energy and Environment

Saransh Tiwari and Sanjeet Singh delve into the crucial balance between industrial and manufacturing sector expansion and environmental concerns. The chapter focuses on the use of Data Envelopment Analysis (DEA) as an approach to assess sustainability, providing a roadmap for researchers and practitioners. The key principles and methodology of DEA in environmental evaluation are discussed, fostering future investigations into its effects on energy and environmental problems.

Chapter 5: Generation Expansion Planning Using Renewable Energy Resources

Authored by Suriya Ponnambalam and M.K. Ilampoornan, this chapter analyzes generation expansion planning using renewable resources. The discussion includes scenarios with and without renewable power plants, assessing their contribution to electricity production and the reduction of greenhouse gas emissions. The chapter explores the role of hydro, biomass, solar, and wind power plants in the expansion planning model.

Chapter 6: Green Building Certificates in Building Projects

Vasco Barata, Jorge Gomes, and Mario Romão focus on the building sector's role in sustainable innovation. The chapter studies the influence of Green Building Certificates (GBC) on collaborative processes in building projects, emphasizing the importance of green project champions in leadership roles. The work draws insights from a multiple case study and comparisons with academic references.

Chapter 7: Machine Learning-Based Load Forecast for Energy Markets

Mauparna Nandan and KS Sastry Musti address time series-based forecasting using machine learning for precise load prediction in power distribution networks. The chapter applies the Gradient Boosting algorithm to forecast energy usage in 10-minute intervals, presenting past and present data trends and accuracy parameters such as Mean Absolute Error (MAE), Mean Squared Error (MSE), and Root Mean Squared Error (RMSE).

Chapter 8: Modeling of Anti-Islanding Protection for Solar Energy-based Microgrid

Dharmbir Prasad, Rudra Singh, Anis Hakim, Ranadip Roy, and Md. Khan explore the effective management of renewable energy resources, focusing on anti-islanding protection in integrated systems. The chapter models a grid-connected PV system in PSCAD software, examining its operation and controlling aspects. The discussion includes the analysis of system capacity sizing for a 5.03 kWp microgrid using PVsyst tools.

Chapter 9: Need of 21st Century – A Sustainable Environment Together a Smart Environment

Shrikant Tiwari, Chidambaram N, and Amit Tyagi discuss the increasing need for a sustainable environment in the 21st century. The chapter emphasizes the importance of adopting sustainable practices and implementing smart technologies to create an eco-friendly living environment. It explores the integration of advanced technologies such as IoT, AI, and data analytics in various aspects of environmental management.

Chapter 10: Revolution Towards DC Microgrids – A History

Authored by Rinki Maurya, Priya Sharma, Ashutosh Singh, and Surya Prakash, this chapter reviews the technical and economic benefits of DC microgrids. The focus is on their advantages over AC microgrids, including improved energy efficiency and easier integration of renewable energy sources. The analysis suggests that transitioning to DC microgrids is a promising approach for addressing 21st-century energy challenges.

Chapter 11: Sustainable Energy Planning and Management Using Metaheuristic Algorithms and Computational Modeling

Latifa Dekhici, Khaled Guerraiche, and Karima Belmabrouk highlight the crucial role of sustainable energy planning and management in achieving a reliable and affordable energy supply. The chapter explores the use of metaheuristic algorithms and computational modeling techniques to optimize renewable energy systems. Case studies on solar power forecasting and wind turbine design are presented to illustrate the application of computational modeling in energy planning.

Chapter 12: Sustainable Technology Accessing the Software Product Line (SPL) via Model-Based Mutation

Muskan Singh, Sudhanshu Srivastava, and Shelly Garg address the challenges of testing the quality of test suites in the context of software product lines (SPL). The chapter focuses on the capabilities of test suites to detect errors in feature models (FMs), emphasizing dissimilar test suites' effectiveness in detecting errors compared to similar test suites. The experiments conducted demonstrate the ability of dissimilar test suites to identify errors in FMs.

Chapter 13: System Modeling and Functional Analysis of Renewable Energies for Management of Smart Grids – Analysis of Renewable Energies for Management of Smart Grids

Authored by Hamza Wertani, this chapter provides an overview of system modeling for managing Smart Grids based on renewable energies. The discussion includes functional analysis techniques such as SADT (Structured Analysis, Design Technique) and SA-RT (Structured Analysis Real Time), presenting a hybrid system and detailing its functional and operational control-command applications.

IN SUMMARY

As we conclude this edited reference book, we reflect on the diverse and insightful contributions that have come together to form a comprehensive exploration of contemporary issues in renewable energy, sustainable environments, and operational research optimization methods.

The chapters within this volume, authored by experts in their respective fields, have provided a multifaceted understanding of the challenges and opportunities in the intersection of technology, sustainability, and energy planning. From algorithms solving intuitionistic fuzzy transportation problems to the application of artificial intelligence in business sustainability, each chapter has added a unique perspective to the evolving landscape of clean technology.

We witnessed discussions on the transition from fossil fuels to entirely renewable energy sources, emphasizing the urgency of establishing a decarbonized energy supply system. The integration of Data Envelopment Analysis (DEA) into environmental evaluation, the analysis of generation expansion planning with renewable resources, and the role of Green Building Certificates in building projects showcased the interdisciplinary nature of our exploration.

Chapters on machine learning-based load forecasting, anti-islanding protection for solar energy-based microgrids, and the history and promise of DC microgrids underscored the role of technology in shaping the future of energy systems. The need for a sustainable environment and the integration of smart technologies, discussed in detail in another chapter, shed light on the crucial steps needed to mitigate environmental challenges.

The application of metaheuristic algorithms and computational modeling in sustainable energy planning and management was a focal point, with case studies demonstrating their effectiveness in optimizing renewable energy systems. Lastly, the exploration of sustainable technology through model-based mutation in software product lines addressed the challenges in maintaining the quality of test suites.

As editors, we are grateful for the dedication and expertise demonstrated by the contributing authors. The diverse audience, ranging from academics and professionals to policymakers and environmental experts, will find valuable insights within these pages. We hope this handbook serves as a roadmap for further research and innovation in the pursuit of a sustainable and environmentally conscious future.

Embark on a journey through the wealth of knowledge presented in this handbook, exploring the cutting-edge advancements and interdisciplinary approaches of *Operational Research* that pave the way for a cleaner, greener, and more sustainable world.

Joshua Thomas
University of Wollongong Malaysia, Malaysia

Gerhard Wilhelm Weber
Poznan University of Technology, Poland

Roman Rodríguez Aguilar
Universidad Panamericana, Mexico

Elias Munapo
North West University, South Africa

Pandian Vasant
Ton Duc Thang University, Vietnam

Chapter 1
Algorithms and Software Packages for Solving Transportation Problems With Intuitionistic Fuzzy Numbers

P. Senthil Kumar
https://orcid.org/0000-0003-4317-1021
Amity School of Engineering and Technology, Amity University, Bengaluru, India

ABSTRACT

To determine the optimal value and solution for balanced and unbalanced intuitionistic fuzzy transportation problems (UBIFTPs), two different approaches are proposed in this chapter. Additionally, the parameters of the proposed problems are considered to be triangular intuitionistic fuzzy numbers (TIFNs). Two efficient methods, namely method-I (named linear programming method) and method-II (named PSK method), are presented. Both of them are used to solve the proposed problems. The ideas of these two methods are illustrated with the help of simple examples, and their relevant computer programming is presented. From software such as RStudio, LINGO, RGui, and MATLAB, the obtained solutions for the proposed problems are compared and analyzed with the solutions obtained by the proposed methodologies and already existing methodologies. The unique or superior results, discussions, advantages, and disadvantages of the proposed methods are all presented. Finally, the author's future work is mentioned.

INTRODUCTION OF TRADITIONAL TRANSPORTATION PROBLEM

In many of the real life situations, there is a need to transport the homogeneous product from various sources (e.g., factories) to different destinations (e.g., retail stores or storage centres) and the main aim of the decision-maker (DM) is to find how much quantity of the product from which source to which destination should be supplied so that all the supply points (e.g., a_i, $i=1,2,\ldots,m$) are fully used and demands (b_j, $j=1,2,\ldots,n$) of all the destinations are fulfilled as well as total transportation cost (TTC) is minimum or total transportation profit (TTP) is maximum.

DOI: 10.4018/978-1-6684-9130-0.ch001

TPs play a vital role in industry and network communication for reducing cost and improving services. In today's very high competitive market, the pressure on organizations to find better ways to create and deliver items and services to customers becomes stronger. How and when to send the items to the customers in the quantities which they want in a cost-effective manner becomes more challenging. The transportation models provide a powerful framework to meet these challenges. They ensure the efficient movement and timely availability of finished goods and/or raw materials.

Resource allocation is a toughest job, used to assign the available resources in an economic way. When the resources to be allocated are scarce, a well-planned action is necessary for decision-makers (DMs) to attain the optimum utility. If the supplying sources and the receiving agents are finite, the best pattern of the allocation to get the minimum cost/the best plan with the maximum return, whichever may be applicable to the problem, is to be found out. Those types of problems are called 'Allocation Problems' and are divided into 2 categories, namely: 1. Transportation Problems (TPs) and 2. Assignment Problems (APs). These categories of problems are studied in Operations Research/Optimization.

During 2nd World War, Britain was having very less military resources; so, there was an urgent need to allocate resources to the various military operations and to the activities within each operation in an effective manner. So, the British military executives called upon a team of scientist to apply scientific method to study the strategic and tactical problems related to 'land' and 'air' defence of the country. As the team was dealing with research of military operations, the work of this team of scientists was named as Operations Research (OR).

TP is the special case of linear programming problem (LPP) where the objective is to transport various quantities of a single homogeneous product that are initially stored at various origins, to different destinations in such a way that maximize the total transportation profit or minimize the TTC.

The traditional TP deals with the transportation a certain commodity from each of m origins (O_i or S_i, $i=1,2,3,\ldots,m$) to any of n destinations (D_j or W_j, $j=1,2,3,\ldots,n$). Assume that O_i's or S_i's are factories with respect capacities $a_1,a_2,a_3,\ldots,a_m$ and D's or W_j's are warehouses with required levels of demands $b_1,b_2,b_3,\ldots,b_n$. For the transport of a unit of the given commodity from O_i to D_j a cost c_{ij} is assigned with $c_{ij}\geq 0$, $\forall i,j$. Hence, one must determine the amounts y_{ij} to be transported from all the sources ($S_1,S_2,S_3,\ldots,S_m$) to all the destinations ($D_1,D_2,D_3,\ldots,D_n$) in such a way that the total cost is minimized.

Typically, the traditional TP can be written as follows.

$$\text{Minimize } Z=\sum_{i=1}^{m}\sum_{j=1}^{n}c_{ij}\times y_{ij}$$

$$s.t\begin{cases}\sum_{j=1}^{n}y_{ij}\leq a_i\ (i=1,2,\ldots,m)\\ \sum_{i=1}^{m}y_{ij}\geq b_j\ (j=1,2,\ldots,n)\end{cases}$$

$y_{ij}\geq 0$ ($i=1,2,\ldots,m$ and $j=1,2,\ldots,n$)

If we assume that the profit (p_{ij}, i=1,2,…,m and j=1,2,…,n) instead of cost in the above model then we can write the maximization problem as follows.

$$\text{Maximize } Z = \sum_{i=1}^{m}\sum_{j=1}^{n} p_{ij} \times y_{ij} \text{ subject to above constraints.}$$

The general mathematical model of the basic TP was introduced by Hitchcock (1941). It is one of the subclasses of non-linear programming problem (NLPP) where, constraints are all either equality/inequality.

The traditional TP is a subclass of LPP. Here, constraints are all equality type (=). It is widely used in the areas of employment scheduling, aggregate planning, inventory control, communication network, and personnel assignment, etc. Depending on the nature of the cost function, the TP can be divided into two categories, namely: linear TP and non-linear TP.

In traditional TP, our aim is to minimize the total cost (or maximizes the total profit) of transporting a commodity which is available at some sources and is required at various destinations. It has the following parameters, namely: 1. Supply, 2. Demand and 3. Cost/profit. Further, they are explained as follows.

1. the quantity of item available at source is referred as supply,
2. the quantity of item required at destination is referred as demand and
3. the cost/profit is referred as the cost/profit of transporting a single item from the supply point to demand point

Therefore, in a traditional TP, we have 3 parameters.

In TP, if the parameters (namely, cost, demand and supply) are all known precisely then the efficient algorithms have been developed for solving TPs in literature. But, in our day by day life, the transportation costs (TCs), supplies and demands are all not fixed crisp numbers which are in the form of uncertainty and hesitation due to several uncontrollable factors. Therefore, to counter this situation, in this book chapter, the author presents 2 different methods, namely: method-I and method-II. In this book chapter, formulation of the numerical models of both the balanced and UBIFTPs are presented. In this balanced and UBIFTPs, the objective function considered to be linear with intuitionistic fuzzy transportation cost (IFTC). |||ly, the constraints and non negative restrictions are all assumed as linear with intuitionistic fuzzy numbers (IFNs).

Organization of the Chapter

The remainder of the study is structured as follows.

1. Section 2 presents the historical background of different types of transportation problems.
2. In Section 3, preliminaries and notions have been presented.
3. In Section 4, the ranking procedure of IFN (e.g., TIFN) and its ordering principles have been discussed.
4. In Section 5, the generalized mathematical model of the fully intuitionistic fuzzy transportation problem (FIFTP) has been discussed. That is, this section provides LPP form of FIFTP.

5. In Section 6, two methods have been discussed, namely: method-I (named linear programming method) and method-II (named PSK method).
6. In Section 7, numerical examples and its relevant computer programming, results and comparative analysis have been discussed.
7. In Section 8, both the discussions and valid conclusions have been presented.
8. Finally, the author's future work has been presented in Section 9.

HISTORICAL BACKGROUND OF DIFFERENT TYPES OF TRANSPORTATION PROBLEMS

This section describes the literature review of traditional, fuzzy and intuitionistic fuzzy transportation problem (IFTP). Initially, the author presents the TP. In the middle of this section, fuzzy transportation problem (FTP) and other relevant problems (for example, fuzzy linear programming problem (FLPP), fuzzy assignment problem (FAP), etc.) have been discussed. 3rd, the concept of intuitionistic fuzzy optimization problems (IFOPs) has been discussed. Lastly, the necessity for studying optimization problems with intuitionistic fuzzy parameters has been discussed also.

In the history of TP, Hitchcock (1941) designed the basic transportation model first. Next, several researchers (e.g., Koopmans, 1949; Charnes & Cooper, 1954; Dantzig, 1963) have been introduced new ideas/methods to solve the TPs. Ye et al. (2010, 2012) gave 'a note on the connection between the Primal-Dual and the A* algorithm'. Jain et al. (2020) studied an analytical solution of TP from Modified Distribution (MODI) Approach. Kalonda and Omekanda (2020) presented Kruskal's Algorithm, Vogel's approximation and modified distribution methods for the design of optimal electrical networks in the democratic republic of Congo. Ahmed and Ahmed (2020) presented a comparative study between new proposed technique (LPAM) and the other existing techniques to find IBFS for TP. Kumari (2020) developed zero accomplishment method for finding an optimal more-for-less solution of TP with mixed constraints. Recently, several researchers (Zeeshan et al., 2022; Akpofure & Jude, 2022; Thanoon, 2022; Rochayata et al., 2022; Sam'an et al., 2022; Elaibi & Hasanain, 2022; Bama & Sumathi, 2022; Akbar et al., 2023; Sonia & Chidambaran, 2023; Mohammed, 2023; Ackora-Prah et al., 2023) have solved TPs with ordinary numbers. But, they are not solved TPs with fuzzy or IFNs.

In traditional TPs, supplies, demands and costs of transportation are all fixed crisp numbers. According to this, we can decide that the DM can predict transportation cost (TC) or profit accurately. In contrast, in real world TPs, the costs are in uncertainty and hesitation due to several inescapable factors, for example, market conditions, variation in rates of fuel, traffic jams, weather in hilly areas etc. In such situations, the DM can't predict transportation cost exactly. That's why, to counter these uncertainties Atanassov (1983), introduced the intuitionistic fuzzy set (IFS) which is more consistent than the fuzzy set proposed by Zadeh (1965). In case of TP, with the help of IFS theory, the DM can decide about the level of acceptance and non-acceptance of the transportation profit/cost. So, the IFS theory is very useful and popular in TP, decision-making, AP, planning, LPP, manufacturing, NLPP, and scheduling.

As there is an uncertainty in the parameters of a crisp TP, several academicians/researchers (e.g., Kumar, 2010; Mohideen & Kumar, 2010a; 2010b; Samuel, 2012; Kumar, 2016a; 2016b; Kumar, 2017a; Sam'an, Hariyanto, & Surarso, 2018; Ngastiti & Surarso, 2018; Crama & Pironet, 2019; Malini, 2019; Srivastava & Bisht, 2019; Kaur, Kacprzyk, & Kumar, 2020; Khalifa et al., 2020; Kalyani & Nagarani, 2020; Mathur & Srivastava, 2020; Mandal, Kumar, & Burnwal, 2020; Sam'an, Surarso, & Irwanto, 2020;

Dymova, 2020; Mishra, Panda, & Hota, 2020; Emam et al., 2022; Deshmukh et al., 2022; Niksirat, 2022; Raj et al., 2022; Giri et al., 2023; Kumar, 2023a) have solved TPs under fuzzy (uncertain) environments.

Although in our daily life, the TC/profit, demands and supply are all not a fixed crisp numbers. They are all in the form of uncertainty as well as hesitation due to some uncontrollable factors such as errors in data, and lack of good communications. According to this, as there is an uncertainty and hesitation in the parameters of TPs, several academicians/researchers (e.g., Hussain & Kumar, 2012a, 2012b, 2012c, 2013; Kumar & Hussain, 2014a, 2015, 2016a; Kumar, 2017b; Kumar, 2018a, 2018b, 2018c, 2018d, 2018e; Chhibber, Srivastava, & Bisht, 2019; Kumar, 2019a, 2019b; Bharati, 2019; Sarma, Das, & Bera, 2020; Kumar, 2020a; Sharma, 2020; El Alaoui, 2020; Traneva & Tranev, 2020, 2021; Haque et al. 2022; Jansi Rani, Manivannan & Dhanasekar, 2023) have solved TPs with intuitionistic fuzzy parameters. Recently, Jansi Rani et al., (2023) developed branch and bound technique for solving IFTP. So, several academicians/researchers have solved TPs under intuitionistic fuzzy environments successfully. Although none of the academicians/researchers have solved IFTPs using the proposed methods in this book chapter. Furthermore, none of them have compared their obtained solutions with software solution methods for IFTPs except the academician Kumar (2021).

TPs are special cases of LPPs. In this book chapter, method-I is based on linear programming approach. In general, several papers are available in the literature for solving fuzzy and intuitionistic fuzzy linear programming problems (IFLPPs). Some of them are specified here (Ghanbari et al., 2019; Suneela & Chakraverty, 2019; Javanmard, & Nehi, 2019; Mahmoudi & Nasseri, 2019; Ozkok, 2019; Das, Mandal, & Behera, 2019; Wang & Peng, 2019; Chopra & Saxena, 2019; El Alaoui, 2021; Pourofoghi et al., 2021; Ali et al., 2021). The most important advantage of the method-I is that the crisp optimal value and solutions of the given IFTPs are all equal to (=) the optimal value and solutions of their respective equivalent crisp TPs.

As there is an uncertainty and hesitation in the parameters of real-life problems (e.g., APs, TPs, etc.), several authors or researchers (for instance, Atanassov, 1995; Angelov, 1997; Li, Nan, & Zhang, 2010; Kumar & Hussain, 2014b, 2014c, 2016b, 2016c) have solved various problems using the concept of IFS. Lately, Kumar (2019b; 2020b,c,d, 2021, 2022, 2024a,b) developed new and efficient methodologies to solve intuitionistic fuzzy assignment problems (IFAPs) and intuitionistic fuzzy solid assignment problems (IFSAPs) under 2 different situations. Thus, due to various applications of IFS theory, several researchers (or authors) have solved either optimization problems or other real life oriented problems under intuitionistic fuzzy environments.

Fundamentally, IFTPs are TPs but their parameters are IFNs. Mainly, the objective for solving IFTPs is to determine the optimum schedule that minimizes the total intuitionistic fuzzy transportation cost (TIFTC) or maximizes the total intuitionistic fuzzy transportation profit (TIFTP) while satisfying the intuitionistic fuzzy supply ($\tilde{a}_i^I$) and intuitionistic fuzzy demand ($\tilde{b}_j^I$) limits.

In this book chapter, to determine the optimal value and solution of the balanced and unbalanced intuitionistic fuzzy transportation problems (UBIFTPs), computationally simple and single stage methods, namely: method-I and method-II have been discussed. The ordering procedure of Varghese and Kuriakose is used to convert the balanced and UBIFTPs to a crisp one so that the traditional methods may be applied to solve both of them. The occupied cell of the crisp unbalanced transportation problem (CUBTP) that we obtained is as same as the occupied cell of the unbalanced intuitionistic fuzzy transportation problem (UBIFTP), but the value of occupied cell of the UBIFTP is the maximum possible value of {supply, demand}. III^{ly}, the above said statement is always applicable for both the crisp balanced

TP and crisp UBIFTP. From these ideas, we conclude that the solution methods are differing from the crisp UBTP to UBIFTP in allocation steps only. |||ly, we can also conclude that the solution methods are differing from the crisp balanced TP to the BIFTP in allocation steps only. So, simple methodologies are proposed to determine the optimal value of the given IFTP in the form of TIFNs in the proposed book chapter. Mainly, to solve any IFTP, the proposed methodologies are much easier to apply as compared to some other existing methodologies. Furthermore, solving TP with intuitionistic fuzzy parameters using MATLAB, RStudio, and RGui is new in the literature.

PRELIMINARIES AND NOTIONS

The basic definitions and notations of Intuitionistic Fuzzy Set (IFS), (α, β)–cuts of IFS A on X, intuitionistic fuzzy number (IFN), TIFN, arithmetic operations on TIFN are all taken from the articles mentioned here in (Atanassov, 1999; Guha & Chakraborty, 2010; Varghese & Kuriakose, 2012; Kumar, 2023b).

Definition (Atanassov, 1999)

Let *X* be the universe of discourse. An IFS *A* in *X* is an object having the form:

$$A = \{\langle x, \mu_A(x), \vartheta_A(x)\rangle / x \in X\},$$

where, the functions $\vartheta A(x)$, $\mu A(x)$: $X \to [0,1]$ define respectively, the degree of non –membership and degree of membership of the element x *in* X *to* the set A, which is a subset of X, and for each element x∈*X*, $0 \le \mu A(x) + \vartheta A(x) \le 1$ (The sum of both membership & non-membership values should always lie between 0 and 1).

Furthermore, we have,

$$\pi A(x) = 1 - \mu A(x) - \vartheta A(x).$$

Usually, $\pi A(x)$ is called the IFS index/hesitation margin of x in *A*. $\pi A(x)$ is *the* degree of indeterminacy of x∈X to *the* IFS A and $\pi A(x) \in [0,1]$, i.e., $\pi A(x)$: $X \to [0,1]$ *and* $0 \le \pi A(x) \le 1$ for *each* x∈X. $\pi A(x)$ *expresses* the lack of knowledge of whether x belongs *to* IFS A/not.

For example, let us consider A be an I*FS* with $\mu A(x)=0.3$ & $\vartheta A(x)=0.2$

$$\Rightarrow \pi A(x) = 1 - (0.3+0.2) = 0.5$$

It can be interpreted as follows.

1. the degree that the object x∈A is 0.*3*.
2. the degree that the object x∉A is 0.2.
3. the degree of hesitancy is 0.5.

Definition

An intuitionistic fuzzy subset $A=\{\langle x,\mu_A(x),\vartheta_A(x)\rangle / x\in X\}$ of the real line $\mathbb{R}$ is called an IFN if the following holds.

1. There exists $m\in\mathbb{R}$, $\mu A(_m) = 1$ and $\vartheta A(m_) = 0$, where m indicates the mean value of A.
2. μA i$_s$ a continuous mapping from $\mathbb{R}\to[0,1]$ and $\forall\ x\in\mathbb{R}$, the relation $0\leq \mu A(_x) + \vartheta A(x_) \leq 1$ holds.

The membership function of IFS A is denoted by $\mu A\ (_x)$ and defined by

$$\left.\begin{cases}0, & \text{for } -\infty<x\leq m-\alpha,\\ f^1(x), & \text{for } x\in[m-\alpha,m],\\ 1, & \text{for } x=m,\\ g^1(x), & \text{for } x\in[m,m+\beta],\\ 0, & \text{for } m+\beta\leq x<\infty\end{cases}\right\}=\mu_A(x)$$

where $f^1(x)$ and $g^1(x)$ are strictly increasing and decreasing function in $[m-\alpha,m]$ and $[m,m+\beta]$ respectively.
The non-membership function of IFS A is denoted by $\vartheta A(_x)$ and defined by

$$\left.\begin{cases}1, & \text{for } -\infty<x\leq m-\alpha',\\ f^2(x), & \text{for } x\in[m-\alpha',m];0\leq f^1(x)+f^2(x)\leq 1,\\ 0, & \text{for } x=m,\\ g^2(x), & \text{for } x\in[m,m+\beta'];0\leq g^1(x)+g^2(x)\leq 1\\ 1, & \text{for } m+\beta'\leq x<\infty\end{cases}\right\}=\vartheta_A(x)$$

Here,

m- is the mean value of A.
α, β- are called the left and right spreads of membership function $\mu A_(x)$, respectively.
α', β'- are represents the left and right spreads of non-membership function $\vartheta A_(x)$, respectively.

Figuratively, the IFN ($\tilde{A}^I$) is represented as $A_{IFN} = (m;\alpha,\beta;\alpha',\beta')$.

Definition

A TIFN $\left(\tilde{A}^I\right)$ is an IFS in $\mathbb{R}$. . Then, it has the following membership function (denoted by $\mu A_(x)$) and non-membership function (denoted by $^{\vartheta}A(_x)$).

$$\begin{cases} 0, & \text{for } x < a^1, \\ \dfrac{x - a^1}{a^2 - a^1}, & \text{for } a^1 \le x \le a^2, \\ 1, & \text{for } x = a^2 \\ \dfrac{a^3 - x}{a^3 - a^2}, & \text{for } a^2 \le x \le a^3, \\ 0, & \text{for } x > a^3, \end{cases} = \mu_A(x)$$

$$\begin{cases} 1, & \text{for } x < a^{1'}, \\ \dfrac{a^2 - x}{a^2 - a^{1'}}, & \text{for } a^{1'} \le x \le a^2, \\ 0, & \text{for } x = a^2, \\ \dfrac{x - a^2}{a^{3'} - a^2}, & \text{for } a^2 \le x \le a^{3'} \\ 1, & \text{for } x > a^{3'} \end{cases} = \vartheta_A(x)$$

where $a^{1'} \le a^1 \le a^2 \le a^3 \le a^{3'}$ and $\vartheta A(x)$, $\mu A(x) \le 0.5$ for $\vartheta A(x) = \mu A(x)$ $\forall x \in \mathbb{R}$.

This TIFN is denoted by $\tilde{A}^I = \left(a^1, a^2, a^3; a^{1'}, a^2, a^{3'}\right)$.

$(\boldsymbol{\alpha},\boldsymbol{\beta})$–Cuts of IFS and Arithmetic Operations on TIFN

By realizing the importance/necessity of the arithmetic operations on TIFNs, some of the academicians/researchers (for instance, Shaw & Roy, 2012; Mahapatra & Roy, 2013; Dutta & Saikia, 2019) have studied arithmetic operations on TIFNs. Generally, operations on TIFNs play a vital role in optimization and decision-making.

Both the basic definitions of arithmetic operations on TIFNs based on (α, β)-cuts and ordering principles are taken from the articles mentioned here in (Varghese & Kuriakose, 2012; Mahapatra & Roy, 2013; Kumar & Hussain, 2016a).

Definition

Let A be an IFS on X. Then, (α,β)–cuts of A is

$$\left\{x \in X : \mu_{\tilde{A}^I}(x) \ge \alpha, \vartheta_{\tilde{A}^I}(x) \le \beta; \alpha, \beta \in [0,1]\right\}.$$

Definition

Let $\tilde{A}^I = \left(a^1, a^2, a^3; a^{1'}, a^2, a^{3'}\right)$ and $\tilde{B}^I = \left(b^1, b^2, b^3; b^{1'}, b^2, b^{3'}\right)$ be any 2 arbitrary TIFNs then the arithmetic operations between them are as follows.

$$\tilde{A}^I \oplus \tilde{B}^I = \left(a^1 + b^1, a^2 + b^2, a^3 + b^3; a^{1'} + b^{1'}, a^2 + b^2, a^{3'} + b^{3'}\right)$$

$$\tilde{A}^I \ominus B^I = \left(a^1 - b^3, a^2 - b^2, a^3 - b^1; a^{1'} - b^{3'}, a^2 - b^2, a^{3'} - b^{1'}\right)$$

$$\tilde{A}^I \otimes \tilde{B}^I = \left(a^1\mathcal{R}\left(\tilde{B}^I\right), a^2\mathcal{R}\left(\tilde{B}^I\right), a^3\mathcal{R}\left(\tilde{B}^I\right); a^{1'}\mathcal{R}\left(\tilde{B}^I\right), a^2\mathcal{R}\left(\tilde{B}^I\right), a^{3'}\mathcal{R}\left(\tilde{B}^I\right)\right)$$

if $\mathcal{R}\left(\tilde{A}^I\right), \mathcal{R}\left(\tilde{B}^I\right) \geq 0$

Note:

1. $c\tilde{A}^I = \left(ca^1, ca^2, ca^3; ca^{1'}, ca^2, ca^{3'}\right)$, for $c \geq 0$
2. $c\tilde{A}^I = \left(ca^3, ca^2, ca^1; ca^{3'}, ca^2, ca^{1'}\right)$, for $c < 0$

RANKING METHODOLOGY OF TIFNs

The definition of IFN was introduced by Burillo et al. (1994). Later, they have defined important as well as useful properties of IFNs. Further, many academicians/researchers (e.g., Grzegorzewski, 2003a; Grzegorzewski, 2003b; Mitchell, 2004; Ban, 2008; Mahapatra & Roy, 2009; Guha & Chakraborty 2010; Nehi, 2010; Varghese & Kuriakose, 2012; Shabani & Jamkhaneh, 2014; Darehmiraki, 2019; Tarmudi & Rahman, 2020; Chutia et al., 2021; Chutia, 2021; Iqbal & Yaqoob, 2023) have studied IFNs. Recently, very few academicians/researchers have analyzed important and useful properties of IFNs in the literature. So, ranking of alternatives in intuitionistic fuzzy environment plays a salient role in decision-making.

If $\tilde{A}^I = \left(a^1, a^2, a^3; a^{1'}, a^2, a^{3'}\right)$ is an arbitrary TIFN, its equivalent crisp number is denoted by $\mathcal{R}(\tilde{A}^I)$ and is defined by the authors Varghese and Kuriakose in 2012 as follows:

$$\mathcal{R}\left(\tilde{A}^I\right) = \frac{1}{3}\left[\frac{\left(a^{3'} - a^{1'}\right)\left(a^2 - 2a^{3'} - 2a^{1'}\right) + \left(a^3 - a^1\right)\left(a^1 + a^2 + a^3\right) + 3\left(a^{3'^2} - a^{1'^2}\right)}{a^{3'} - a^{1'} + a^3 - a^1}\right]$$

From this, we obtain the following biconditional logical statements.

$$\mathcal{R}(\tilde{A}^I) > (\tilde{B}^I) \Leftrightarrow \tilde{A}^I \succ \tilde{B}^I$$

$$\mathcal{R}(\tilde{A}^I) < (\tilde{B}^I) \Leftrightarrow \tilde{A}^I \prec \tilde{B}^I$$

$$\mathcal{R}(\tilde{A}^I) = (\tilde{B}^I) \Leftrightarrow \tilde{A}^I \approx \tilde{B}^I$$

If the Varghese and Kuriakose's formula doesn't give any finite value for the given TIFNs then we can use the following accuracy function to find the rank of them.

$$f\left(\tilde{A}^I\right) = \frac{\left(a^1 + 2a^2 + a^3\right) + \left(a^{1'} + 2a^2 + a^{3'}\right)}{8}.$$

From this, we obtain the following biconditional logical statements.

$$f(\tilde{A}^I) > f(\tilde{B}^I) \Leftrightarrow \tilde{A}^I \succ \tilde{B}^I$$

$$f(\tilde{A}^I) < f(\tilde{B}^I) \Leftrightarrow \tilde{A}^I \prec \tilde{B}^I$$

$$f(\tilde{A}^I) = f(\tilde{B}^I) \Leftrightarrow \tilde{A}^I \approx \tilde{B}^I$$

Theorem: 1 The ranking function Ʀ: IF(R)→ R is a linear function.

Proof: Let $\tilde{A}^I = \left(\phi^1, \phi^2, \phi^3; \phi^{1'}, \phi^2, \phi^{3'}\right)$ and $\tilde{B}^I = \left(\Psi^1, \Psi^2, \Psi^3; \Psi^{1'}, \Psi^2, \Psi^{3'}\right)$ be any 2 arbitrary TIFNs. Then, ∀c≥0, we have,

$$\mathcal{R}\left(c\tilde{A}^I \oplus \tilde{B}^I\right) = \mathcal{R}\left[c\left(\left(\phi^1, \phi^2, \phi^3; \phi^{1'}, \phi^2, \phi^{3'}\right)\right) \oplus \left(\Psi^1, \Psi^2, \Psi^3; \Psi^{1'}, \Psi^2, \Psi^{3'}\right)\right]$$

$$\because \mathrm{c}\tilde{A}^I = c\left[\left(\phi^1, \phi^2, \phi^3; \phi^{1'}, \phi^2, \phi^{3'}\right)\right]$$

$$= \left(c\phi^1, c\phi^2, c\phi^3; c\phi^{1'}, c\phi^2, c\phi^{3'}\right), \forall c \geq 0$$

$$\mathcal{R}\left(c\tilde{A}^I \oplus \tilde{B}^I\right) = \mathcal{R}\left[\left(c\phi^1, c\phi^2, c\phi^3; c\phi^{1'}, c\phi^2, c\phi^{3'}\right) \oplus \left(\Psi^1, \Psi^2, \Psi^3; \Psi^{1'}, \Psi^2, \Psi^{3'}\right)\right]$$

$$c\tilde{A}^I \oplus \tilde{B}^I = \left[\left(c\phi^1 + \Psi^1, c\phi^2 + \Psi^2, c\phi^3 + \Psi^3; c\phi^{1'} + \Psi^{1'}, c\phi^2 + \Psi^2, c\phi^{3'} + \Psi^{3'}\right)\right]$$

$$\because\ \tilde{A}^{I} \oplus \tilde{B}^{I} = \left(\phi^{1}+\Psi^{1},\phi^{2}+\Psi^{2},\phi^{3}+\Psi^{3};\phi^{1'}+\Psi^{1'},\phi^{2}+\Psi^{2},\phi^{3'}+\Psi^{3'}\right)$$

$$\mathcal{R}\left(c\tilde{A}^{I}+\tilde{B}^{I}\right)=\frac{\left[c\phi^{1}+\Psi^{1}+2\left(c\phi^{2}+\Psi^{2}\right)+c\phi^{3}+\Psi^{3}\right]+\left[c\phi^{1'}+\Psi^{1'}+2\left(c\phi^{2}+\Psi^{2}\right)+c\phi^{3'}+\Psi^{3'}\right]}{8}$$

$$\because\ \mathcal{R}\left(\tilde{A}^{I}\right)=\left(\frac{\left(\phi^{1}+2\phi^{2}+\phi^{3}\right)+\left(\phi^{1'}+2\phi^{2}+\phi^{3'}\right)}{8}\right)$$

$$=\frac{\left[c\phi^{1}+\Psi^{1}+2c\phi^{2}+2\Psi^{2}+c\phi^{3}+\Psi^{3}\right]+\left[c\phi^{1'}+\Psi^{1'}+2c\phi^{2}+2\Psi^{2}+c\phi^{3'}+\Psi^{3'}\right]}{8}$$

$$=\left(\frac{\left[\left(c\phi^{1}+2c\phi^{2}+c\phi^{3}\right)+\left(c\phi^{1'}+2c\phi^{2}+c\phi^{3'}\right)\right]+\left(\Psi^{1}+2\Psi^{2}+\Psi^{3}\right)+\left(\Psi^{1'}+2\Psi^{2}+\Psi^{3'}\right)}{8}\right)$$

$$=c\left(\frac{\left(\phi^{1}+2\phi^{2}+\phi^{3}\right)+\left(\phi^{1'}+2\phi^{2}+\phi^{3'}\right)}{8}\right)+\left(\frac{\left(\Psi^{1}+2\Psi^{2}+\Psi^{3}\right)+\left(\Psi^{1'}+2\Psi^{2}+\Psi^{3'}\right)}{8}\right)$$

$\mathcal{R}\left(c\tilde{A}^{I}\oplus\tilde{B}^{I}\right)=c\mathcal{R}\left(\tilde{A}^{I}\right)+\mathcal{R}\left(\tilde{B}^{I}\right)\Rightarrow$ the ranking function $\mathcal{R}$ is a linear function.

Note. We can prove this theorem for all $c<0$.

Definition

The ordering ($\preccurlyeq$ and $\succcurlyeq$) between any 2 arbitrary TIFNs $\tilde{A}^{I}$ and $\tilde{B}^{I}$ are defined as follows.

1. $\tilde{A}^{I} \preceq \tilde{B}^{I}$ iff $\tilde{A}^{I} \prec \tilde{B}^{I}$ or $\tilde{A}^{I} \approx \tilde{B}^{I}$
2. $\tilde{A}^{I} \succeq \tilde{B}^{I}$ iff $\tilde{A}^{I} \succ \tilde{B}^{I}$ or $\tilde{A}^{I} \approx \tilde{B}^{I}$

Definition

Assume that $\{\tilde{A}_{i}^{I}, i=1,2,\ldots,n\}$ is a set of TIFNs. In this set, if

1. $\forall$i, $\mathcal{R}(\tilde{A}_{k}^{I}) \le \mathcal{R}(\tilde{A}_{i}^{I})$ then the TIFN $\tilde{A}_{k}^{I}$ is the min of $\{\tilde{A}_{i}^{I}, i=1,2,\ldots,n\}$
2. $\forall$i, $\mathcal{R}(\tilde{A}_{t}^{I}) \ge \mathcal{R}(\tilde{A}_{i}^{I})$ then the TIFN $\tilde{A}_{t}^{I}$ is the max of $\{\tilde{A}_{i}^{I}, i=1,2,\ldots,n\}$

Definition

A TIFN $\tilde{A}^I = \left(a^1, a^2, a^3; a^{1'}, a^2, a^{3'}\right)$ is said to be $\tilde{0}^I$ iff $\mathcal{R}\left(\tilde{A}^I\right) = 0$. Similarly, $\tilde{A}^I \succeq \tilde{0}^I$ is said to be non-ve TIFN iff $\mathcal{R}\left(\tilde{A}^I\right) \geq 0$.

THE PROPOSED MODEL

The TCs of many real world applications are in IFNs. Consider the $m \times n$ TP (i.e., a TP with m -rows or origins, and n columns or destinations). Let

$$\tilde{c}_{ij}^I \left(= \left(c_{ij}^1, c_{ij}^2, c_{ij}^3; c_{ij}^{1'}, c_{ij}^2, c_{ij}^{3'}\right)\right)$$

be the cost of transporting one unit of the product from i[th] origin to j[th] destination,

$$\tilde{a}_i^I \left(= \left(a_i^1, a_i^2, a_i^3; a_i^{1'}, a_i^2, a_i^{3'}\right)\right)$$

be the quantity of commodity available at origin i,

$$\tilde{b}_j^I \left(= \left(b_j^1, b_j^2, b_j^3; b_j^{1'}, b_j^2, b_j^{3'}\right)\right)$$

the quantity of commodity needed at destination j,

$$\tilde{y}_{ij}^I \left(= \left(y_{ij}^1, y_{ij}^2, y_{ij}^3; y_{ij}^{1'}, y_{ij}^2, y_{ij}^{3'}\right)\right)$$

is the quantity transported from i[th] origin to j[th] destination, so as to minimize the TIFTC.

Typically, the FIFTP can be written as follows.

$$(P_1)\ \text{Min}\ \tilde{Z}^I = \sum_{i=1}^{m}\sum_{j=1}^{n} \tilde{c}_{ij}^I \otimes \tilde{y}_{ij}^I$$

$$s.t \begin{cases} \sum_{j=1}^{n} \tilde{y}_{ij}^I \approx \tilde{a}_i^I \ (i = 1, 2, \ldots, m) \\ \sum_{i=1}^{m} \tilde{y}_{ij}^I \approx \tilde{b}_j^I \ (j = 1, 2, \ldots, n) \end{cases} \quad (1)$$

$\tilde{y}_{ij}^{I} \succeq \tilde{0}^{I}$ (i=1,2,…,m and j=1,2,…,n)

Here, m and n denote the number of supply and demand points respectively.

The constraints of the FIFTP are given in Equation (1). Tabular form for the FIFTP is presented in Table 1.

Table 1. m×n FIFTP

Origin or Source (S_i)	**Destination (D_j)**				**Intuitionistic Fuzzy Availability $\tilde{a}_i^I$**
	D_1	D_2	…	D_n	
S_1	$\tilde{c}_{11}^I$	$\tilde{c}_{12}^I$	…	$\tilde{c}_{1n}^I$	$\tilde{a}_1^I$
S_2	$\tilde{c}_{21}^I$	$\tilde{c}_{22}^I$	…	$\tilde{c}_{2n}^I$	$\tilde{a}_2^I$
⋮	⋮	⋮	…	⋮	⋮
S_m	$\tilde{c}_{m1}^I$	$\tilde{c}_{m2}^I$	…	$\tilde{c}_{mn}^I$	$\tilde{a}_m^I$
Intuitionistic fuzzy demand $\tilde{b}_j^I$	$\tilde{b}_1^I$	$\tilde{b}_2^I$	…	$\tilde{b}_n^I$	$\sum_{i=1}^{m}\tilde{a}_i^I = \sum_{j=1}^{n}\tilde{b}_j^I$

THE PROPOSED METHODOLOGIES FOR SOLVING A BALANCED AND UBIFTPs

The list of steps for the method-I is as follows.

Method-I

The optimal solution of the FIFTP (P_1) is the set of non-negative integers (say $\left\{\tilde{y}_{ij}^I\right\}$), which satisfies the below characteristics.

- $\sum_{j=1}^{n}\tilde{y}_{ij}^I \approx \tilde{a}_i^I$ (i=1,2,…,m) and $\sum_{i=1}^{m}\tilde{y}_{ij}^I \approx \tilde{b}_j^I$ (j=1,2,…,n);
- If ∃ any set of non-negative integers (say $\left\{\tilde{y}_{ij}^{I'}\right\}$) such that

$\sum_{j=1}^{n}\tilde{y}_{ij}^{I'} \approx \tilde{a}_i^I$ (i=1,2,…,m) and $\sum_{i=1}^{m}\tilde{y}_{ij}^{I'} \approx \tilde{b}_j^I$ (j=1,2,…,n) then

$$\mathcal{R}\left(\sum_{i=1}^{m}\sum_{j=1}^{n}\tilde{c}_{ij}^{I}\otimes\tilde{y}_{ij}^{I}\right)\le\mathcal{R}\left(\sum_{i=1}^{m}\sum_{j=1}^{n}\tilde{c}_{ij}^{I}\otimes\tilde{y}_{ij}^{I'}\right).$$

1. Form the intuitionistic fuzzy transportation table (IFTT) for the given data. i.e., to make a TP with intuitionistic fuzzy parameters listed in Table 1.
2. Now, modify IFTP obtained in step-1 of the method-I to the crisp linear programming problem (CLPP). That is, to change the IFTP to the following model.

Min/max $Z=\sum_{i=1}^{m}\sum_{j=1}^{n}\mathcal{R}\left(\tilde{c}_{ij}^{I}\right)\otimes\mathcal{R}\left(\tilde{y}_{ij}^{I}\right)$ [If $\tilde{c}_{ij}^{I}$ is cost, then minimize the objective function, whereas if $\tilde{c}_{ij}^{I}$ is profit, then maximize the objective function].

Subject to the given constraints (see Equation (1)).

3. Now, solve CLPP, obtained in step-2 of the method-I, to find out the optimal solution $\left\{\tilde{y}_{ij}^{I}\right\}$ by using Varghese and Kuriakose ranking index relevant to minimum total intuitionistic fuzzy transportation cost (MTIFTC) $\sum_{i=1}^{m}\sum_{j=1}^{n}\mathcal{R}\left(\tilde{c}_{ij}^{I}\right)\otimes\mathcal{R}(\tilde{y}_{ij}^{I})$.

Note 1: For solving the CLPP, we can use any software package (for example, LINGO, RStudio, MATLAB, RGui and TORA).

Method-II

The steps for the method-II are as follows.

1. Let us consider a TP with IFNs listed in Table 1.
2. 1st, check whether $\sum_{i=1}^{m}\tilde{a}_{i}^{I}=\sum_{j=1}^{n}\tilde{b}_{j}^{I}$ or $\sum_{i=1}^{m}\tilde{a}_{i}^{I}\neq\sum_{j=1}^{n}\tilde{b}_{j}^{I}$. If $\sum_{i=1}^{m}\tilde{a}_{i}^{I}=\sum_{j=1}^{n}\tilde{b}_{j}^{I}$ then go to step 3. If not, introduce a dummy row or column having all its cost elements as 0 or $\tilde{0}^{I}$ and $\tilde{a}_{i}^{I}$ or $\tilde{b}_{j}^{I}$ as the '+ve' difference ('+ve' difference is identified or calculated by using ranking index) of $\tilde{a}_{i}^{I}$ and $\tilde{b}_{j}^{I}$.
3. After using step-2 of this method, now apply the ranking method of Varghese and Kuriakose to calculate the ranking index of all TIFNs.
4. Now construct the crisp or standard TP using step-3 of this method. Next, solve the same standard TP using any one or all of the standard methods (for example, MODI, Zero Point, Improved Zero Point, and so forth) or software packages (for example, RStudio, RGui, etc.). Then, we will get the optimal allocation and optimal objective value of the standard TP. Generally, this optimal allotted cell in standard transportation table is named as occupied cell.
5. After using step-4 of this method, now, ensure if one or more rows and/or columns having exactly one occupied cell, then, allot the maximum possible value in that kind of cells first, and adjust the corresponding $\tilde{a}_{i}^{I}$ or $\tilde{b}_{j}^{I}$ with a '+ve' difference of $\tilde{a}_{i}^{I}$ and $\tilde{b}_{j}^{I}$. Otherwise, if all the rows and/or

columns having more than one occupied cells then select a cell in the γ- row and δ- column of the transportation table (TT) whose cost is maximum (note that if the maximum cost is more than one, i.e., a tie occurs then select arbitrarily) and examine which one of the cell is minimum cost (note that if the minimum cost is more than one, i.e., a tie occurs then select at random) among all the occupied cells in that row and column then allot the maximum possible value to that kind of cells. Like this, we should proceed with the selected row and column entirely. If the entire row and column of the occupied cells having fully allotted then select the next maximum cost of the transportation table and check which one of the cell is minimum cost among all the occupied cells in that row and column then allot the maximum possible value to that kind of cells. Like this, we should repeat this process until all $\tilde{a}_i^I$ (supply points) are fully used and all $\tilde{b}_j^I$ (demand points) are fully received. Exactly, this step will provide the optimum intuitionistic fuzzy allotment table.

6. To determine the optimal objective value of the FIFTP, substitute the values of the decision variables ($\tilde{y}_{ij}^I$) in the objective function ($\tilde{Z}^I = \sum_{i=1}^{m}\sum_{j=1}^{n}\tilde{c}_{ij}^I \otimes \tilde{y}_{ij}^I$).

Theorem: The feasible solutions of all of the type- II IFTPs are also the feasible solutions of their respective equivalent crisp TPs, it is said to be the PSK theorem.

Theorem: The optimal solutions of all of the type- II IFTPs are also the optimal solutions of their respective equivalent crisp TPs, it is also called the PSK theorem.

Remark 1: Allocate the maximum possible value (maximum possible value means the Min {supply, demand}) in the IFTP to the most preferable column or row having exactly one occupied cell.

Remark 2: The proposed methodologies are applicable iff the existing software are efficient.

Remark 3: If any UBIFAP (unbalanced intuitionistic fuzzy assignment problem) is possible to be solved using the PSK method then any UBIFTP is also possible to be solved by the PSK method. Vice versa is also true.

The proposed methodologies can be illustrated with the help of following examples.

ILLUSTRATIVE EXAMPLES

Example 1: *Type- II IFTP*

Let us consider the following simple numerical example.

A firm (named as Sashwin) has 3 factories- P, Q, and R. These factories are producing the same kind of fridges in 3 different places- Krishnagiri, Tiruppur and Madurai. The owner of this firm is Sashwin. He would like to transport fridges from 3 different factories to 3 different districts/cities namely, Tiruvarur (U), Virudhunagar (V) and Namakkal (W). All the factories are connected to all the districts/cities by roads and the fridges are all transported by heavy goods vehicles. The supply and demand of the fridges are all well known crisp numbers. But the TC is not known exactly due to the variations in rates of gas/diesel/petrol/traffic jams/nature of the road/weather in hilly areas, etc. So, the crisp demand, crisp supply and uncertain TC (that is, IFTC) per unit (Rs. in 100's) for fridges from 3-different factories to 3-different districts/cities are presented in Table 2 from the recent experience.

Table 2. 3×3 type- II IFTP

Districts → Factories ↓	Districts/Cities			Supply a_i
	Tiruvarur U	Virudhunagar V	Namakkal W	
Krishnagiri P	(7,21,29; 2,21,34)	(7,20,57; 3,20,61)	(12,25,56; 8,25,60)	18
Tiruppur Q	(8,9,16; 2,9,22)	(4,12,35; 1,12,38)	(6,14,28; 3,14,31)	12
Madurai R	(5,9,22; 2,9,25)	(10,15,20; 5,15,25)	(4,16,19; 1,16,22)	4
Demand b_j	8	16	10	

Find out the optimal allocation to minimize the IFTC for the type- II IFTP.
Method 1 *type- II IFTP: LPP method* (Example:1)

1. 1. The intuitionistic fuzzy linear programming problem (IFLPP) formulation of IFTP, chosen in example-1 is as follows.

Min [(7,21,29; 2,21,34) × y_{11}+ (7,20,57; 3,20,61) × y_{12}+ (12,25,56; 8,25,60) × y_{13}+ (8, 9, 16; 2, 9, 22) × y_{21}+ (4,12,35; 1,12,38) × y_{22}+ (6,14,28; 3,14,31) × y_{23}+ (5, 9, 22; 2,9,25) × y_{31}+ (10,15,20; 5,15,25) × y_{32}+ (4,16,19; 1,16,22) × y_{33}]

$$s.t.\begin{cases} y_{11}+y_{12}+y_{13}=18, \\ y_{21}+y_{22}+y_{23}=12, \\ y_{31}+y_{32}+y_{33}=4, \\ y_{11}+y_{21}+y_{31}=8, \\ y_{12}+y_{22}+y_{32}=16, \\ y_{13}+y_{23}+y_{33}=10. \end{cases}$$

$y_{ij}\geq 0$ (for all i=1,2,3 and j=1,2,3).

2. 2. By using step 2 of the method-I, the formulated IFLPP is converted to the following CLPP.

Min [ℝ(7,21,29; 2,21,34) × y_{11}+ ℝ(7,20,57; 3,20,61) × y_{12}+ ℝ(12,25,56; 8,25,60) × y_{13}+ ℝ(8, 9, 16; 2,9,22) × y_{21}+ ℝ(4,12,35; 1,12,38) × y_{22}+ ℝ(6,14,28; 3,14,31) × y_{23}+ ℝ(5, 9, 22; 2,9,25) × y_{31}+ ℝ(10,15,20; 5,15,25) × y_{32}+ ℝ(4,16,19; 1,16,22) × y_{33}]

$$s.t.\begin{cases} y_{11}+y_{12}+y_{13}=18, \\ y_{21}+y_{22}+y_{23}=12, \\ y_{31}+y_{32}+y_{33}=4, \\ y_{11}+y_{21}+y_{31}=8, \\ y_{12}+y_{22}+y_{32}=16, \\ y_{13}+y_{23}+y_{33}=10. \end{cases}$$

$y_{ij}\geq 0$ ($\forall$ i=1,2,3 and j=1,2,3).

To find out the values of $\mathcal{R}\left(\tilde{c}_{ij}^{I}\right)$, ($i$=1,2,3 and j=1,2,3) using definitions of Section 3 and Section 4, we obtain the following.

$$\left\{\mathcal{R}\left(\tilde{c}_{11}^{I}\right)=19, \mathcal{R}\left(\tilde{c}_{12}^{I}\right)=28, \mathcal{R}\left(\tilde{c}_{13}^{I}\right)=31, \mathcal{R}\left(\tilde{c}_{21}^{I}\right)=11, \mathcal{R}\left(\tilde{c}_{22}^{I}\right)=17,\right.$$
$$\left.\mathcal{R}\left(\tilde{c}_{23}^{I}\right)=16, \mathcal{R}\left(\tilde{c}_{31}^{I}\right)=12, \mathcal{R}\left(\tilde{c}_{32}^{I}\right)=15, \mathcal{R}\left(\tilde{c}_{33}^{I}\right)=13\right\}$$

Using the values of $\mathcal{R}\left(\tilde{c}_{ij}^{I}\right)$, the CLPP obtained in step-2, can be written as the following mathematical model:

Min $[19\times y_{11}+28\times y_{12}+31\times y_{13}+11\times y_{21}+17\times y_{22}+16\times y_{23}+12\times y_{31}+15\times y_{32}+13\times y_{33}]$

$$s.t.\begin{cases} y_{11}+y_{12}+y_{13}=18, \\ y_{21}+y_{22}+y_{23}=12, \\ y_{31}+y_{32}+y_{33}=4, \\ y_{11}+y_{21}+y_{31}=8, \\ y_{12}+y_{22}+y_{32}=16, \\ y_{13}+y_{23}+y_{33}=10. \end{cases}$$

$y_{ij}\geq 0$ (i=1,2,3 and j=1,2,3).

3. 3. After solving the CLPP obtained in Step 2, we obtain the following optimal value and solution.

Min $\tilde{Z}_{LPP}^{I}$= (202, 588, 1256; 74, 588, 1384).

$$\begin{bmatrix} y_{11} \\ y_{12} \\ y_{22} \\ y_{23} \\ y_{33} \end{bmatrix} = \begin{bmatrix} 8 \\ 10 \\ 6 \\ 6 \\ 4 \end{bmatrix}$$

Varghese and Kuriakose ranking index relevant to MTIFTC of this problem is Rs. 682.

Hence, from the obtained objective value and solutions, we conclude that both the method-I and method-II are give the same optimal objective value and solutions.

The optimal solution and objective value of type- II IFTP, chosen in example-1, can be obtain using the below steps of the PSK Method (i.e., method-II).

Method 2 *type- II IFTP: PSK method* (Example: 1)

Solution:

Now, by using step 2 of the method-II, we obtain $\sum_{j=1}^{n} b_j = 34 = \sum_{i=1}^{m} a_i$.

This is the balanced condition for the traditional TP. That is, this implies that the given problem is a BTP. |||[ly], the same condition is satisfied in type- II IFTP also. Therefore, we state that the given problem is a balanced type- II IFTP.

Next, by using step 3 of the method-II (i.e., to calculate the ranking index of all TIFNs using the suitable ranking procedure), we obtain the revised transportation table (TT) as shown in Table 3.

Table 3. The equivalent defuzzified problem of Example 1

Factories ↓	Districts			Supply a_i
	U	V	W	
P	19	28	31	18
Q	11	17	16	12
R	12	15	13	4
Demand b_j	8	16	10	

After using step 3 of the method-II we obtain the optimal allocation of the reduced problem of example-1 as shown in Table 4.

Table 4. The optimum table of defuzzified problem of Example 1

Factories ↓	**Districts**			**Supply** a_i
	U	**V**	**W**	
P	19 (→0) **(8)**	28(→0) **(10)**	31	18
Q	11	17(→0) **(6)**	16 (→0)**(6)**	12
R	12	15	13(→0) **(4)**	4
Demand b_j	8	16	10	

∴ the minimum objective value (MOV) Z_{LPP} = [(19×8)+(28×10)+(17×6)+(16×6)+ (13×4)]=682.

After using step 4 of the method-II now, using step 5, we obtain the optimal allocation directly to the given FIFTP as presented in Table 5.

Table 5. Optimum table for 3×3 type- II IFTP

Retail Stores → **Factories ↓**	**Districts**			**Supply** a_i
	Tiruvarur **U**	**Virudhunagar** **V**	**Namakkal** **W**	
Krishnagiri P	**8** (7,21,29; 2,21,34)	**10** (7,20,57; 3,20,61)	(12,25,56; 8,25,60)	18
Tiruppur Q	(8,9,16; 2,9,22)	**6** (4,12,35; 1,12,38)	**6** (6,14,28; 3,14,31)	12
Madurai R	(5,9,22; 2,9,25)	(10,15,20; 5,15,25)	**4** (4,16,19; 1,16,22)	4
Demand b_j	8	16	10	

Finally, we obtain the optimal solution to the type- II IFTP as given below.

$$\begin{bmatrix} y_{11} \\ y_{12} \\ y_{22} \\ y_{23} \\ y_{33} \end{bmatrix} = \begin{bmatrix} 8 \\ 10 \\ 6 \\ 6 \\ 4 \end{bmatrix}$$

Now, using step 6 of the method-II, we obtain the MTIFTC as follows.

Min $\tilde{Z}^{I}$ = [(7, 21, 29; 2, 21, 34)×8]+[(7, 20, 57; 3, 20, 61)×10]+[(4, 12, 35; 1, 12, 38) ×6]+[(6, 14, 28; 3, 14, 31)×6]+[(4, 16, 19; 1, 16, 22)×4]

= [(56, 168, 232; 16, 168, 272)+(70, 200, 570; 30, 200, 610)+ (24, 72, 210; 6, 72, 228)+ (36, 84, 168; 18, 84, 186)+ (16, 64, 76; 4, 64, 88)]

Min $\tilde{Z}^{I}_{\mathrm{PSK}}$ = (202, 588, 1256; 74, 588, 1384)

$\Re\left(\tilde{Z}^{1}\right) = \Re(202,588,1256; 74,588,1384)=682$

RStudio, RGui, MATLAB, and LINGO codes for this problem and their outputs are all given in the Appendix.

Figures 1 and 2 show the RStudio and RGui codes of example 1 with their solutions (see Appendix).

Figures 3 and 4 show the MATLAB code of example 1 with its solution (see Appendix).

Figures 5-8 show the different types of lingo matrix pictures relevant to example 1 (see Appendix).

The authors claim to illustrate their ideas with real-life applications. However, in their first example, apart from a typical cover story, the TIFN has been used in the relevant TP without any explanation about the necessity of introducing it. Hence, a better explanation is given in Example 2 about the necessity of introducing TIFN by the authors.

Example 2: *Unbalanced FIFTP (UBFIFTP)*

If demands, supplies and costs of the TP are all well-known, we will get an exact optimal value as well as exact optimal solutions for that problem. Contrarily, if any one and/or all of the parameters are in FNs, that is, if any one and/or all of the parameters are not in clear then we will get an optimal value as well as optimal solutions for this problem in terms of FNs. However, from the FNs we can find out the following:

- the boundary (extreme values) of the optimal TTC,
- 100% expected optimal TTC

and

- the acceptance level of TTC.

Although, from the FNs we cannot find out the following:

- the hesitation index/hesitation margin

and

- non-acceptance (rejection) level of TTC.

For this demerit in FNs, the author introduced the TIFN to solve TP. If we solve the TPs using TIFN, we can find out the following:

- The boundary/extreme values of TTC.

- 100% expected optimal TTC.
- Level of acceptance/level of satisfaction of TTC.
- Level of rejection/level of non-acceptance of TTC.
- Hesitation index/hesitation margin of TTC.

If the author not give the justification about the necessity of introducing TIFN for his/her example then it will make a question (?) to the practical value of his/her proposed methods. For this reason, the author gave the justification about the necessity of introducing TIFN in the present example.

Consider a TP having the parameters listed in Table 6, that is, an UBFIFTP.

Table 6. 3×3 UBFIFTP

Source (S_i)	**Destination** (D_j)			**Intuitionistic Fuzzy Availability** $\tilde{a}_i^I$
	D_1	D_2	D_3	
S_1	(2,4,7; 1,4,11)	(3,13,14; 2,13,15)	(5,15,16; 4,15,17)	(8,10,13; 7,10,17)
S_2	(2,4,7; 1,4,11)	(8,18,19; 7,18,20)	(15,17,20; 14,17,24)	(2,4,7; 1,4,11)
S_3	(1,3,6; 0,3,10)	(9,19,20; 8,19,21)	(19,21,24; 18,21,28)	(2,12,13; 1,12,14)
Intuitionistic fuzzy demand $\tilde{b}_j^I$	(3,8,10; 1,8,12)	(10,12,15; 9,12,19)	(7,9,12; 6,9,16)	

Now, using step 2 of the method-II, we obtain the following.

$$\sum_{i=1}^{m}\tilde{a}_i^I = (12,26,33;\ 9,26,42) = \text{Ʀ}(12,26,33;\ 9,26,43);$$

$$\sum_{j=1}^{n}\tilde{b}_j^I = (20,29,37;\ 16,29,47) = \text{Ʀ}(20,29,37;\ 16,29,47) = 30.$$

Clearly, we note that, $\sum_{i=1}^{m}\tilde{a}_i^I \not\approx \sum_{j=1}^{n}\tilde{b}_j^I$.

This implies that the balanced condition is not satisfied. From this, we can say that, the given problem is an UBFIFTP.

Table 7 shows the BFIFTP relevant to Table 6.

Table 7. The 4×3 BFIFTP relevant to example 2

Source (S_i)	Destination (D_j)			Intuitionistic Fuzzy Availability $\tilde{a}_i^I$
	D_1	D_2	D_3	
S_1	(2,4,7; 1,4,11)	(3,13,14; 2,13,15)	(5,15,16; 4,15,17)	(8,10,13; 7,10,17)
S_2	(2,4,7; 1,4,11)	(8,18,19; 7,18,20)	(7,16,24; 0,16,41)	(2,4,7; 1,4,11)
S_3	(1,3,6; 0,3,10)	(9,19,20; 8,19,21)	(19,21,24; 18,21,28)	(2,12,13; 1,12,14)
S_4	(0,0,0; 0,0,0)	(0,0,0; 0,0,0)	(0,0,0; 0,0,0)	(-13,3,25; -26,3,38)
Intuitionistic fuzzy demand $\tilde{b}_j^I$	(3,8,10; 1,8,12)	(10,12,15; 9,12,19)	(7,9,12; 6,9,16)	

Now, using step 3 of the method-II (i.e., to calculate the ranking index of all TIFNs using the suitable ranking procedure), we obtain the revised crisp TT as shown in Table 8.

Table 8. The equivalent defuzzified problem of Example 2

Source (S_i)	Destination (D_j)			Availability a_i
	D_1	D_2	D_3	
S_1	5	10	12	11
S_2	5	15	18	5
S_3	4	16	22	9
Demand b_j	7	13	10	

We can change an UBTP to a balanced transportation problem (BTP) using the traditional method of TP. Therefore, the UBTP in Table 8 has been changed to a BTP in Table 9.

Table 9. BTP relevant to the defuzzified problem of Example 2

Source (S_i)	Destination (D_j)			Availability a_i
	D_1	D_2	D_3	
S_1	5	10	12	11
S_2	5	15	18	5
S_3	4	16	22	9
S_4	0	0	0	5
Demand b_j	7	13	10	30

After using step 4 of the method-II, we obtain the optimal allocation of the crisp TP as shown in Table 10.

Table 10. The optimum table of the BTP relevant to the defuzzified problem of Example 2

Source (S_i)	Destination (D_j)			Availability a_i
	D_1	D_2	D_3	
S_1	5	10(→0) **(6)**	12(→0) **(5)**	11
S_2	5	15(→0) **(5)**	18	5
S_3	4(→0) **(7)**	16(→0) **(2)**	22	9
S_4	0	0	0(→0) **(5)**	5
Demand b_j	7	13	10	30

∴ the MOV Z_{LPP} = [(10×6)+(12×5)+(15×5)+(4×7)+(16×2)+(0×5)]
=255

After using step 4 of the method-II now, using step 5, we obtain the optimal allocation directly to the UBIFTP as presented in Table 11.

Table 11. Optimum table for 4×3 UBIFTP relevant to Example 2

Source (S_i)	Destination (D_j)			Intuitionistic Fuzzy Availability $\tilde{a}_i^I$
	D_1	D_2	D_3	
S_1	(2,4,7; 1,4,11)	**(-17,4,31; -35,4,49)** (3,13,14; 2,13,15)	**(-18,6,25; -32,6,42)** (5,15,16; 4,15,17)	(8,10,13; 7,10,17)
S_2	(2,4,7; 1,4,11)	**(2,4,7; 1,4,11)** (8,18,19; 7,18,20)	(7,16,24; 0,16,41)	(2,4,7; 1,4,11)
S_3	**(3,8,10; 1,8,12)** (1,3,6; 0,3,10)	**(-8,4,10; -11,4,13)** (9,19,20; 8,19,21)	(19,21,24; 18,21,28)	(2,12,13; 1,12,14)
S_4	(0,0,0; 0,0,0)	(0,0,0; 0,0,0)	**(-13,3,25; -26,3,38)** (0,0,0; 0,0,0)	(-13,3,25; -26,3,38)
Intuitionistic fuzzy demand $\tilde{b}_j^I$	(3,8,10; 1,8,12)	(10,12,15; 9,12,19)	(7,9,12; 6,9,16)	

Thus, the obtained intuitionistic fuzzy optimal solution (IFOS) in terms of TIFN is given below.

$$\begin{bmatrix} \tilde{y}^I_{12} \\ \tilde{y}^I_{13} \\ \tilde{y}^I_{22} \\ \tilde{y}^I_{31} \\ \tilde{y}^I_{32} \\ \tilde{y}^I_{44} \end{bmatrix} = \begin{bmatrix} (-17,4,31;-35,4,49) \\ (-18,6,25;-32,6,42) \\ (2,4,7;1,4,11) \\ (3,8,10;1,8,12) \\ (-8,4,10;-11,4,13) \\ (-13,3,25;-26,3,38) \end{bmatrix}$$

Finally, we obtain the MTIFTC using step 6 of the method-II.

Min $\tilde{Z}^I_{\text{PSK}}$ =[(3,13,14; 2,13,15)⊗ (-17,4,31; -35,4,49)]+[(5,15,16; 4,15,17) ⊗ (-18,6,25; -32,6,42)]+[(8,18,19; 7,18,20) ⊗ (2,4,7; 1,4,11)]+[(1,3,6; 0,3,10) ⊗ (3,8,10; 1,8,12)]+ [(9,19,20; 8,19,21) ⊗ (-8,4,10; -11,4,13)]+[0⊗ (-13,3,25; -26,3,38)]

Min $\tilde{Z}^I_{\text{PSK}}$ =[(3,13,14; 2,13,15) ⊗Ꞧ (-17,4,31; -35,4,49)]+[(5,15,16; 4,15,17) ⊗Ꞧ (-18,6,25; -32,6,42)]+[(8,18,19; 7,18,20) ⊗Ꞧ (2,4,7; 1,4,11)]+[(1,3,6; 0,3,10) ⊗Ꞧ (3,8,10; 1,8,12)]+[(9,19,20; 8,19,21) ⊗Ꞧ (-8,4,10; -11,4,13)]+[(0,0,0; 0,0,0) ⊗Ꞧ (-13,3,25; -26,3,38)]

Min $\tilde{Z}^I_{\text{PSK}}$ =[(3,13,14; 2,13,15)× (6)]+[(5,15,16; 4,15,17) × (5)]+[(8,18,19; 7,18,20) × (5)]+[(1,3,6; 0,3,10) × (7)]+[(9,19,20; 8,19,21) × (2)]+[(0,0,0; 0,0,0) × (5)]

Min $\tilde{Z}^I_{\text{PSK}}$ =[(18,78,84; 12,78,90)+(25,75,80; 20,75,85)+(40,90,95;35,90,100)+ (7,21,42; 0,21,70)+(18,38,40; 16,38,42)+ (0,0,0; 0,0,0)]

Min $\tilde{Z}^I_{\text{PSK}}$ =(108,302,341; 83,302,387)

$Ꞧ\left(\tilde{Z}^I_{\text{PSK}}\right)$=Ꞧ(108,302,341; 83,302,387) ≈ Rs.255

RStudio, RGui, MATLAB, and LINGO codes for this problem and their outputs are all given in the Appendix.

Figures 9 and 10 show the RStudio and RGui codes of example 2 with their solutions (see Appendix).

Figure 11 shows the MATLAB code with the solution to Example 2 (see Appendix).

Figure 12 shows the lingo input of Example 2 (see Appendix).

Figure 13 shows the lingo solver status of example 2 (see Appendix).

Figures 14–17 show the different types of Lingo matrix pictures relevant to example 2 (see Appendix).

From the presented examples it has been shown that the total optimal cost obtained by the method-I and II remains same as that of obtained by converting the TIFTC by applying the ranking method of Varghese and Kuriakose (2012). A comparative study of the method-I and II with some software packages are presented in Table 12 and 13.

Table 12. Comparative analysis-I

Theoretical and Software Solution Methods →		Theoretical Solution				Software Solution			
		Zero Point Method	VAM	MODI	Improved Zero Point Method	Matlab	RGui	Lingo	RStudio
Varghese and Kuriakose ranking method	Problem No. 1	682	682	682	682	682	682	682	682
	Problem No. 2	255	255	255	255	255	255	255	255

Table 13. Comparative analysis-II

Theoretical and Software Solution Methods →	Theoretical Solution			Software Solution			
	Intuitionistic Fuzzy Zero Point Method	Proposed Method-I and II	Improved Intuitionistic Fuzzy Zero Point Method	Matlab	RGui	Lingo	RStudio
Problem No. 1	(202,588,1256; 74,588,1384) 682	(202,588,1256; 74,588,1384) 682	(202,588,1256; 74,588,1384) 682	(202,588,1256; 74,588,1384) 682	(202,588,1256; 74,588,1384) 682	(202,588,1256; 74,588,1384) 682	(202,588,1256; 74,588,1384) 682
Problem No. 2	(108,302,341; 83,302,387) 255	(108,302,341; 83,302,387) 255	(108,302,341; 83,302,387) 255	(108,302,341; 83,302,387) 255	(108,302,341; 83,302,387) 255	(108,302,341; 83,302,387) 255	(108,302,341; 83,302,387) 255
Varghese and Kuriakose ranking method							

DISCUSSIONS OF OBTAINED RESULTS, AND CONCLUSION

The MTIFTC of problem-2 is given in equation 2.

$$\tilde{Z}^{I} = (108,302,341;\ 83,302,387) \quad (2)$$

Figure 18 shows the graphical representation of IFTC (see Appendix).
The obtained result on example-2 i.e., equation (2) can be explained as follows (see Figure 18).

- The minimum TTC lies in [108,341] i.e., $108 \leq c \leq 341$. Here, c is the minimum total transportation cost (MTTC),
- 100% i.e., hundred percent expects are in favour that the MTTC is 302 as $m_{\tilde{Z}^{I}}(c) = 1$. Here, c=302,
- Assume that n and m be non-membership and membership values respectively at TCc. Then,

We have,

- 100.*m* percent ('.' denotes usual multiplication operation) expects are in favour that the TC is *c*
- 100.*n* percent expects are in opposing that the TC is *c*
- But 100.(1–*m*–*n*) percent are in confusion that the TC is *c*

Values of non-membership (i.e., $n_{\tilde{Z}^I}(c)$) and membership (i.e., $m_{\tilde{Z}^I}(c)$) at different values of *c* (transportation cost) can be determined using the following equations (see equations (3) and (4)):

$$n_{\tilde{Z}^I}(c)=\begin{cases}1, & \text{for } c<83,\\ \dfrac{302-c}{219}, & \text{for } 83\le c\le 302,\\ 0, & \text{for } c=302,\\ \dfrac{c-302}{85}, & \text{for } 302\le c\le 387,\\ 1, & \text{for } c>387.\end{cases} \tag{3}$$

$$m_{\tilde{Z}^I}(c)=\begin{cases}0, & \text{for } c<108,\\ \dfrac{c-108}{194}, & \text{for } 108\le c\le 302,\\ 1, & \text{for } c=302,\\ \dfrac{341-c}{39}, & \text{for } 302\le c\le 341,\\ 0, & \text{for } c>341.\end{cases} \tag{4}$$

Generally in mathematics, we have 2 sub-branches namely, Pure Mathematics and Applied Mathematics. Particularly, we can say that both the 'Pure Mathematics' and 'Applied Mathematics' are two sides of the same coin. If the coin has only tail it is not valid. Vice-versa, if the coin has only head then also it is not valid. It comprises both. Without pure mathematics we can't develop any applications. |||ly, if there is no applied mathematics then there is no need of pure mathematic, that is, there is no use of pure mathematics. Therefore, due to the importance of applied mathematics, the author studied the solution of real life problems in this book chapter.

This book chapter proposes 2 innovative and simple methodologies for solving the linear IFTP. Both of the methodologies are based on the idea of ranking the TIFNs involved and replacing them with the corresponding real values. In this way, the IFTP is converted to a traditional TP, which is solved by using already existing software for this purpose. Commonly, the following question will arise in this case: What is in practical application(s) the reason of forming the IFTP 1st and not proceeding directly to a traditional TP? Due to raising the logical question by the learned reviewers, the author is going to give the explanation to their question, which is given below.

IFTP, FTP and TP are all used to transport the finished goods/raw materials from various supply points to different demand points with the lowest cost or highest profit. This is the general concept for all IFTP, FTP and TP. The traditional TP generally called as TP. In IFTP and FTP, all the parameters are considered as IFNs and fuzzy numbers (FNs), whereas in TP the parameters are all considered as crisp numbers. In TP, the optimal value (that is, objective function value) and optimal solution both are well-known because, the parameters are all well-known crisp numbers. In IFTP, the objective function value and optimal/optimum solution both are in terms of uncertain as well as hesitation. Since the parameters used in this case are all in terms of uncertain as well as hesitation that is, IFNs. $|||^{ly}$ in FTP, the optimal solution and objective function value both are uncertain that is, not clear. Since the parameters used in this case are uncertain (fuzzy) numbers.

In this book chapter, the author discussed TP with IFNs that is, the author solved the IFTP. If we discuss the IFTP then first we need its model. In addition, to distinguish and understand the IFTP, FTP and TP, we need a separate model for that. Therefore, the author formulated IFTP first.

The problems solved in this book chapter are not a traditional TP. Therefore, we cannot solve this problem directly to as a traditional TP. Since the parameters used in this problems are in IFNs. Moreover, the proposed methodologies are based on the idea of ranking the TIFNs involved and replacing them with the corresponding real values. So, first we need to formulate IFTP, second we need to transform the formulated IFTP to crisp TP. Next, we need to solve the obtained crisp TP by using any one of the proposed methodologies.

Merits of the Proposed Method-I and II Are as Follows

The proposed methods in this book chapter can help the DMs in the logistics related issues of real life problems by aiding them in the decision making process and providing an optimal/appropriate solution in a simple, efficient and effective manner.

A DM has the below advantages by using the proposed methodologies.

- The obtained optimal value of the linear UBIFTP is always non -ve (-ve denotes 'negative') TIFN (either 0 or +ve TIFN (+ve denotes 'positive')) that is, there is no -ve part in the obtained TIFN.
- The proposed methods in this book chapter are 'not only easy to understand but also computationally very simple'.
- Those who are using proposed methods, they will get the appropriate/optimal solution for an unbalanced linear intuitionistic fuzzy and unbalanced linear fuzzy optimization problems.
- The proposed methods in this book chapter always give the appropriate or optimal solution to all the types of linear IFTP.
- The proposed methods give an alternative idea to solve the variety of linear optimization problems under fuzzy and/or intuitionistic fuzzy environment.
- The proposed methods in this book chapter are directly or indirectly reduce to the DMs computation time.
- The minimization and maximization linear IFTPs can be easily solved with the help of the proposed methods.
- The systematic approaches are presented in this book chapter.
- The solution/solutions obtained by the proposed methods are all always reliable, efficient and appropriate/optimal.
- The proposed methods in this book chapter are also called the student-friendly methods since its computation steps are very less.

- The proposed methods in this book chapter are all helpful to solve the real life related transportation problems.
- From the computational point of view, the proposed methods are all have significantly fewer calculations when compared to all the existing methods.
- The researcher scholars/practitioners/DMs can be easily verified, whether their obtained solution is optimal or not. Because, the proposed methods always depends on the CLPP or software such as RStudio, LINGO, RGui, MATLAB, and TORA.
- Not only 'linear balanced' but also 'linear unbalanced IFTPs' can be easily solved by using the proposed methods.
- Everyone will get the good amount of knowledge or skill in 'IFSs' if they study this book chapter properly. If everyone has deep knowledge or skill in IFSs then they can effectively solve transportation related problems under uncertain environment.
- If anybody wants to solve the problems related to 'Management Science' or 'Applied Operations Research', they can study the proposed methods.
- The proposed methods in this book chapter are the backbone to study or solve the IFSTPs. An extension of IFTP is called an IFSTP.
- Both the methodologies are logically designed. Therefore, these methods are not working in a trial and error basis. The trial and error is one of the problem solving methods in which multiple attempts are made to reach a solution. Moreover, these methodologies will provide optimal/appropriate solution always.
- If the problem has an alternative optimal solution, the proposed methodologies will give an alternative optimal solution. If the problem has more than one optimal solution, it is called alternative optimal solution/multiple optimal solutions. We cannot find out multiple optimal solutions to the problem by using the existing methodologies.
- The proposed methodologies both are not depending on $n+m-1$ non -ve allotted entries that is, initial basic feasible solutions (IBFSs) $\Rightarrow$ there is no need to find out IBFSs. But, the already existing methodologies are always depending on IBFSs.

Demerits of the Proposed Method-I and II are as follows:

- Particularly, the proposed methods in this book chapter are depending on the CLPP or software like RGui or LINGO or RStudio or TORA or MATLAB. Without knowing the CLPP or already mentioned software packages no one can solve the problem by using the proposed methods.
- Nobody can solve both the balanced linear IFTPs and linear UBIFTPs if the already specified software are corrupted or not working properly. Further, the proposed methodologies will helpful to solve 'linear optimization problems with IFNs' but not the 'non-linear optimization problems with IFNs'.
- Either we should aware of the LPP or software before we study the proposed methods (some examples for software: RGui, LINGO, RStudio, TORA and MATLAB).
- The solution/solutions obtained by the proposed methods are optimal / efficient only if the corresponding software is efficient. In lieu, if the corresponding software is not efficient then the solution/solutions obtained by the proposed methods are also not an optimal/efficient.
- The proposed methodologies both are based on changing TIFNs to real values. If we change TIFNs to real values, we will lose some helpful information.

At the end of this chapter the author presented a list of 20 advantages/merits about using his/her proposed 2 methods for solving the IFTP. From his/her proposed advantages, there is no need to think all of them are general and/or abstract because some of the merits are unique and are all clearly explained. If we read deeply in the advantages about using his proposed 2 methods then we can understand; what are the advantages of his methods with respect to the already existing ones, apart from his simple statement that they involve a lighter computational burden. Additionally, the author has listed 5 demerits about his proposed methods. In general, TIFNs are 6-tuple numbers. A 6-tuple is a tuple that consists of six elements and is known as sextuple. When we use the Pandian's approach to solve the TP with TIFN, we should split the given problem to 5 sub-problems (because, the middle value of both the non-membership and membership function in TIFNs are equal (otherwise, we should split 6 sub-problems)). Afterwards, we should solve all of them independently whereas if we use the proposed methodologies to solve the same problem, we no need to split the given problem to 5 sub-problems. Furthermore, solving a single problem is enough to obtain the optimal solution if we use the proposed methodologies. Because, the proposed methodologies are based on the idea of ranking the TIFNs involved and replacing them with the corresponding real values. Therefore, the proposed methodologies have a lighter computational burden. Furthermore, the proposed methodologies are having very less number of computational steps. Therefore, the author has proved his given statement in a simple and easy manner i.e., he has proved his proposed methodologies are having a lighter computational burden.

In general, a simple and better explanation must be given on why the use of IFNs is more suitable than the use of FNs/other generalizations of them, when the TP is characterized by uncertainty. If we give it, it will be create more interest to study or solve the TP with IFNs to the readers. In addition, the readers will understand why we need to study/solve the TP by using IFNs, i.e., they will understand the greatness of studying TP using IFNs. Therefore, the author has given the following explanation in this regard. The use of IFNs is more suitable than the use of FNs/other generalizations of them, when the TP is characterized by uncertainty. Because, the FNs/other generalizations of them (except IFNs) are not helpful to find the level of non-acceptance and hesitation index (hesitation margin) of the optimal transportation cost whereas in IFNs we can find both the 'level of non-acceptance' and 'hesitation margin' of the optimal transportation cost. So, the use of IFNs is more suitable than the use of FNs/other generalizations of them, when the TP is characterized by uncertainty.

Through by the author, a stronger argument has been given about the advantage of the proposed methodologies with respect to the already existing ones for solving the IFTP. Further, an explanation is given in the example-2 about the necessity of introducing TIFN in first place instead of real coefficient.

The listed advantages of the proposed methodologies with respect to other already existing methodologies are not in general and abstract. So, the question/questions will not arise about the importance of the proposed methodologies. In the demerits, the statement 'corresponding software tool is efficient' denotes the software that related to solve the conventional TP.

From this study, we can conclude that FTPs, IFTPs, BIFTPs and UBIFTPs which can be solved by the recent methods (Hussain & Kumar, 2012a, 2012b; Pandian, 2014; Priyadharshini & Anju, 2020; Dinagar & Keerthivasan, 2020; Sharma, 2020; El Alaoui, 2020) can also be solved by the proposed methods. On the other hand, it is much easier to apply the proposed methods as compared to all the recent methods. The solution and objective value of BIFTPs and/or UBIFTPs obtained by the proposed methods remains always '+ve', i.e., there is no '-ve' part in the obtained TIFN. Therefore, the proposed methods are physically meaningful and computationally very simple when compared to all other recent methods.

For approaching the proposed methods in mathematical term the author has solved 2 numerical examples. The methodologies discussed in this book chapter, will guide to the DMs in the field of logistics, which are related to practical problems of life. Furthermore, it will provide optimal solution(s) in a very simple and effective manner. The techniques to formulate both the balanced and UBIFTPs as a minimization (or maximization) problem can be used to solve the other types of transportation models or issues.

In this book chapter, the balanced and UBIFTPs have been solved by proposed methods in a simple and easy way. The stated results in this book chapter have been checked by using RStudio, MATLAB, LINGO and RGui software. The result analysis between the proposed methods and the already existing methods with software solution is given in this book chapter. The computer code for the proposed problems and their results/outputs both are clearly presented. Furthermore, the unique result, result analysis, discussions, advantages and disadvantages of the proposed methods are all given.

The author's perception on this book chapter is, it is started with good introduction and it contains necessary background description i.e., it provides sufficient background information and literature review. The organization/logical structure of this book chapter is generally good. The given references in this chapter are sufficient, appropriate and up-to-date. Particularly, the required number of references is added in this chapter, it is very good. The merits/advantages presented in this chapter are all not generic. Anyhow, out of 20 merits/advantages some of them are generic. So, there is no need of more clarity for all the described advantages. The justifications about why the TIFN should be introduced to solve TP are given in the example. These kinds of justifications are given for the understanding purpose and the importance of knowing the concept of TIFN. Moreover, it will be helpful for the readers/practitioners to get more ideas about IFN, and to read, understand and apply. Finally, Figure 2 is given with a clear explanation.

The constructive comments and fruitful suggestions to this chapter have been given by the learned reviewers. In a simple and effective manner, the author answered all of the questions raised by the reviewers. The author has been implemented all of the reviewers recommendations in his final and printable book chapter. Further, this book chapter is a double-blind peer reviewed. Therefore, these are all strengthening this chapter directly.

The author's perception is:

Every book chapter printed on papers is not a good book chapter; conversely, any book chapter printed on papers if it is used to solve the public issues then it is called a good book chapter. So, the author has solved social issues in this book chapter. Therefore, as said above, this chapter considered as a good book chapter. The author expectantly this chapter will be helpful for the business executives, academicians, practitioners, and research scholars (e.g., PhD and master students) after its printing/publication.

FUTURE RESEARCH ON OPTIMIZATION WITH IFSs AND/OR IFNs

The author would like to develop the python programming code for the proposed problems in this chapter because; Python is an easily readable computer programming language. Also, the author has planned to solve the non linear optimization problems using IFSs and/or IFNs. His key objective is to solve the public problems of the people using IFSs and/or IFNs. Additionally, he would like to introduce a new method to solve non linear optimization problems under fuzzy and/or intuitionistic fuzzy environment. In the very near future, the proposed methods may be modified to solve IFSAPs, MIFSAPs, IFSTPs and MIFSTPs.

ACKNOWLEDGMENT

The author is thankful to the referees for their valuable suggestions and extensive comments which have helped to improve the chapter. The author is thankful to the editors for their valuable suggestions and comments on this work. The author would also like to acknowledge for his motivation and kind support.

REFERENCES

Ackora-Prah, J., Acheson, V., Owusu-Ansah, E., & Nkrumah, S. K. (2023). A proposed method for finding initial solutions to transportation problems. *Pakistan Journal of Statistics and Operation Research, 19*(1), 63–75. doi:10.18187/pjsor.v19i1.4196

Ahmed, A. A., & Ahmed, A. A. (2020). A comparative study between new proposed Technique (LPAM) and the other existing techniques to Find IBFS for transportation problem. *Journal of Pure & Applied Sciences, 19*(5), 40–45.

Akbar, Y. R., Zain, I., Septianingsih, R., & Nuraini, P. (2023). The transportation method for efficient cost of shipping goods. *International Journal of Economics, Management, Business, and Social Science, 3*(1), 67–76.

Akpofure, N., & Jude, O. (2022). An explicit and efficient self-developed algorithm for IBFS of a transportation problem. *Asian Journal of Multidimensional Research, 11*(11), 58–72.

Ali, A., Ahmadini, H., & Ahmad, F. (2021). Solving intuitionistic fuzzy multiobjective linear programming problem under neutrosophic environment. *AIMS Mathematics, 6*(5), 4556–4580. doi:10.3934/math.2021269

Angelov, P. P. (1997). Optimization in an intuitionistic fuzzy environment. *Fuzzy Sets and Systems, 86*(3), 299–306.

Atanassov, K.T. (1983). *Intuitionistic fuzzy sets, VII ITKR's Session, Sofia, June 1983*. Deposed in Central Sci. - Techn. Library of Bulg. Acad. of Sci., 1697/84.

Atanassov, K. T. (1995). Ideas for intuitionistic fuzzy equations, inequalities and optimization. *Notes on Intuitionistic Fuzzy Sets, 1*(1), 17–24.

Atanassov, K.T. (1999). Intuitionistic Fuzzy Sets. *Studies in Fuzziness and Soft Computing*. doi:10.1007/978-3-7908-1870-3

Bama, C. S., & Sumathi, P. (2022, November). A new proposal for getting transportation cost using redundancy. In. AIP Conference Proceedings: Vol. 2516. *No. 1* (p. 320011). AIP Publishing LLC. doi:10.1063/5.0108708

Ban, A. (2008). Trapezoidal approximations of intuitionistic fuzzy numbers expressed by value, ambiguity, width and weighted expected value. *Notes on Intuitionistic Fuzzy Sets, 14*(1), 38–47.

Bharati, S. K. (2019). Detailed study of uncertainty and hesitation in transportation problem. *Life Cycle Reliability and Safety Engineering, 8*(4), 357–364. doi:10.1007/s41872-019-00095-y

Burillo, P., Bustince, H., & Mohedano, V. (1994). Some definitions of intuitionistic fuzzy number first properties. *Proceedings of the First Workshop on Fuzzy Based Expert System*, 53-55.

Charnes, A., & Cooper, W. W. (1954). The stepping- stone method for explaining linear programming calculation in transportation problem. *Management Science*, *1*(1), 49–69.

Chhibber, D., Srivastava, P. K., & Bisht, D. C. (2019). Average duo triangle ranking technique to solve fully and type-2 intuitionistic fuzzy transportation problem. *Nonlinear Studies*, *26*(3), 487–504.

Chopra, R., & Saxena, R. R. (2019). Linear programming problems with coefficients as trapezoidal intuitionistic fuzzy numbers. *Advances in Fuzzy Mathematics*, *14*(1), 41–52.

Chutia, R. (2021). Ordering intuitionistic fuzzy numbers by a convex combination of values and multiple of ambiguity inclusion functions with ambiguities of membership and nonmembership functions. *International Journal of Intelligent Systems*, *36*(10), 5785–5815. doi:10.1002/int.22530

Chutia, R., Saikia, S., & Gogoi, M. K. (2021). A theoretical approach to ranking of parametric fuzzy numbers using value and left–right ambiguity. *Mathematical Sciences*, 1–17. doi:10.1007/s40096-021-00422-4

Crama, Y., & Pironet, T. (2019). Vehicle allocation problem with uncertain transportation requests over a multi-period rolling horizon. *Logistics Research*, *12*(1), 1–22. doi:10.23773/2019_1

Dantzig, G. B. (1963). *Linear programming and extensions*. Princeton Univ. Press.

Darehmiraki, M. (2019). A novel parametric ranking method for intuitionistic fuzzy numbers. *Iranian Journal of Fuzzy Systems*, *16*(1), 129–143. doi:10.22111/IJFS.2019.4489

Das, S. K., Mandal, T., & Behera, D. (2019). A new approach for solving fully fuzzy linear programming problem. *International Journal of Mathematics in Operational Research*, *15*(3), 296–309. doi:10.1504/ijmor.2019.102074

Deshmukh, A., Mhaske, A. S., Jadhav, A., & Bondar, K. L. (2022). New methods for ranking of trapezoidal fuzzy to obtained optimum solution to fuzzy transportation problem. *Jisuanji Jicheng Zhizao Xitong*, *28*(11), 257–264.

Dinagar, D. S., & Keerthivasan, R. (2020). Solving transportation problem with modern zero suffix method under fuzzy environment. In AIP Conference Proceedings (Vol. 2261, No. 1, p. 030060). AIP Publishing LLC. doi:10.1063/5.0016998

Dutta, P., & Saikia, B. (2019). Arithmetic operations on normal semi elliptic intuitionistic fuzzy numbers and their application in decision-making. *Granular Computing*, 1-17. doi:10.1007/s41066-019-00175-5

Dymova, L. (2020). A numerical solution of fully fuzzy distribution problem. In *International Conference on Artificial Intelligence and Soft Computing* (pp. 305-314). Springer. doi:10.1007/978-3-030-61401-0_29

El Alaoui, M. (2020). Intuitionistic fully fuzzy balanced transportation problem. *Notes on Intuitionistic Fuzzy Sets*, *26*(1), 69–80. doi:10.7546/nifs.2020.26.1.69-80

El Alaoui, M. (2021). Type 2 fully fuzzy linear programming. *International Journal of Fuzzy System Applications*, *10*(4), 37–56. doi:10.4018/ijfsa.2021100103

Elaibi, W. M., & Hasanain, W. S. (2022). Optimality criterion for three proposed methods to unbalanced transportation models. *Italian Journal of Pure and Applied Mathematics, 48*, 67-79.

Emam, O., Haggag, R. M. Y., & Nabil, N. (2022). Using a fuzzy genetic algorithm for solving transportation logistics problems. *Journal of Business and Management Studies*, *4*(4), 195–211. doi:10.32996/jbms.2022.4.4.19

Ghanbari, R., Ghorbani-Moghadam, K., Mahdavi-Amiri, N., & De Baets, B. (2019). Fuzzy linear programming problems: Models and solutions. *Soft Computing*, 1–31. doi:10.1007/s00500-019-04519-w

Giri, P. K., Maiti, M. K., & Maiti, M. (2023). *Profit maximization fuzzy 4D-TP with budget constraint for breakable substitute items: a swarm based optimization approach.* OPSEARCH. doi:10.1007/s12597-023-00621-8

Grzegorzewski, P. (2003a). Distance and orderings in a family of intuitionistic fuzzy numbers. *Proceedings of the Third Conference of the European Society for Fuzzy Logic and Technology*, 223-227.

Grzegorzewski, P. (2003b). Intuitionistic fuzzy numbers. *Proceedings of the IFSA 2003 World Congress.*

Guha, D., & Chakraborty, D. (2010). A theoretical development of distance measure for intuitionistic fuzzy numbers. *International Journal of Mathematics and Mathematical Sciences.*

Haque, S., Bhurjee, A. K., & Kumar, P. (2022). Multi-objective non-linear solid transportation problem with fixed charge, budget constraints under uncertain environments. *Systems Science & Control Engineering*, *10*(1), 899–909. doi:10.1080/21642583.2022.2137707

Hitchcock, F. L. (1941). The distribution of a product from several sources to numerous localities. *Journal of Mathematics and Physics*, *20*(2), 224–230. doi:10.1002/sapm1941201224

Hussain, R. J., & Kumar, P. S. (2012a). The transportation problem in an intuitionistic fuzzy environment. *International Journal of Mathematics Research*, *4*(4), 411–420.

Hussain, R. J., & Kumar, P. S. (2012b). Algorithmic approach for solving intuitionistic fuzzy transportation problem. *Applied Mathematical Sciences*, *6*(77-80), 3981–3989.

Hussain, R. J., & Kumar, P. S. (2012c). The transportation problem with the aid of triangular intuitionistic fuzzy numbers. *Proceedings in International Conference on Mathematical Modeling and Applied Soft Computing (MMASC-2012)*, 819-825.

Hussain, R. J., & Kumar, P. S. (2013). An optimal more-for-less solution of mixed constraints intuitionistic fuzzy transportation problems. *International Journal of Contemporary Mathematical Sciences*, *8*(12), 565–576. doi:10.12988/ijcms.2013.13056

Iqbal, S., & Yaqoob, N. (2023). Ranking of linear Diophantine fuzzy numbers using circumcenter of centroids. *AIMS Mathematics*, *8*(4), 9840–9861. doi:10.3934/math.2023497

Jain, V., Sethi, P., Verma, R., Arya, S., & Chawla, C. (2020). An analytical solution of transportation problem from modified distribution (MODI) approach. *Wesleyan Journal of Research*, *13*(16), 43–55.

Jansi Rani, J., Dhanasekar, S., Micheal, D. R., & Manivannan, A. (2023). On solving fully intuitionistic fuzzy transportation problem via branch and bound technique. *Journal of Intelligent & Fuzzy Systems,* 1–11. doi:10.3233/JIFS-221345

Jansi Rani, J., Manivannan, A., & Dhanasekar, S. (2023). Interval valued intuitionistic fuzzy diagonal optimal algorithm to solve transportation problems. *International Journal of Fuzzy Systems*, 1–15. doi:10.1007/s40815-022-01446-1

Javanmard, M., & Nehi, H. M. (2019). A solving method for fuzzy linear programming problem with interval type-2 fuzzy numbers. *International Journal of Fuzzy Systems*, *21*(3), 882–891. doi:10.1007/s40815-018-0591-3

Kalonda, P. O., & Omekanda, A. M. (2020). *Kruskal's Algorithm, Vogel's Approximation and Modified Distribution Methods for the Design of Optimal Electrical Networks in the Democratic Republic of Congo. In 2020 IEEE PES/IAS PowerAfrica*. IEEE. doi:10.1109/PowerAfrica49420.2020.9219801

Kalyani, S., & Nagarani, S. (2020). A fully fuzzy transportation problem with hexagonal fuzzy number. In AIP Conference Proceedings (Vol. 2261, No. 1, p. 030078). AIP Publishing LLC. doi:10.1063/5.0016903

Kaur, A., Kacprzyk, J., & Kumar, A. (2020). New methods for solving the fully fuzzy transportation problems with the LR flat fuzzy numbers. In *Fuzzy Transportation and Transshipment Problems* (pp. 81–101). Springer. doi:10.1007/978-3-030-26676-9_5

Khalifa, A. E. W., Kumar, P., & Alharbi, M. G. (2020). Enhancement of capacitated transportation problem in fuzzy environment. *Advances in Fuzzy Systems*, *2020*, 1–9. doi:10.1155/2020/8893976

Koopmans, T. C. (1949). Optimum utilization of the transportation system. *Econometrica, 17*, 136-146. http://www.jstor.org/stable/1907301

Kumar, P. S. (2010). *A comparative study on transportation problem in fuzzy environment* [M.Phil thesis]. Jamal Mohamed College, Tiruchirappalli, India.

Kumar, P. S. (2016a). PSK method for solving type-1 and type-3 fuzzy transportation problems. *International Journal of Fuzzy System Applications*, *5*(4), 121–146. doi:10.4018/IJFSA.2016100106

Kumar, P. S. (2016b). A simple method for solving type-2 and type-4 fuzzy transportation problems. *International Journal of Fuzzy Logic and Intelligent Systems*, *16*(4), 225–237. doi:10.5391/IJFIS.2016.16.4.225

Kumar, P.S. (2017a). PSK method for solving type-1 and type-3 fuzzy transportation problems. *Fuzzy Systems,* 367–392. doi:10.4018/978-1-5225-1908-9.ch017

Kumar, P. S. (2017b). *Algorithmic approach for solving allocation problems under intuitionistic fuzzy environment* [PhD thesis]. Jamal Mohamed College, Tiruchirappalli, India.

Kumar, P. S. (2018a). PSK method for solving intuitionistic fuzzy solid transportation problems. *International Journal of Fuzzy System Applications*, *7*(4), 62–99. doi:10.4018/IJFSA.2018100104

Kumar, P. S. (2018b). A note on 'a new approach for solving intuitionistic fuzzy transportation problem of type-2'. *International Journal of Logistics Systems and Management*, *29*(1), 102–129. doi:10.1504/IJLSM.2018.088586

Kumar, P. S. (2018c). Linear programming approach for solving balanced and unbalanced intuitionistic fuzzy transportation problems. *International Journal of Operations Research and Information Systems*, *9*(2), 73–100. doi:10.4018/IJORIS.2018040104

Kumar, P. S. (2018d). Search for an optimal solution to vague traffic problems using the PSK method. In *Handbook of Research on Investigations in Artificial Life Research and Development* (pp. 219–257). IGI Global. doi:10.4018/978-1-5225-5396-0.ch011

Kumar, P. S. (2018e). A simple and efficient algorithm for solving type-1 intuitionistic fuzzy solid transportation problems. *International Journal of Operations Research and Information Systems*, *9*(3), 90–122. doi:10.4018/IJORIS.2018070105

Kumar, P. S. (2019a). PSK method for solving mixed and type-4 intuitionistic fuzzy solid transportation problems. *International Journal of Operations Research and Information Systems*, *10*(2), 20–53. doi:10.4018/ijoris.2019040102

Kumar, P. S. (2019b). Intuitionistic fuzzy solid assignment problems: A software-based approach. *International Journal of System Assurance Engineering and Management*, *10*(4), 661–675. doi:10.1007/s13198-019-00794-w

Kumar, P. S. (2020a). Intuitionistic fuzzy zero point method for solving type-2 intuitionistic fuzzy transportation problem. *International Journal of Operational Research*, *37*(3), 418–451. doi:10.1504/IJOR.2020.105446

Kumar, P. S. (2020b). The PSK method for solving fully intuitionistic fuzzy assignment problems with some software tools. In *Theoretical and Applied Mathematics in International Business* (pp. 149–202). IGI Global. doi:10.4018/978-1-5225-8458-2.ch009

Kumar, P. S. (2020c). Developing a new approach to solve solid assignment problems under intuitionistic fuzzy environment. *International Journal of Fuzzy System Applications*, *9*(1), 1–34. doi:10.4018/IJFSA.2020010101

Kumar, P. S. (2020d). Algorithms for solving the optimization problems using fuzzy and intuitionistic fuzzy set. *International Journal of System Assurance Engineering and Management*, *11*(1), 189–222. doi:10.1007/s13198-019-00941-3

Kumar, P. S. (2021). Finding the solution of balanced and unbalanced intuitionistic fuzzy transportation problems by using different methods with some software packages. In Handbook of Research on Applied AI for International Business and Marketing Applications (pp. 278-320). doi:10.4018/978-1-7998-5077-9.ch015

Kumar, P. S. (2022). Computationally simple and efficient method for solving real-life mixed intuitionistic fuzzy 3D assignment problems. *International Journal of Software Science and Computational Intelligence*. Advance online publication. doi:10.4018/ijssci.291715

Kumar, P. S. (2023a). The PSK method: A new and efficient approach to solving fuzzy transportation problems. In *Transport and Logistics Planning and Optimization* (pp. 149–197). IGI Global. doi:10.4018/978-1-6684-8474-6.ch007

Kumar, P. S. (2023b). The theory and applications of the software-based PSK method for solving intuitionistic fuzzy solid transportation problems. In M. Habib (Ed.), *Perspectives and Considerations on the Evolution of Smart Systems* (pp. 137–186). IGI Global. doi:10.4018/978-1-6684-7684-0.ch007

Kumar, P. S. (2024a). AI-Driven Decision Support System for Intuitionistic Fuzzy Assignment Problems. In T. Nguyen & N. Vo (Eds.), *Using Traditional Design Methods to Enhance AI-Driven Decision Making* (pp. 352–398). IGI Global. doi:10.4018/979-8-3693-0639-0.ch016

Kumar, P. S. (2024b). Theory and applications of the software-based PSK method for solving intuitionistic fuzzy solid assignment problems. In *Applications of New Technology in Operations and Supply Chain Management*. IGI Global.

Kumar, P. S., & Hussain, R. J. (2014a). A systematic approach for solving mixed intuitionistic fuzzy transportation problems. *International Journal of Pure and Applied Mathematics*, *92*(2), 181–190. doi:10.12732/IJPAM.V92I2.4

Kumar, P. S., & Hussain, R. J. (2014b). A method for finding an optimal solution of an assignment problem under mixed intuitionistic fuzzy environment. In *Proceedings in International Conference on Mathematical Sciences (ICMS-2014)* (pp. 417-421). Academic Press.

Kumar, P. S., & Hussain, R. J. (2014c). New algorithm for solving mixed intuitionistic fuzzy assignment problem. *Elixir Appl. Math*, *73*, 25971–25977.

Kumar, P. S., & Hussain, R. J. (2015). A method for solving unbalanced intuitionistic fuzzy transportation problems. *Notes on Intuitionistic Fuzzy Sets*, *21*(3), 54–65.

Kumar, P. S., & Hussain, R. J. (2016a). Computationally simple approach for solving fully intuitionistic fuzzy real life transportation problems. *International Journal of System Assurance Engineering and Management, 7*(S1), 90-101. doi:10.1007/s13198-014-0334-2

Kumar, P. S., & Hussain, R. J. (2016b). A simple method for solving fully intuitionistic fuzzy real life assignment problem. *International Journal of Operations Research and Information Systems*, *7*(2), 39–61. doi:10.4018/IJORIS.2016040103

Kumar, P. S., & Hussain, R. J. (2016c). An algorithm for solving unbalanced intuitionistic fuzzy assignment problem using triangular intuitionistic fuzzy number. *The Journal of Fuzzy Mathematics*, *24*(2), 289–302.

Kumari, N. (2020). Zero accomplishment method for finding an optimal more-for-less solution of transportation problem with mixed constraints. *International Journal of Mathematics Trends and Technology*, *66*(8), 144–149.

Li, D. F., Nan, J. X., & Zhang, M. J. (2010). A ranking method of triangular intuitionistic fuzzy numbers and application to decision making. *International Journal of Computational Intelligence Systems*, *3*(5), 522–530. doi:10.1080/18756891.2010.9727719

Mahapatra, G. S., & Roy, T. K. (2009). Reliability evaluations using triangular intuitionistic fuzzy numbers, arithmetic operations. *International Scholarly and Scientific Research & Innovation*, *3*(2), 422–429.

Mahapatra, G. S., & Roy, T. K. (2013). Intuitionistic fuzzy number and its arithmetic operation with application on system failure. *Journal of Uncertain Systems*, *7*(2), 92–107.

Mahmoudi, F., & Nasseri, S. H. (2019). A new approach to solve fully fuzzy linear programming problem. *Journal of Applied Research on Industrial Engineering*, *6*(2), 139–149. doi:10.22105/JARIE.2019.183391.1090

Malini, P. (2019). A new ranking technique on heptagonal fuzzy numbers to solve fuzzy transportation problem. *International Journal of Mathematics in Operational Research*, *15*(3), 364–371. doi:10.1504/IJMOR.2019.102078

Mandal, M. K., Kumar, A., & Burnwal, A. P. (2020). A fuzzy approach for with quadratic from of achievement function for transportation problem. *Journal of Information and Computational Science*, *10*(9), 316–326.

Mathur, N., & Srivastava, P. K. (2020). An inventive approach to optimize fuzzy transportation problem. *International Journal of Mathematical, Engineering and Management Sciences*, *5*(5), 985–994. doi:10.33889/IJMEMS.2020.5.5.075

Mishra, M., Panda, D., & Hota, S. (2020). Unbalanced transportation problems with multiple fuzzy goals. *Materials Today: Proceedings*. Advance online publication. doi:10.1016/j.matpr.2020.09.362

Mitchell, H. B. (2004). Ranking intuitionistic fuzzy numbers. *International Journal of Uncertainty, Fuzziness and Knowledge-based Systems*, *12*(3), 377–386.

Mohammed, B. F. (2023). (Tc) Technique for finding optimal solution to transportation problem. *Mathematical Theory and Modeling*, *13*(1), 30–38.

Mohideen, S. I., & Kumar, P. S. (2010a). A comparative study on transportation problem in fuzzy environment. *International Journal of Mathematics Research*, *2*(1), 151–158.

Mohideen, S. I., & Kumar, P. S. (2010b). A comparative study on transportation problem in fuzzy environment. *International Conference on Emerging Trends in Mathematics and Computer Applications (ICETMCA2010)*, 226-229.

Nehi, H. M. (2010). A new ranking method for intuitionistic fuzzy numbers. *International Journal of Fuzzy Systems*, *12*(1), 80–86.

Ngastiti, P. T. B., & Surarso, B. (2018). Zero point and zero suffix methods with robust ranking for solving fully fuzzy transportation problems. *Journal of Physics: Conference Series*, *1022*(1), 012005.

Niksirat, M. (2022). A new approach to solve fully fuzzy multi-objective transportation problem. *Fuzzy Information and Engineering*, *14*(4), 456–467. doi:10.1080/16168658.2022.2152836

Ozkok, B. A. (2019). Finding fuzzy optimal and approximate fuzzy optimal solutionof fully fuzzy linear programming problems with trapezoidal fuzzy numbers. *Journal of Intelligent & Fuzzy Systems*, *36*(2), 1389–1400. doi:10.3233/JIFS-18016

Pandian, P. (2014). Realistic method for solving fully intuitionistic fuzzy transportation problems. [AMS]. *Applied Mathematical Sciences*, *8*(113), 5633–5639.

Pourofoghi, F., Salokolaei, D. D., & Abraham, A. (2021, August). Designing transportation problem under grey linear programming based on sensitivity analysis. In *International Conference on Intelligent and Fuzzy Systems* (pp. 666-675). Springer. doi:10.1007/978-3-030-85626-7_78

Priyadharshini, P., & Anju, A. (2020). Type-2 duality trapezoidal fuzzy fractional transportation problem using fuzzy optimization techniques. In AIP Conference Proceedings (Vol. 2261, No. 1, p. 030058). AIP Publishing LLC. doi:10.1063/5.0016927

Raj, M. E. A., Sivaraman, G., & Vishnukumar, P. (2022). A novel kind of arithmetic operations on trapezoidal fuzzy numbers and its applications to optimize the transportation cost. *International Journal of Fuzzy Systems*, 1–8. doi:10.1007/s40815-022-01425-6

Rochayata, K. S. B., Widodasih, R. W. K., & Gunardi, H. (2022). Optimization of transportation costs with the implementation of the Milk-Run method. *Enrichment: Journal of Management*, *12*(5), 3662–3669.

Sam'an, M., Dasril, Y., & Bujang, N. B., & Farikhin. (2022). Improved total difference method (ITDM): A new approach to solving transportation problem based on modifications of total difference method 1 and integration of total ratio cost matrix. *International Journal of Computing Science and Mathematics*, *16*(1), 13–23. doi:10.1504/IJCSM.2022.10051689

Sam'an, M., Hariyanto, S., & Surarso, B. (2018). Optimal solution of full fuzzy transportation problems using total integral ranking. *Journal of Physics: Conference Series*, *983*(1), 012075.

Sam'an, M., Surarso, B., & Irwanto, B. (2020). Solving of fuzzy transportation problem using fuzzy analytical hierarchy process (AHP). In *The 2nd International Seminar on Science and Technology (ISSTEC 2019)* (pp. 10-15). Atlantis Press.

Samuel, A. E. (2012). Improved zero point method (IZPM) for the transportation problems. *Applied Mathematical Sciences*, *6*(109), 5421–5426.

Sarma, D., Das, A., & Bera, U. K. (2020). Generalized type-2 intuitionistic fuzzy approaches for allocation and redistribution of resources in the disaster operation. *Studies in Computational Intelligence*, *863*, 327–341. doi:10.1007/978-3-030-34152-7_25

Shabani, A., & Jamkhaneh, E. B. (2014). A new generalized intuitionistic fuzzy number. *Journal of Fuzzy Set Valued Analysis*, *24*, 1–10. doi:10.5899/2014/JFSVA-00199

Sharma, M. K. (2020). Zero-point maximum allocation method for solving intuitionistic fuzzy transportation problem. *International Journal of Applied and Computational Mathematics*, *6*(4), 1–11. doi:10.1007/s40819-020-00867-6

Shaw, A. K., & Roy, T. K. (2012). Some arithmetic operations on triangular intuitionistic fuzzy number and its application on reliability evaluation. *International Journal of Fuzzy Mathematics and Systems*, *2*(4), 363–382.

Sonia, N. A., & Chidambaran, S. (2023). Time paradox in transportation problems. *International Journal of Operational Research*, *45*(4), 540–552. doi:10.1504/IJOR.2022.128397

Srivastava, P. K., & Bisht, D. C. (2019). An efficient fuzzy minimum demand supply approach to solve fully fuzzy transportation problem. *Mathematics in Engineering, Science and Aerospace*, *10*(2), 253–269.

Suneela, S., & Chakraverty, S. (2019). New ranking function for fuzzy linear programming problem and system of linear equations. *Journal of Information and Optimization Sciences*, *40*(1), 141–156. do i:10.1080/02522667.2018.1450196

Tarmudi, Z., & Rahman, N. A. (2020). Diverse ranking approach in MCDM based on trapezoidal intuitionistic fuzzy numbers. In *Proceedings of the Tenth International Conference on Soft Computing and Pattern Recognition (SoCPaR 2018).* Springer. doi:10.1007/978-3-030-17065-3_2

Thanoon, F. H. (2022). Proposed method for solving transportation problems. *Wasit Journal of Pure Sciences, 1*(3), 9-14. doi:10.31185/WJPS.62

Traneva, V., & Tranev, S. (2020). Intuitionistic fuzzy transportation problem by zero point method. *Proceedings of the 2020 Federated Conference on Computer Science and Information Systems, 21*, 349–358. doi:10.15439/2020F61

Traneva, V., & Tranev, S. (2021). An intuitionistic fuzzy zero suffix method for solving the transportation problem. In I. Dimov & S. Fidanova (Eds.), *Advances in High Performance Computing. HPC 2019. Studies in Computational Intelligence* (Vol. 902). Springer. doi:10.1007/978-3-030-55347-0_7

Varghese, A., & Kuriakose, S. (2012). Centroid of an intuitionistic fuzzy number. *Notes on Intuitionistic Fuzzy Sets*, *18*(1), 19–24.

Wang, G., & Peng, J. (2019). Fuzzy optimal solution of fuzzy number linear programming problems. *International Journal of Fuzzy Systems*, *21*(3), 865–881. doi:10.1007/s40815-018-0594-0

Ye, X., Han, S., & Lin, A. (2010). A note on the connection between the Primal-Dual and the A* algorithm. *International Journal of Operations Research and Information Systems*, *1*(1), 73–85. doi:10.4018/JORIS.2010101305

Ye, X., Han, S., & Lin, A. (2012). A note on the connection between the Primal-Dual and the A* algorithm. In J. Wang (Ed.), *Innovations in Information Systems for Business Functionality and Operations Management* (pp. 170–181). IGI Global. doi:10.4018/978-1-4666-0933-4.ch010

Zadeh, L. A. (1965). Fuzzy sets. *Information and Control*, *8*(3), 338–353. doi:10.1016/s0019-9958(65)90241-x

Zeeshan, A, M., Subash, A., & C S, A. (2022). Model employing the VAM-MODI method for risk management in the transportation of commodities. *International Journal of Research and Review*, *9*(11), 96–99. doi:10.52403/IJRR.20221116

APPENDIX

The first example's RStudio code and output are:

```
#Ctrl+L — Clear the Console
> library(lpSolve)
> C=c(19,28,31,11,17,16,12,15,13)
> CM=matrix(C,3,3, byrow=T)
> CM
## Not run:
     [,1] [,2] [,3]
[1,]   19   28   31
[2,]   11   17   16
[3,]   12   15   13
## End(Not run)
> row.signs <- rep ("=", 3)
> row.rhs <- c(18,12,4)
> col.signs <- rep ("=", 3)
> col.rhs <- c (8,16,10)
> require(lpSolve)
> sol=lp.transport(CM,direction = "min", row.signs, row.rhs, col.signs, col.
rhs)
> sol
## Not run: Success: the objective function is 682
> lp.transport(CM,direction = "min", row.signs, row.rhs, col.signs, col.
rhs)$solution
## Not run:
     [,1] [,2] [,3]
[1,]    8   10    0
[2,]    0    6    6
[3,]    0    0    4
## End (Not run)
```

Figure 1. The RStudio code and solution of Example 1

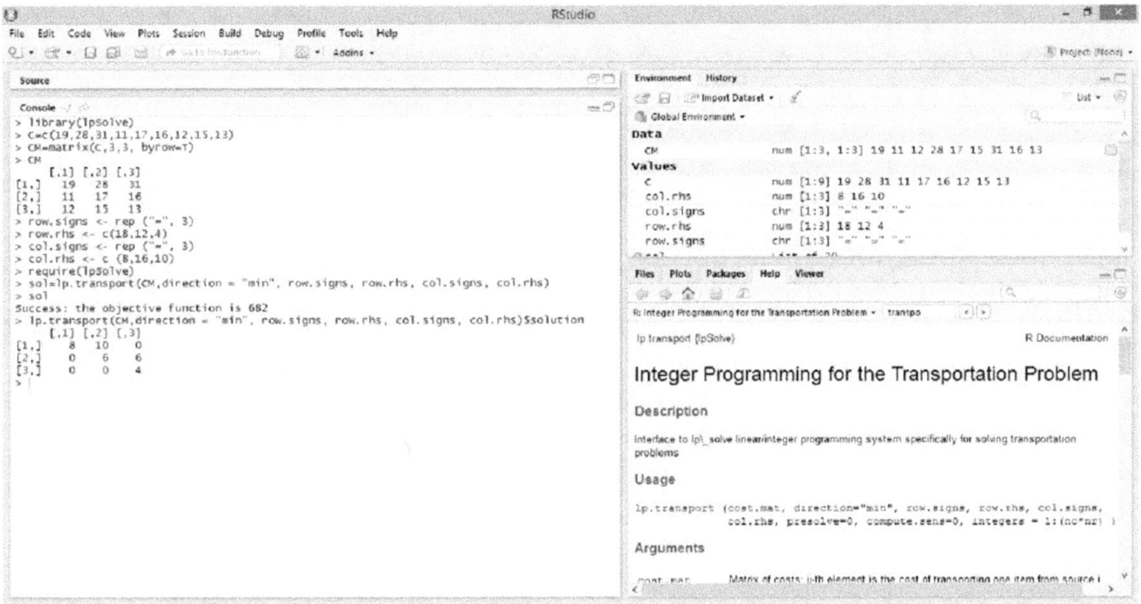

The first example's RGui code and output are:

```
> library(lpSolve)
> C=c(19,28,31,11,17,16,12,15,13)
> CM=matrix(C,3,3, byrow=T)
> CM
     [,1] [,2] [,3]
[1,]   19   28   31
[2,]   11   17   16
[3,]   12   15   13
> col.signs <- rep ("=", 3)
> row.signs <- rep ("=", 3)
> col.rhs <- c (8,16,10)
> row.rhs <- c(18,12,4)
> sol=lp.transport(CM,direction = "min", row.signs, row.rhs, col.signs, col.
rhs)
> sol
Success: the objective function is 682
> sol$solution
     [,1] [,2] [,3]
[1,]    8   10    0
[2,]    0    6    6
[3,]    0    0    4
```

Figure 2. The RGui code and solution of Example 1

The first example's MATLAB code and output are:

```
%% clean screen ('clc' stands for 'clean screen').
%% The syntax 'clc' is used to clear all input and output from the command
window in MATLAB.
>> clc;
%% Coefficient of the objective function.
>> f=[19 28 31 11 17 16 12 15 13];
%% Left hand-side (LHS) coefficient of the < or = constraints.
>> A=[];
%% Right-hand side (RHS) value of the < or = constraints.
>> b=[];
%% LHS coefficient of the equality constraints.
>> Aeq= [1 1 1 0 0 0 0 0 0;
         1 0 0 1 0 0 1 0 0;
         0 0 0 1 1 1 0 0 0;
         0 1 0 0 1 0 0 1 0;
         0 0 0 0 0 0 1 1 1;
         0 0 1 0 0 1 0 0 1];
%% RHS value of the equality constraints.
>> beq= [18 8 12 16 4 10];
%% Lower bound of the variable.
>> lb=[0 0 0 0 0 0 0 0 0];
```

```
%% Upper bound of the variable.
>> ub=[];
%% 'linprog' stands for 'linear programming problem (LPP)'. Also, this syntax
is used to solve the same LPP.
>> [y,f]= linprog(f,A,b,Aeq,beq,lb,ub);
Optimization terminated.
%% The syntax 'y.'' is used to displays the values of the decision variables.
>> y.'
ans =
    8.0000    10.0000     0.0000     0.0000     6.0000     6.0000     0.0000
0.0000    4.0000
%% The syntax 'f.'' is used to displays the objective value.
>> f.'
f =
  682.0000
```

Figure 3. The MATLAB code of Example 1

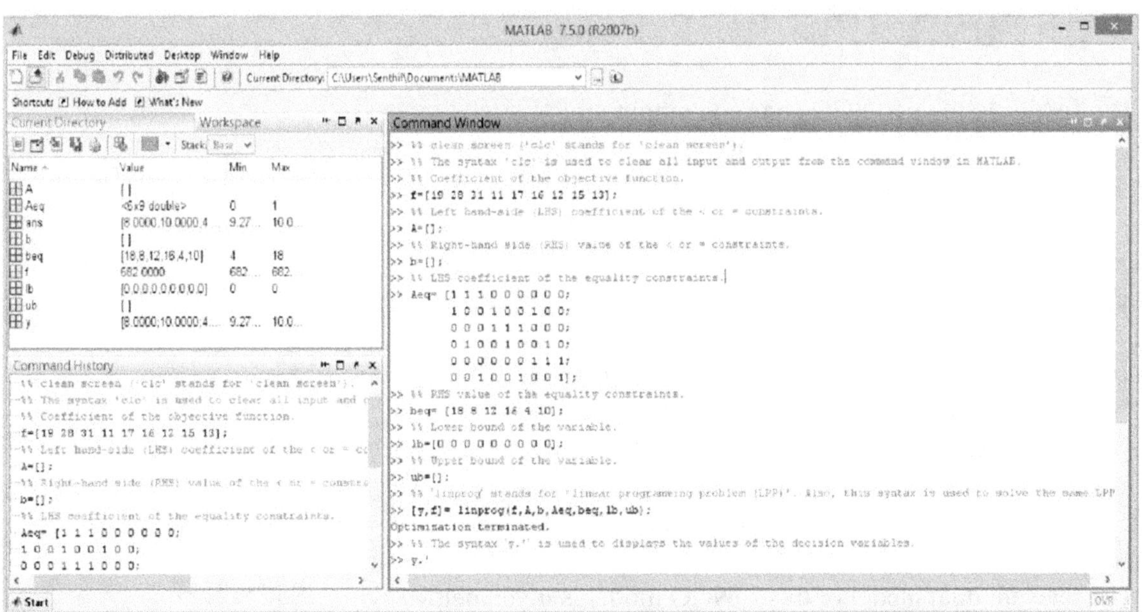

Figure 4. The MATLAB solution of Example 1

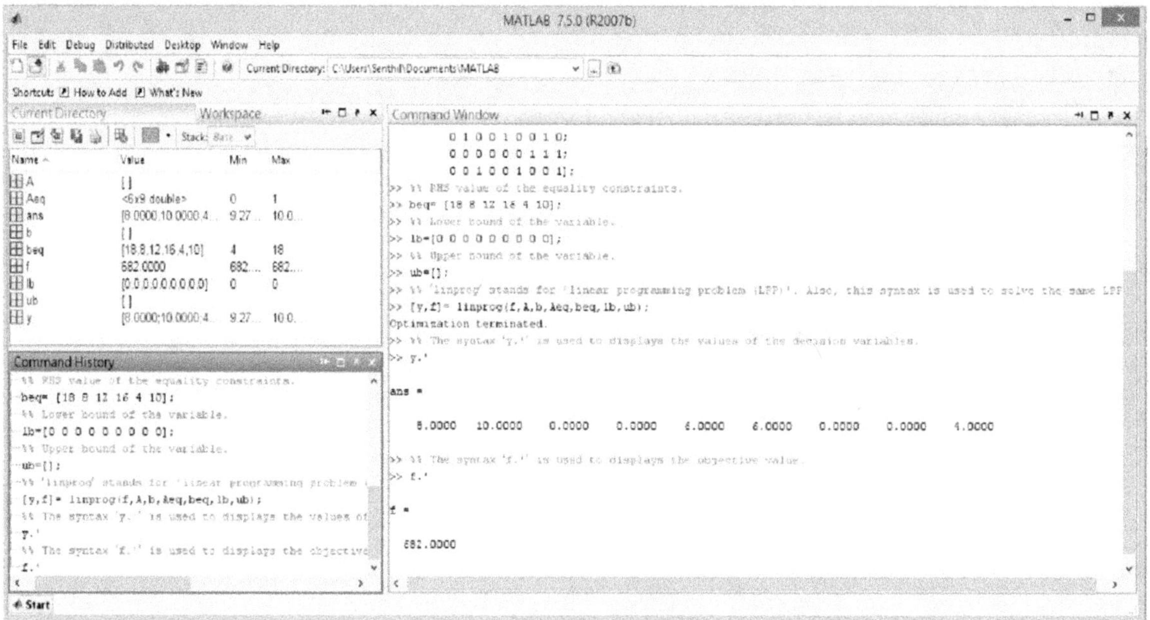

The first example's LINGO code and output are:

```
!Objective function;
Min=19*y11+28*y12+31*y13+11*y21+17*y22+16*y23+12*y31+15*y32+13*y33;
!s.t;
!Supply constraints (row restrictions);
y11+y12+y13= 18; y21+y22+y23=12; y31+y32+y33=4;
!Demand constraints (column restrictions);
y11+y21+y31= 8; y12+y22+y32=16; y13+y23+y33=10;
!Non-ve restrictions, that is, it shows that all decision variables should be
> or = zero;
y11>=0; y12>=0; y13>=0; y21>=0; y22>=0; y23>=0; y31>=0; y32>=0; y33>=0;
```

The relevant dual model of this LINGO model is given below.

```
MODEL:
  MAX= 18 * _2 + 12 * _3 + 4 * _4 + 8 * _5 + 16 * _6 + 10 * _7;
 [Y11] _2 + _5 + _8 <= 19;
 [Y12] _2 + _6 + _9 <= 28;
 [Y13] _2 + _7 + _10 <= 31;
 [Y21] _3 + _5 + _11 <= 11;
 [Y22] _3 + _6 + _12 <= 17;
 [Y23] _3 + _7 + _13 <= 16;
```

```
[Y31] _4 + _5 + _14 <= 12;
[Y32] _4 + _6 + _15 <= 15;
[Y33] _4 + _7 + _16 <= 13;
@FREE( _2); @FREE( _3); @FREE( _4); @FREE( _5); @FREE( _6); @FREE( _7);
END
```

Figure 5. Unpermuted matrix picture relevant to Example 1

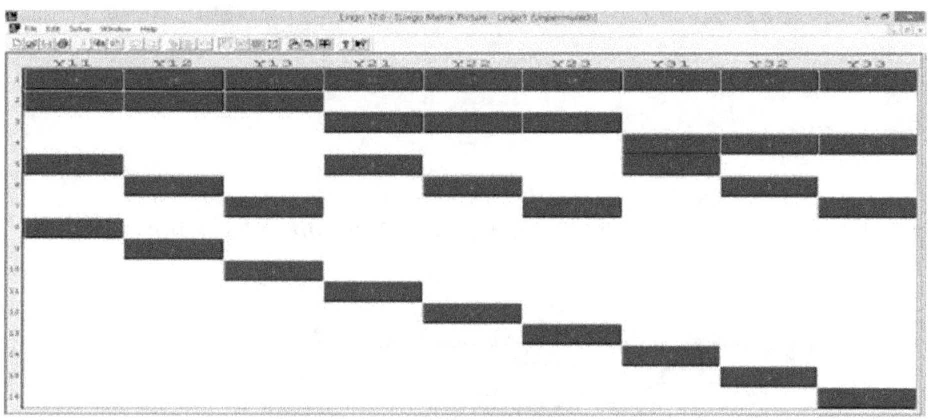

Figure 6. Lower triangular matrix picture relevant to Example 1

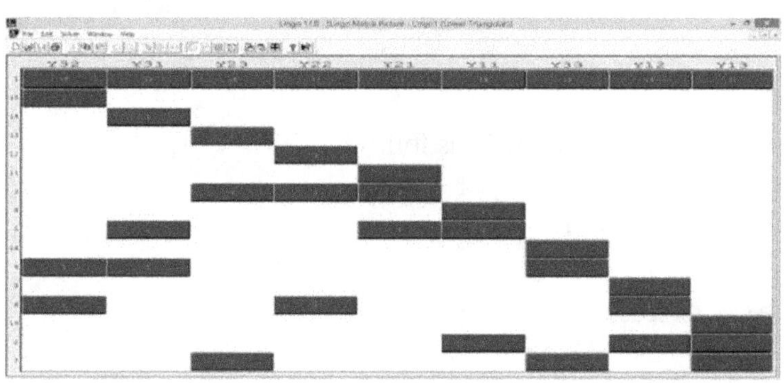

Figure 7. Block triangular (GP1-3 blocks) matrix picture relevant to Example 1

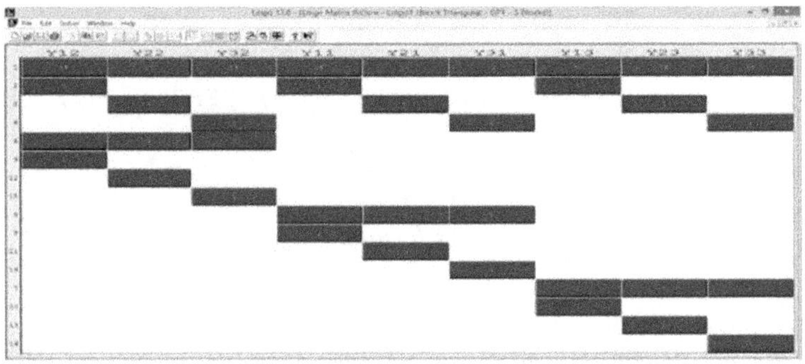

Figure 8. Block triangular (GP2-3 blocks) matrix picture relevant to Example 1

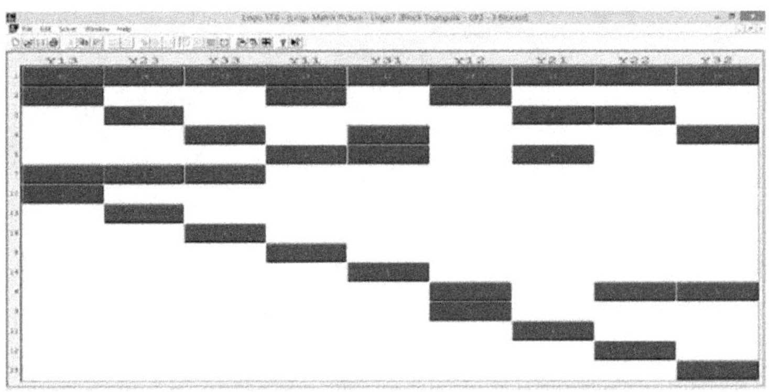

The LINGO output of the first example is as follows.

```
Global optimal solution found.
  Objective value:                          682.00
  Infeasibilities:                          0.00
  Total solver iterations:                  5
  Elapsed runtime seconds:                  0.03
  Model Class:                              LP
  Total variables:                         9
  Nonlinear variables:                     0
  Integer variables:                   0
  Total constraints:                       16
  Nonlinear constraints:               0
  Total non-zeros:                     36
  Nonlinear non-zeros:                 0
```

```
Variable     Value           Reduced Cost

Y11          8.00               0.00
Y12          10.0               0.00
Y13          0.00               4.00
Y21          0.00               3.00
Y22          6.00               0.00
Y23          6.00               0.00
Y31          0.00               7.00
Y32          0.00               1.00
Y33          4.00               0.00

Row    Slack or Surplus         Dual Price

1      682.00                     -1.00
2        0.00                    -19.00
3        0.00                     -8.00
4        0.00                     -5.00
5        0.00               0.00
6        0.00                     -9.00
7        0.00                     -8.00
8        8.00               0.00
9        10.0               0.00
10       0.00               0.00
11       0.00               0.00
12       6.00                    0.00
13       6.00                    0.00
14       0.00                    0.00
15       0.00                    0.00
16       4.00                    0.00
```

The LINGO model statistics of the first example is as follows.

```
Rows=        16 Vars=         9  ( all are linear)
  Integer vars=        0 Binary vars=        0
  Nonzeros=        42 Const nonz=        27(        27 are +- 1) Density= 0.263
  Smallest and largest elements in abs value=     1.00000         31.0000
  No. < :        0 No. =:        6 No. > :        9 Obj= MIN, GUBs <=        9
  Single cols=        0
```

The second example's RStudio code and output are:

```
#Ctrl+L – Clear the Console
library(lpSolve)
A=c(5,10,12,5,15,18,4,16,22,0,0,0)
AA=matrix(A,4,3, byrow=T)
AA
     [,1] [,2] [,3]
[1,]    5   10   12
[2,]    5   15   18
[3,]    4   16   22
[4,]    0    0    0

r=c(11,5,9,5)
c=c(7,13,10)
rs=c("=","=","=","=")
cs=c("=","=","=")
require(lpSolve)
sol=lp.transport(AA,direction = "min",rs,r,cs,c)
sol
Success: the objective function is 255
sol$solution

     [,1] [,2] [,3]
[1,]    0    6    5
[2,]    0    5    0
[3,]    7    2    0
[4,]    0    0    5
```

Figure 9. The RStudio code with solution of Example 2

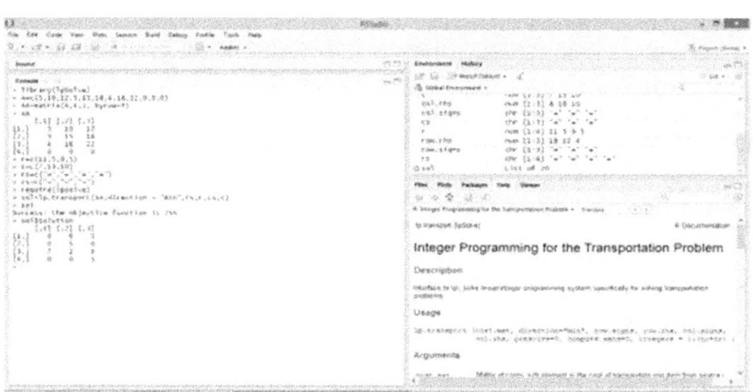

The second example's RGui code and output are:

```
> library(lpSolve)
> A=c(5,10,12,5,15,18,4,16,22,0,0,0)
> AA=matrix(A,4,3, byrow=T)
> AA
      [,1] [,2] [,3]
[1,]     5   10   12
[2,]     5   15   18
[3,]     4   16   22
[4,]     0    0    0
> r=c(11,5,9,5)
> rs=c("=","=","=","=")
> c=c(7,13,10)
> cs=c("=","=","=")
> sol=lp.transport(AA,direction = "min",rs,r,cs,c)
> sol
Success: the objective function is 255
> lp.transport(AA,direction = "min",rs,r,cs,c)$solution
 [,1] [,2] [,3]
[1,]     0    6    5
[2,]     0    5    0
[3,]     7    2    0
[4,]     0    0    5
```

Figure 10. The RGui code with solution of Example 2

The second example's MATLAB code and output are:

```
>> clear all;
>> f=[5 10 12 5 15 18 4 16 22 0 0 0];
>> A=[];
>> b=[];
>> Aeq= [1 1 1 0 0 0 0 0 0 0 0 0;
```

```
         0 0 0 1 1 1 0 0 0 0 0 0;
         0 0 0 0 0 0 1 1 1 0 0 0;
         0 0 0 0 0 0 0 0 0 1 1 1;
         1 0 0 1 0 0 1 0 0 1 0 0;
         0 1 0 0 1 0 0 1 0 0 1 0;
         0 0 1 0 0 1 0 0 1 0 0 1];
>> beq= [11 5 9 5 7 13 10];
>> lb=[0 0 0 0 0 0 0 0 0 0 0 0];
>> ub=[];
>> [y,f]= linprog(f,A,b,Aeq,beq,lb,ub);
Optimization terminated.
>> disp(y);
    0.0000
    6.0000
    5.0000
    0.0000
    5.0000
    0.0000
    7.0000
    2.0000
    0.0000
    0.0000
    0.0000
    5.0000
>> disp(f);
  255.0000
```

Figure 11. The MATLAB code with the solution of Example 2

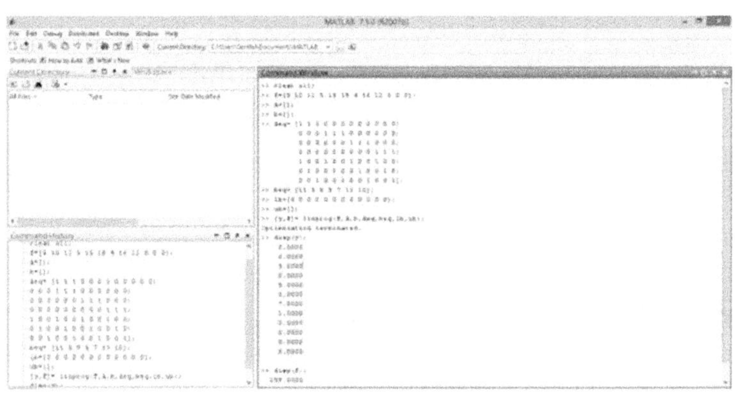

The LINGO model/code of the second example is as follows.

```
Min=5*y11+10*y12+12*y13+5*y21+15*y22+18*y23+4*y31+16*y32+22*y33+0*y41+0*y42+0
*y43;
y11+y12+y13=11;y21+y22+y23=5;y31+y32+y33=9;y41+y42+y43=5;
y11+y21+y31+y41=7;y12+y22+y32+y42=13;y13+y23+y33+y43=10;
y11>=0;y12>=0;y13>=0;y21>=0;y22>=0;y23>=0;y31>=0;y32>=0;y33>=0;y41>=0;y42>=0
;y43>=0;
```

The LINGO model statistics of the second example is as follows.

```
Rows=       20 Vars=       12   ( all are linear)
  Integer vars=        0 Binary vars=        0
  Nonzeros=        52 Const nonz=        36(        36 are +- 1) Density= 0.200
  Smallest and largest elements in abs value=     1.00000        22.0000
  No. < :         0 No. =:         7 No. > :        12 Obj= MIN, GUBs <=        12
  Single cols=        0
```

The relevant dual model of the above LINGO model is as follows.

```
MODEL:
  MAX= 11 * _2 + 5 * _3 + 9 * _4 + 5 * _5 + 7 * _6 + 13 *
       _7 + 10 * _8;
 [Y11] _2 + _6 + _9 <= 5;
 [Y12] _2 + _7 + _10 <= 10;
 [Y13] _2 + _8 + _11 <= 12;
 [Y21] _3 + _6 + _12 <= 5;
 [Y22] _3 + _7 + _13 <= 15;
 [Y23] _3 + _8 + _14 <= 18;
 [Y31] _4 + _6 + _15 <= 4;
 [Y32] _4 + _7 + _16 <= 16;
 [Y33] _4 + _8 + _17 <= 22;
 [Y41] _5 + _6 + _18 <= 0;
 [Y42] _5 + _7 + _19 <= 0;
 [Y43] _5 + _8 + _20 <= 0;
 @FREE( _2); @FREE( _3); @FREE( _4); @FREE( _5); @FREE( _6); @FREE( _7);
 @FREE( _8);
 END
```

Figure 12. The lingo input of Example 2

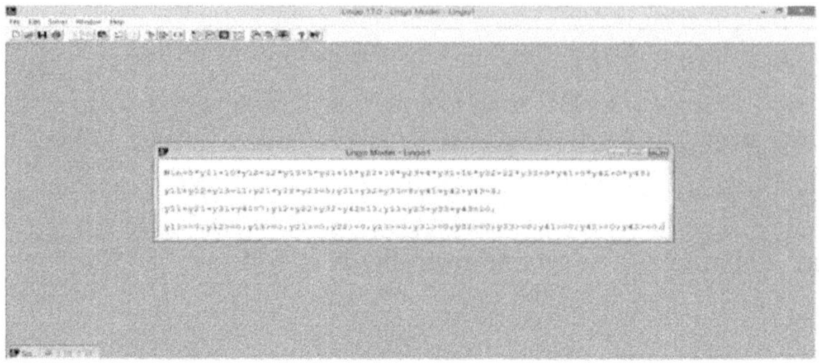

The LINGO output of example 2 is:

```
Global optimal solution found.
  Objective value:                              255.0000
  Infeasibilities:                              0.000000
  Total solver iterations:                             9
  Elapsed runtime seconds:                          0.03

  Model Class:                                        LP

  Total variables:                    12
  Nonlinear variables:                 0
  Integer variables:                   0

  Total constraints:                  20
  Nonlinear constraints:               0

  Total nonzeros:                     45
  Nonlinear nonzeros:                  0

Variable      Value           Reduced Cost

Y11        0.000000            7.000000
Y12        6.000000            0.000000
Y13        5.000000            0.000000
Y21        0.000000            2.000000
Y22        5.000000            0.000000
Y23        0.000000            1.000000
Y31        7.000000            0.000000
Y32        2.000000            0.000000
Y33        0.000000            4.000000
```

Y41	0.000000	14.00000
Y42	0.000000	2.000000
Y43	5.000000	0.000000

Row	Slack or Surplus	Dual Price
1	255.0000	-1.000000
2	0.000000	2.000000
3	0.000000	-3.000000
4	0.000000	-4.000000
5	0.000000	14.00000
6	0.000000	0.000000
7	0.000000	-12.00000
8	0.000000	-14.00000
9	0.000000	0.000000
10	6.000000	0.000000
11	5.000000	0.000000
12	0.000000	0.000000
13	5.000000	0.000000
14	0.000000	0.000000
15	7.000000	0.000000
16	2.000000	0.000000
17	0.000000	0.000000
18	0.000000	0.000000
19	0.000000	0.000000
20	5.000000	0.000000

Figure 13. The lingo solver status of Example 2

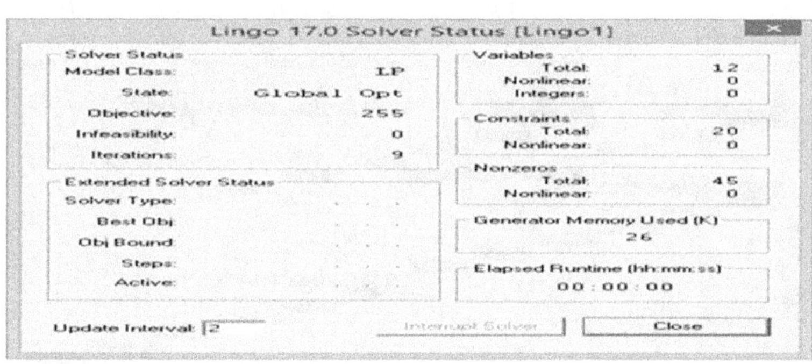

Figure 14. Unpermuted matrix picture relevant to Example 2

Figure 15. Lower triangular matrix picture relevant to Example 2

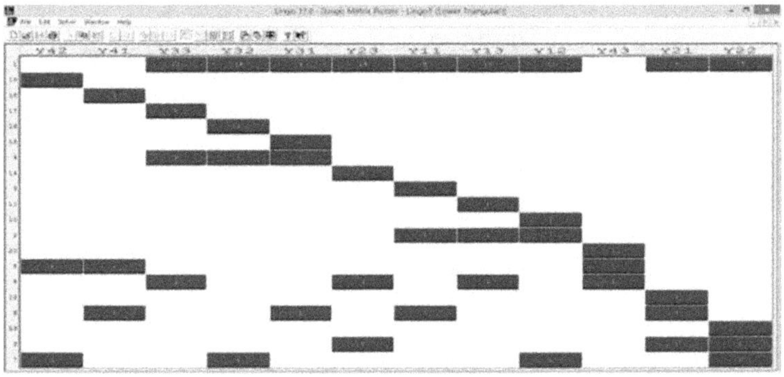

Figure 16. Block triangular (GP1-4 blocks) matrix picture relevant to Example 2

Figure 17. Block triangular (GP2-4 blocks) matrix picture relevant to Example 2

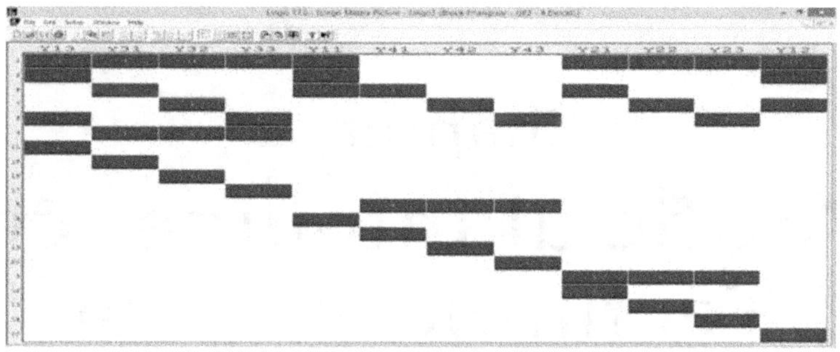

Figure 18. Graphical form of FIFTC

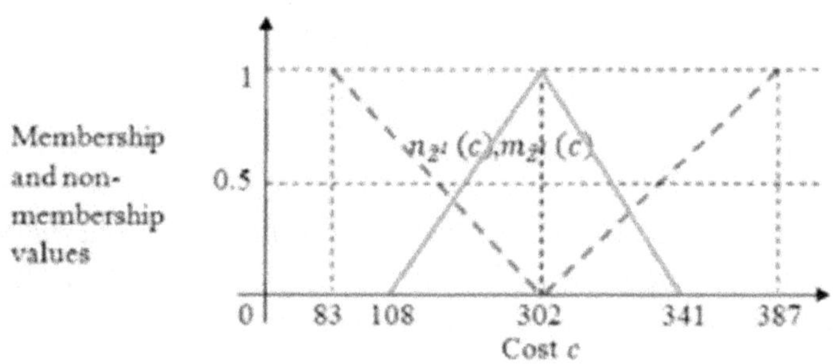

Chapter 2
Artificial Intelligence Technologies for Sustainability in Business

Yesim Can Saglam
https://orcid.org/0000-0003-0528-6135
Gebze Technical University, Turkey

ABSTRACT

In a fiercely competitive environment, modern technologies have attracted the tremendous attention of businesses for existence. As a consequence of this development, research areas including supply chain management, industrial engineering, operation management, and so on have been influenced by the effect of these cutting-edge technologies. Leading-edge technologies such as artificial intelligence are assisting to remodel the operational management and production systems of businesses. In the literature, increasing consensus and investigations exist that new technologies can provide a variety of advantages for performing and developing sustainability issues. However, empirical insights on how the advantages of artificial intelligence can enhance business sustainability (in terms of economic, environmental, and social sustainability) are relatively scarce. Therefore, this chapter investigates the advantages of artificial intelligence on business sustainability from the angle of three aspects, namely environmental, economic, and social sustainability.

INTRODUCTION

Due to the various significant problems caused by climate change, environmental sustainability has become a critical issue for business operations worldwide. Various environmental policies are intended to develop resource capacity as well as efficiency while limiting the utilization of non-beneficial proposals that might have an unfavorable effect on the environment (Kar et al., 2022). Endeavors to respond to difficulties of resource scarcity, economic benefits or environmental effects, or amalgamations of these have been done by industries, governments, and societies around the world (Lieder and Rashid, 2016). Besides regulatory pressure, various internal and external factors such as market changes, pressure from

DOI: 10.4018/978-1-6684-9130-0.ch002

stakeholders, and corporate strategies have driven businesses to increasingly measure and evaluate their environmental as well as social impacts. Most of these interests focus on adverse effects on the environment and society like quantifying the diminishing of material and energy utilization, or the avoidance of waste as well as pollution (Dijkstra-Silva et al., 2022). Researchers have focused their efforts on the enhancement of sustainable activities by adopting these three pillars of sustainability in recent years. They have unearthed that the enhancement of sustainable activities necessitates a balance of social as well as economic growth (Gupta et al., 2023).

In general, sustainability has been evaluated in three aspects namely, social, environmental, and economic sustainability (Yildiz Çankaya and Sezen, 2019; Kar et al., 2022; Gupta et al., 2023). Environmental sustainability describes the conservation of natural entities like climate stability, minerals, and many others which the human race cannot exist without of these. To conserve natural resources, the improvement of novice and sustainable technologies as well as strategies has a critical role. Social sustainability, on the other hand, can be evaluated as a tool for achieving environmental as well as economic sustainability purposes. Social sustainability comprises indirect actions like providing the economic and political privileges of communities or individuals embarking on societal activities such as constructing social awareness administration structures or developing higher standards for human life (Yusuf et al., 2013). Finally, economic sustainability is an essential subject for businesses and demonstrates that effective and efficient exploitation of resources is significant to prohibit inefficacious outcomes of resource utilization as well as safeguarding existence and profitability (Durmaz et al., 2021). To summarize from the general point of view, sustainability can be evaluated as meeting today's requirements without sacrificing the capability of future generations to meet their own requirements (Yildiz Çankaya and Sezen, 2019).

Artificial intelligence, which was first introduced by Dartmouth College in the UK in 1956, can be defined as "a system's ability to interpret external data correctly, to learn from such data, and to use those learnings to achieve specific goals and tasks through flexible adaption" (Kaplan and Haenlein, 2019, p.17). It has developed to enable machines to think and behave like humans. To realize its function, Artificial intelligence has intelligence, learning, and reasoning capabilities (Kar et al., 2022). To provide a general view of the subject of artificial intelligence, three core concepts of artificial intelligence should be explained: (1) neural networks, (2) machine learning; and (3) deep learning with further enhancements like data mining or natural language processing (Heins, 2022). Artificial intelligence technologies have revolutionized many industries from the side of various aspects (Perucica and Andjelkovic, 2022). It has attracted huge interest from enterprises and governments thanks to the advantages which enable the way various industries can plan the services they provide (Sciarretta et al., 2022). The World Economic Forum characterized artificial intelligence as the core point of the world's extant technological revolution since, with supplementary novel technologies, it has laid the foundations for business developments that enhance the supply chain, streamline the delivery of products as well as service and facilitate the generation of new business solutions (Heins, 2022). It is not a just only concept, but rather a set of techniques for computers to simulate human mentality. Thereby, computers can behave just like humans, such as problem-solving, learning, and reasoning (Sciarretta et al., 2022).

Artificial intelligence technology can be beneficial to pursue sustainability issues. It can develop decision-making processes in human vulnerability, climate change, and biodiversity loss applications. For instance, natural or controlled forest fires have a tremendous effect on the destruction of biodiversity. Monitoring and foreseeing forest fire risk has a critical role in preventing undesirable conditions. Artificial intelligence technologies have been utilized to support the forecast of forest fires. In this way, artificial intelligence technologies improved for pursuing and forecasting forest situations may also be utilized to assist in combat and control fires in advance of the disasters (Salas et al., 2022). According to researchers, although artificial intelligence has attracted considerable attention in various fields, scarce investigations have been conducted on the profound environmental impact of artificial intelligence (Perucica and Andjelkovic, 2022). Therefore, this chapter's purpose is to shed light on how artificial intelligent address and facilitate the issue of sustainability. It will provide an overview features of artificial intelligence and then discuss how artificial intelligence technologies can enhance the environmental, economic, and social performance of businesses.

ARTIFICIAL INTELLIGENCE

The digitalization of enterprise operations is significant for conventional businesses to compete in the digital economy epoch. In this sense, the adoption of developed applications and systems like Artificial Intelligence, concurrently with the improvement and maturity of associated digital abilities and competency, has been crucial for the digital transformation of the companies (Tsolakis et al., 2022). In 1956, artificial intelligence was first raised at the Dartmouth workshop promoted by the American mathematician John McCarthy as a scientific discipline. Up to the present, no universally approved description of it has been agreed upon (Perucica and Andjelkovic, 2022). Therefore, in the literature, many descriptions of artificial intelligence exist. Prevalently used definition of it as follows "Artificial Intelligence is the field of computer science dedicated to solving cognitive problems commonly associated with human intelligence, such as learning, problem-solving, and pattern recognition" (Goralski and Tan, 2022: p.2). The utilization of artificial intelligence has appeared as a novel and innovative road to understanding problems and creating a plausible solution with the utilization of language via a machine learning process (Bhagat et al., 2022).

Artificial intelligence is a powerful technology application in various fields that are progressively attaining popularity (Andeobu et al., 2022). It is divided into three main classifications: (1) Artificial Narrow Intelligence, which is described as weak, (2) Artificial General Intelligence, which is attributed as strong, and (3) Artificial Super Intelligence, which is the last category (Goralski and Tan, 2022; Nti et al., 2022; Heins, 2022). Artificial Narrow Intelligence is unable to autonomously solve problems in other fields and could be solely implemented in a particular field (Heins, 2022). When Artificial intelligence equals the intelligence of humans, has a self-consciousness awareness, and can separately cope with difficulties and solve problems, and make plans for the future, in the next step it will be passed the classification of Artificial General Intelligence (Goralski and Tan, 2022). At the present time, Artificial Narrow Intelligence has been prevalently used such as applications comprising Google's Alexa, Apple's Siri, and IBMs Watson. This can be applied to various fields and can autonomously solve problems in other fields (Heins, 2022). The last category called Artificial Super Intelligence describes the surpass of the human brain (Goralski and Tan, 2022), where the machine instantaneously solves complex problems (Heins, 2022), but in reality, this is not yet achieved level (Goralski and Tan, 2022). In addition, there are four critical components of artificial intelligence comprising: (1) Expert system which concerns with circumstances under investigation as an expert and renders the ambitioned or anticipated outcomes, (2) Natural language processing that facilitates conversation among peoples and machines in innate language, (3) Heuristic problem-solving for assessing a small accumulation of solutions which comprises some guesswork to explore near-optimal solutions, and (4) Computer vision which generate the competence to identify features and shapes spontaneously (Andeobu et al., 2022).

Artificial intelligence technologies render three core benefits. Firstly, artificial intelligence presents the automation of significant, yet iterative, and time-consuming duties, empowering communities to be interested in and study higher-value issues. Secondly, artificial intelligence demonstrates comprehension that is otherwise trapped in tremendous quantities of unstructured information that formerly necessitated personal administration and analysis like data constituted by photos, videos, written reports, social media posts, business documents, or e-mail messages. Lastly, artificial intelligence can combine thousands of computers and other resources to figure out the most complicated problems. Ultimately, artificial intelligence competencies should be improved to explore ways to alleviate the climate crisis. To achieve this, the meticulous examination is important to determine how artificial intelligence solutions can be integrated with human feelings, cognitions, behavioral responses, and social norms (Nishant et al., 2020).

Among the diverse techniques of artificial intelligence, the most significant is the one called machine learning, which is the capability for computers to obtain novel information and abilities by analyzing data from the actual world and how they interconnect with it. At the present, machine learning is the primary approach applied in artificial intelligence and the motivation for why astounding consequences have been obtained in recent years. It is based on the design of a presumptive model that may be generated from previously known data which after that uses novel data to constitute scenarios not thought at first, learning from someone's experience (Sciarretta et al., 2022). It can be categorized into two kinds: supervised learning, which is the kind explained before, and unsupervised learning. Unsupervised learning commences from uncategorized data and does not trust previously known consequences to fix the algorithm (Miller and Brown, 2018). The accomplishment obtained by machine learning is because of its extensive perspective in determining the most productive model in each circumstance and in processing complicated data. Except that, to guarantee consequences, it requires much larger datasets (Sciarretta et al., 2022).

SUSTAINABILITY

The linear economic model dates back to the industrial revolution and is described as a unidirectional model of production. The mentality of the linear economic model is based on the consumption of the product and then disposed of them (Okorie et al., 2018). In another words, the linear economy depends on "take, produce, consume, and waste", which is not sustainable for both businesses and communities (Kazancoglu et al., 2021). In a linear economic model, product efficiency and lifespan are not prioritized. Scarce resources led to the emergence of another approach called the circular economy. In the approach of circular economy, the 3R principles of "Reduce", "Reuse", and "Recycle" are generally stressed (Okorie et al., 2018). It is strongly associated with the environmental sustainability activities of businesses such as enhanced energy effectiveness, utilization of renewable energy and waste recovery, and gain of recycled or renewable materials in the manufacturing process (Lieder and Rashid, 2016).

The exigence to conserve the environment absent of economic growth and impoverishment alleviation dawned in the 1960s (Sarkis et al., 2011). Previously, attention was predominantly on the chemical and its potential impacts on humans and animals. This sparked the creation of the Environmental Protection Agency in the USA. Supply chains were initially (in the 1970s and 1980s) impassive to sustainability topics (Yusuf et al., 2013). However, recently, sustainability in the supply chain management has attracted a lot of attention from both practitioners and academics, since it holds out the promise of embracing a new production philosophy with the potential to obtain better environmental performance (Fallahpour et al., 2016). Arguably the most impressive definition of sustainability is presented by Brundtland Commission (1987) as follows: "development that meets the needs of the present without compromising the ability of future generations to meet their needs." Nevertheless, this clarification might be too far-fetched from the side of practical life to be implemented (Durmaz et al., 2021). Glavič and Lukman (2007, p. 1883) defined sustainability from the side of production as "creating goods by using processes and systems that are non-polluting, that conserve energy and natural resources in economically viable, safe and healthy ways for employees, communities, and consumers and which are socially and creatively rewarding for all stakeholders for the short- and long-term future". The core purposes of sustainability in the supply chain is to diminish the adverse environmental effects during the manufacturing, purchasing, distribution, and sale of products, besides administrating end-of-life products and waste (Fallahpour et al., 2016). Indeed, the aim of the minimization of waste is not only for environmental concerns but also for economic reasons. In other words, waste indicates a greater economic loss (Sarkis et al., 2011).

To achieve supply chain sustainability, supply chain members particularly should focus on performance indicators based on three dimensions: environmental performance, economic performance, and social performance (Gimenez et al., 2012; Nishant et al., 2020; Durmaz et al., 2021; Nti et al., 2022). Gimenez et al. (2012) defined sustainability as an integration of social, environmental, and economic responsibilities. The side of Economic performance of sustainability is generally well understood. It has been operationalized as manufacturing expenditures at the factory level. Nonetheless, the description of environmental and social performance is not clear (Gimenez et al., 2012). The environmental performance demonstrates the competency to perform external and internal green activities for the natural environment (Yee et al., 2021). Put differently, environmental sustainability signifies to the utilization of energy and other resources of firms leave behind as a consequence of their production process. It is usually associated with pollution reduction, waste decrease, energy efficiency, a diminish in the consumption of harmful materials, emissions reduction, a diminish in the density of environmental accidents, etc. (Gimenez et al., 2012). Social performance refers to the business image in the eyes of the public and client satisfaction (Yee et al., 2021). It indicates that businesses provide equal opportunities, promote diversity, encourage correlativity within and outside society, provide a quality of life, and present democratic processes as well as accountable management structures. In summary, sustainability recommends that businesses not solely necessitate to engage in environmentally as well as socially responsible attitudes but also financial gains should be enhanced in the operation process (Gimenez et al., 2012).

THE ADVANTAGES OF ARTIFICIAL INTELLIGENCE IN SUSTAINABILITY ISSUES

In recent years, researchers have commenced discovering how recent technologies like artificial intelligence can be utilized to improve sustainable development. As a result of this investigation, the concept of sustainable artificial intelligence has emerged. According to van Wynsberghe (2021), sustainable artificial intelligence is "a movement to foster change in the entire lifecycle of AI products (i.e. idea generation, training, re-tuning, implementation, governance) towards greater ecological integrity and social justice". Artificial intelligence technologies have created an unprecedented opportunity for development across all the dimensions of sustainability, which comprise an economic, social, and environmental aspects of (Goralski and Tan, 2022). In fact, addressing environmental sustainability challenges is sophisticated yet with the introduction of artificial intelligence, most of the prevalent environmental problems have transformed easily to clarify by putting the concern of the people in activity (Nti et al., 2022). Previously, firms received periodic updates on climate situations, carbon emissions, or the attitude of employees at supplier facilities for sustainability. Nowadays, new technology and digitization with artificial intelligence have created a different environment where these kinds of incidents are pursued in real-time with data constituted constantly (Sanders et al., 2019).

When rightly implemented, artificial intelligence technologies can provide social value and public economic advantages. For instance, smart environment technologies that are implemented in cities are generally artificial intelligence-driven systems, which come in the establishment of noise prediction, smart traffic lights, air quality prediction, and car traffic prediction. In the same way, the implementation of artificial intelligence technologies is frequently developing processes for gathering, transporting, allocating, and recycling diverse kinds of waste, from medical waste to electronic waste, and to other kinds of solid waste (Andeobu et al., 2022). To assist manufacturers, artificial intelligence can be utilized to predictions of energy and environmental effects (Nabavi-Pelesaraei et al., 2018), to choose the most appropriate sustainable supplier that facilitates the improvement of the manufacturers' sustainability (Fallahpour et al., 2016), for constituting traceability along an output's life cycle (Okorie et al., 2018). Particularly, adopting artificial intelligence technology can expand data collecting, interoperability, and analysis competencies across end-to-end stages of operations to facilitate sustainability, supply network security, and cost-competitive resiliency, thus eventually leading to improve consumers trust. To a greater extent, artificial intelligence implementations may leverage data to notify decision-making in organizational processes and facilitate to the production of data-driven products and services (Tsolakis et al., 2022). In addition, the degradation of the physical environment and the climate crisis are complicated problems that necessitate appropriate innovative and progressed solutions. The actual worth of artificial intelligence hence lies in its capability to encourage and stimulate social and environmental governance, rather than just as a means to diminish poverty, pollution, and resource demolition (Musleh Al-Sartawi et al., 2022). Kar et al. (2022) highlighted that artificial intelligence has the power to accelerate worldwide efforts to protect the environment and maintain resources through discovering diverse factors such as energy emission and reductions, surveilling deforestation, greener transportation networks, crash prediction, forecasting weather, and productive supply chain management. Some significant advantages of artificial intelligence and references have been presented in Table 1.

Table 1. The effect of artificial intelligence on sustainability and related references

The Effect of Artificial Intelligence on Sustainability	References
Green supplier selection	Fallahpour et al. (2016)
Cost reduction by avoiding waste	Tsolakis et al. (2022)
Increasing productivity and efficiency	Tsolakis et al. (2022); Nishant et al. (2020)
Monitoring climate changes	Sanders et al. (2019)
Operation and maintenance	Nishant et al. (2020)
Ease pursuing of labor environment	Sanders et al. (2019)
Facilitates to enhance business image	Bhagat et al. (2022)

Artificial intelligence has many positive effects on sustainability as mentioned. This study has investigated the advantages of artificial intelligence on sustainability into three sides, namely environmental, economic, and social sustainability in the following sub-sections.

ENVIRONMENTAL SUSTAINABILITY AND ARTIFICIAL INTELLIGENCE

The literature on artificial intelligence for environmental sustainability cuts across numerous domains. Critically, to address most of the contemporary global as well as territorial environmental themes, artificial intelligence is utilized in fields such as biodiversity, energy, water, and transportation, while many of such areas have diffused every classification and maintenance on evolving. There has been the practical application of artificial intelligence in transportation and biodiversity in many advanced economies, this embrace but are not restricted to obtaining e-waste by developing routing plan, safeguarding the sea from being soiled, applying to strengthen artificial intelligence autonomous rubbish gathering lorry and wildlife protection for developed biodiversity (Nti et al., 2022). The critical role of artificial intelligence in the administration of solid waste is in the sorting process. It is similar to the role of artificial intelligence in production whereby materials on a conveyor arch (new products or waste), are spontaneously scanned with cameras and evaluated through profound learning algorithms; and renders compound statistics and analytics for increasing recycling ratios (Andeobu et al., 2022). In the energy field, artificial intelligence technologies can facilitate decreasing fuel consumption by enhancing effectiveness in logistics and energy conversion. In addition, artificial intelligence can enhance the effectiveness and reliability earnings of renewable energies, for example, by augmenting weather estimation accuracy as well as fine-tuning energy storage (Tsolakis et al., 2022). There are diverse technologies applied in the administration and recycling of solid waste including; Adaptive neuro-fuzzy inference systems, Artificial immune systems, Non-inferior set estimation, K-means, Wavelet transform, Naive bayes, Random forest, Data mining, Linear regression, Logistic model tree, Support vector machine, Rough sets, Decision trees, Ant colony optimization, Artificial neural network, Goal programming, and Genetic algorithm (Andeobu et al., 2022).

Implementations of artificial intelligence to improve sustainability are multi-faced. They comprise innate language processing as well as image identification to pattern recognition and data analytics. Artificial intelligence algorithms, for example, can be utilized to determine jeopardized genres by collecting and analyzing data like animal footprints (Perucica and Andjelkovic, 2022). Besides, artificial intelligence can be applied to improve "smart energy solutions" that minimize energy depletion and administer energy infertilities. For instance, with regard to electrical systems, artificial intelligence tools can be utilized to expect the development of electricity fees on the grounds of electricity demand and supply projections and thereby assist ease electricity trade (Niet et al., 2021). Researchers also present that the technologies of artificial intelligence and robots will progressively be deployed to protect and administer energy factories. From the angle of the circular economy, artificial intelligence can assist in designing and sustaining circular outputs by, for example, providing better e-waste administration with the assistance of image recognition or augmenting the usability of electronics owing to predictive care, or providing better e-waste administration with the assist of image recognition. other different sustainable applications of artificial intelligence exist such as estimating environmental threats, land-use planning, waste administrating as well as geoscience (Perucica and Andjelkovic, 2022). On the other hand, artificial intelligence and environmental sustainability is extensively classified into four core fields which contain;

(1) sustainable agriculture,
(2) management of waste and pollution,
(3) conservation of environmental assets,
(4) pollution monitoring and pollutant treatment (Nishant et al., 2020; Nti et al., 2022).

ECONOMIC SUSTAINABILITY AND ARTIFICIAL INTELLIGENCE

Money is central to the road capitalist communities have developed and functioned. Artificial intelligence exhibits a chance to leverage equipment and program to gain an understanding of recent ways of arranging manufacturing processes and worldwide supply chains to diminish the energy, water, and territory inputs related to upstream product manufacturing like oil, mining, gas, and utilities while diminishing net supports to carbon emissions. Because of a novel chance to decrease the operational expenses of small and large manufacturing operations, artificial intelligence is legitimated as an investment priority for multiple business performances (Nishant et al., 2020). Artificial intelligence can enhance productivity and efficiency in various industries, thereby leading to economic sustainability (Sanders et al., 2019; Camaréna, 2020; Goralski et al., 2022; Tsolakis et al., 2022).

In the food industry, artificial intelligence technologies have been applied to optimize the reaping and processing of yields, for instance by utilizing drones with cameras as well as machine learning algorithms to identify the decomposition proportion of vegetables (Camaréna, 2020). In addition, artificial intelligence supported by other promoter technologies facilitates the categorizing of food supplies while incessantly monitoring the hygiene grade across operations. Similar advantages can also be accomplishable for the fishing sector on various fronts, from pursuing fish cropping and downstream sectoral operation to providing traceability and transparency across global trade processes (Tsolakis et al., 2022). Artificial intelligence can further assist diminish operational costs associated with personnel errors, failure, labor, and equipment (Cubric, 2020) or expenditures associated with fuel depletion for manufacturing and transportation (Tsolakis et al., 2022).

Except for the advantages at the internal operation's grade, artificial intelligence has also been described to ensure benefits for end-to-end management of the supply chain. Such advantages comprise enhanced customer demand administration and forecasting (Cubric, 2020), augmented transparency in the supply chain (Tsolakis et al., 2022), decision-support on the pricing of goods (Sanders et al., 2019), and knowledge sharing between supply chain partners (Tsolakis et al., 2022). Despite artificial intelligence systems not being the talent to transcribe human intelligence, they can ensure proper outputs that may far alter human efforts. In accounting, it is needed to analyze the capacities and the boundaries of artificial intelligence systems to make them beneficial in coping with accounting and working problems and to identify the education and skills required to permit accountants to lightly control intelligent systems. When accountants begin operating the associated artificial intelligence systems, they will be talented to augment their authority and generate quality financial reports for financiers (Musleh Al-Sartawi et al., 2022). Besides, from the side of manufacturing industries, artificial intelligence renders in-depth tracking and control as well as provides an extensive view of the equipment's performance as well as health (Kar et al., 2022).

SOCIAL SUSTAINABILITY AND ARTIFICIAL INTELLIGENCE

Businesses are progressively interested in better comprehending the labor conditions of employees in plants across their supply chains. In former times, these processes were obtained by indiscriminately dispatching auditors to plants and warehouses for periodic evaluations. The purpose was to obtain information about the social conditions and environment from these desultory visits. Nowadays, businesses across various sectors have rules of management for suppliers that span working states, and the quality of life of their employees. With by use of artificial intelligence technologies, connected sensors as well as devices also track essential sustainability metrics like emissions and resource usage. In addition, sensors may incessantly monitor and stream visual and other information to the main firm, just as the capabilities of humans. Employees are authorized to utilize their smartphones to record working conditions as well as dangers and next share that knowledge worldwide. Some businesses in various sectors are taking the chance to showcase their support for labor conditions. For instance, global retailers such as Walmart, Marks & Spencer, and Adidas are directly and without hesitation contacting individual employees in their suppliers' plants to get a sense of labor conditions and human liberties. For many of these businesses, it is a topic of leading on sustainability by outrightly pioneering suppliers' labor conditions (Sanders et al., 2019).

The effect of artificial intelligence is tremendous and businesses competing globally cannot overlook it. In addition, the recognition of artificial intelligence universally has achieved momentum by leaps and bounds. The utilization of it provides that the decision taken by the business is correct and economically applicable. The algorithms applied by artificial intelligence assist managers diminish costs as well as facilitate clients' increased profitability. When it comes to procurement and consumption-associated decision-making, artificial intelligence is becoming a significant perspective for consumers in today's conditions. Consumers have commenced more interest in a healthy lifestyle that comprises purchasable depletion mentalities along with a protected, sustainable environment. Artificial intelligence has transformed into a tool without which sustainable improvement cannot be obtained. Consumers are considering artificial intelligence so favorably that they are volunteering to accept it entirely because it is making daily life much easier as well as it also assists them in protecting a sustainable lifestyle for a living (Bhagat et al., 2022). For instance, many consumers utilize the application of health in the smartphone for protecting a healthy lifestyle.

Moreover, artificial intelligence may assist to advance more productive forms of employment for the poor in rural fields and improve the sustainable utilization of continental ecosystems in cultivated land to decrease land degradation (Goralski et al., 2022). Artificial intelligence can enhance productivity and applications in the food and farm sector to diminish inequality and poverty among countries. Besides, it can assist in to progress of more inclusive as well as the efficient shape of businesses for the evil in rural fields and enhance sustainable utilization of continental ecosystems in cultivable terrain to diminish land deterioration (Goralski and Tan, 2022).

From a social sustainability aspect, transparency is one of the critical advantages that artificial intelligence could deliver, especially downstream of a supply chain. For instance, artificial intelligence may enlighten clients about making more awareness purchasing decisions towards accountably sourced as well as manufactured products because of the enhanced data processing competencies that authorize following the upstream supply chain to the raw materials phase (Tsolakis et al., 2022). Besides, artificial intelligence-driven applications may enhance social prosperity; for instance, socially promoter robots might diminish the workload of caregivers and improve the health of senior society by providing mobility, cognitive support, and social contacts (Cubric, 2020).

THE CHALLENGES OF ADOPTING ARTIFICIAL INTELLIGENCE TECHNOLOGIES

Despite its benefits, some key challenges in implementing these technologies generally exist in various sectors (Tsolakis et al., 2022). The challenges for artificial technology implementation include technical, legal, ethical, administrative, and socio-economic considerations. A primary technical challenge for performing artificial intelligence in firms' operations is associated with the presence and utilization of data. Data present for business are generally unstructured and hard to share among the supply chain partners. Structuring this data might be very expensive. Moreover, the data applied for a particular incident cannot be generalizable (Cubric, 2020). For instance, problems might stem from datasets, which do not exactly picture actuality, or from overfitting the artificial intelligence algorithm to the educating data set (Tsolakis et al., 2022). A deficiency in systems thinking of the supply chain partners may lead to fundamental challenges in performing such technologies in a complicated industrial ambiance (Caména, 2020). On the other side, a deficiency in educating data can lead to decreased performance of complicated artificial intelligence algorithms (Cubric, 2020). Besides, an absence of standardization of knowledge might lead to complexities in selecting an accurate artificial intelligence solution. There is a new trend toward individualizing firms' digital solutions through internal data architecture (Tsolakis et al., 2022).

Among the ethical challenges, the main one apprehensions the deficiency and disintegration of the arrangement of the utilization of artificial intelligence. Without an explicit legal structure, these technologies cannot be applied since responsibility cannot be described. For instance, in the medical healthcare industry, if an artificial intelligence administrated instrument makes an incorrect selection that impacts a patient's health, who should be held responsible? Besides, the faithlessness of existing systems also constitutes other problems associated with the impartialness of artificial intelligence systems: prejudices in data may lead to injustices based on race, gender, or other factors. What's worse, the systems as well as the data that utilize them might be manipulated and distorted; they can be utilized to misguide results or even for criminal aims (Sciarretta et al., 2020).

Progressively nested and complicated systems are vulnerable to unforeseen shocks, and gradual improvement endogenously, also known as "normal accidents". This describes that domestic defeats may appear surprisingly and fluctuate and boost across network connections (for instance, a territorial food supply chain) and effectuate defeats in the system as a complete, particularly if the ingredient of the system is optimized and administrated appropriately (Galaz et al., 2021). Venomous external assaults can uncover such endogenous vulnerabilities as well, and also the most developed artificial intelligence systems based on profound neural networks are undefended to sabotage (Heaven, 2019). Additionally, the black-box nature of artificial intelligence models is a significant topic in the prevalent implementation of such techniques. In most circumstances, it is problematic to reiterate most artificial intelligence models since they are largely established on large datasets that are reliable, conserved, or confidential in publications. The absence of transparency has stood as one of the critical causes behind the lack of consistency and prosecution in many artificial intelligence applications that improved the administration and recycling of solid waste (Andeobu et al., 2022).

IMPLICATIONS FOR MANAGER

There is an extensive comprehension of sustainable progress in the business through the unification of artificial intelligence and other cutting-edge technologies. However, there are still numerous barriers and challenges to improving this domain in practical life (Gupta et al., 2023). According to researchers (Musleh Al-Sartawi et al., 2022), artificial intelligence technologies will progressively be part of daily life increases in the near future and this increases the question of whether regulation is required and in which form. In addition, regulations necessitate being thought of from various angles. These comprise micro-angles (regulations associated with algorithms and businesses), and macro-angles (regulations associated with employment).

Artificial Intelligence unified into the business will overcome most problems through developing sustainability performance (environmental performance, economic performance, and social performance). Business managers should discover the advantages and benefits of utilizing various artificial intelligence technologies. However, there are also concerns about the justice of the algorithms underlying artificial intelligence. Hence, businesses should follow and monitor the quality of education as well as input data, constitute algorithms that are plausible and convenient, and asses the outcomes for possible prejudice. Moreover, artificial intelligence algorithms require to be accountable: clear, transparent, and comprehensible for decision-makers to consent to the examination. Despite the fact that algorithmic decision-making in principle alludes that artificial intelligence systems could be autonomous, for implementation effectively and ethically suitably, there must be a person in the loop (Leszkiewicz et al. 2022).

Internal social programs develop social and environmental performance but also enhance production costs. The development in environmental performance as a consequence of performing social programs may be clarified as follows: Employer participation and education can encourage to decrease in potentially damaging environmental activities. Moreover, the application of social action programs like projects to support society might also develop the organizational social image and performance (Gimenez et al., 2012). In addition, engineers play a key role in constituting a long-run competitive advantage for the business through propounding creative solutions to an extensive range of social and technological problems. For this reason, it is critical to comprehend how different it is to form theories of sustainable entrepreneurship. The absence of legislation to stimulate sustainable improvement in business, the absence of comprehension of the importance of sustainability, the absence of support from the education ministries, and the absence of collaboration among the educational, environmental, health, agriculture, and so on are whole issues. Consequently, there is a necessity for the improvement of novel perspectives and techniques to unify artificial intelligence and other latest technologies with sustainable development in business (Gupta et al., 2023).

CONCLUSION

Artificial intelligence technologies touch the overall side of the business and the economy, comprising health, environmental management, waste management, finance, welfare, education, manufacturing, and government services (Andeobu et al., 2022). They can remarkably revolutionize the design of various business process since it provides intelligent, automated, and versatile decision-making policies. In addition, automated decision-making as well as estimated analytics with the help of artificial intelligence, in integration with speed advances in technologies like robotics and sensor technology are probably to alter the road communities' and governments' comprehension and reply to climate and ecological alteration. In spite of expanding interest in, and arrangement of artificial intelligence technologies in fields crucial for sustainability, only a few studies have investigated the role of artificial intelligence on sustainability in depth (Galaz et al., 2021). Therefore, the aim of this chapter is to profoundly explore the critical role of artificial intelligence in sustainability.

Frankly speaking, artificial intelligence should not be evaluated as the panacea for every difficulty, because this is usually the case when a promising novel technology is inserted into the market. In fact, if a thorough comprehension of the action is missing, artificial intelligence may spark extra topics. A mutual apprehension is that artificial intelligence will finally replace experts and thus remove the human factor. Nonetheless, for the present, this apprehension is groundless. Because autonomous intelligence is still a long-lasting way from becoming an actuality and will not be performed except the difficulties debated are initially elucidated (Sciarretta et al., 2020). Solutions to economic, social, and environmental problems will move towards advancing human intelligence, as artificial intelligence is developed. This will surround systems that have the capability to acquire knowledge from interactions with the environment and humans, and inform human decision-making to choose and assess options. If followed properly, artificial intelligence provides advantages and benefits that can be as transformative as the industrial revolution (Andeobu et al., 2022).

To sum up, the motivators for artificial intelligence adoption are generally associated with economic sustainability (cost, time, and performance), environmental sustainability (decreasing fuel consumption, protecting the environment, and minimizing energy depletion), and social sustainability (customer and society satisfaction, accuracy in decision-making and predictions, enhancing transparency). In other words, the role of artificial intelligence in economic sustainability describes enhancing the opportunities and incentives for organization development, formation of jobs, procurement of incomes, enhancing agricultural productivity, and so on that enhance possibilities for human progress, particularly for the poor in non-rustic fields and also rural inhabitants as a complete. The role of artificial intelligence in social sustainability implies diminishing inequality and discrimination, providing well work conditions for employees, enhancing school registration or educational gaining among the poor, reducing infant mortality, and so on that advantages the poor, migrants, and minorities. The role of artificial intelligence in environmental sustainability depicts recuperating the effect of climate change, pollution of ecosystems or excessive exploitation, and natural resource depletion, which may deploy the natural environment and livelihoods of people and which will disproportionately influence the poor (Goralski and Tan, 2022). Moreover, the barriers to artificial intelligence adoption are related to economic, technical, legal, ethical, and administrative, such as dependence on non-humans, job security, trust and lack of multiple stakeholder perspectives, lack of knowledge, and safety.

REFERENCES

Andeobu, L., Wibowo, S., & Grandhi, S. (2022). Artificial intelligence applications for sustainable solid waste management practices in Australia: A systematic review. *The Science of the Total Environment, 155389*. Advance online publication. doi:10.1016/j.scitotenv.2022.155389 PMID:35460765

Bhagat, R., Chauhan, V., & Bhagat, P. (2022). Investigating the impact of artificial intelligence on consumer's purchase intention in e-retailing. Foresight. doi:10.1108/FS-10-2021-0218

Camaréna, S. (2020). Artificial intelligence in the design of the transitions to sustainable food systems. *Journal of Cleaner Production*, *271*, 122574. doi:10.1016/j.jclepro.2020.122574

Cubric, M. (2020). Drivers, barriers and social considerations for AI adoption in business and management: A tertiary study. *Technology in Society*, *62*, 101257. doi:10.1016/j.techsoc.2020.101257

Dijkstra-Silva, S., Schaltegger, S., & Beske-Janssen, P. (2022). Understanding positive contributions to sustainability. A systematic review. *Journal of Environmental Management*, *320*, 115802. doi:10.1016/j.jenvman.2022.115802 PMID:35952560

Durmaz, A., Demir, H., & Sezen, B. (2021). The role of negative entropy within supply chain sustainability. *Sustainable Production and Consumption*, *28*, 218–230. doi:10.1016/j.spc.2021.04.014 PMID:34722845

Fallahpour, A., Olugu, E. U., Musa, S. N., Khezrimotlagh, D., & Wong, K. Y. (2016). An integrated model for green supplier selection under fuzzy environment: Application of data envelopment analysis and genetic programming approach. *Neural Computing & Applications*, *27*(3), 707–725. doi:10.1007/s00521-015-1890-3

Galaz, V., Centeno, M. A., Callahan, P. W., Causevic, A., Patterson, T., Brass, I., Baum, S., Farber, D., Fischer, J., Garcia, D., McPhearson, T., Jimenez, D., King, B., Larcey, P., & Levy, K. (2021). Artificial intelligence, systemic risks, and sustainability. *Technology in Society*, *67*, 101741. doi:10.1016/j.techsoc.2021.101741

Gimenez, C., Sierra, V., & Rodon, J. (2012). Sustainable operations: Their impact on the triple bottom line. *International Journal of Production Economics*, *140*(1), 149–159. doi:10.1016/j.ijpe.2012.01.035

Glavič, P., & Lukman, R. (2007). Review of sustainability terms and their definitions. *Journal of Cleaner Production*, *15*(18), 1875–1885. doi:10.1016/j.jclepro.2006.12.006

Goralski, M. A., & Tan, T. K. (2022). Artificial intelligence and poverty alleviation: Emerging innovations and their implications for management education and sustainable development. *International Journal of Management Education*, *20*(3), 100662. doi:10.1016/j.ijme.2022.100662

Gupta, B. B., Gaurav, A., Panigrahi, P. K., & Arya, V. (2023). Analysis of artificial intelligence-based technologies and approaches on sustainable entrepreneurship. *Technological Forecasting and Social Change*, *186*, 122152. doi:10.1016/j.techfore.2022.122152

Heaven, D. (2019). Why deep-learning AIs are so easy to fool. *Nature*, *574*(7777), 163–166. doi:10.1038/d41586-019-03013-5 PMID:31597977

Heins, C. (2022). Artificial intelligence in retail – a systematic literature review. Foresight. doi:10.1108/FS-10-2021-0210

Kaplan, A., & Haenlein, M. (2019). Siri, siri, in my hand: Who's the fairest in the land? On the interpretations, illustrations, and implications of artificial intelligence. *Business Horizons*, *62*(1), 15–25. doi:10.1016/j.bushor.2018.08.004

Kar, A. K., Choudhary, S. K., & Singh, V. K. (2022). How can artificial intelligence impact sustainability: A systematic literature review. *Journal of Cleaner Production*, *134120*. Advance online publication. doi:10.1016/j.jclepro.2022.134120

Kazancoglu, I., Sagnak, M., Kumar Mangla, S., & Kazancoglu, Y. (2021). Circular economy and the policy: A framework for improving the corporate environmental management in supply chains. *Business Strategy and the Environment*, *30*(1), 590–608. doi:10.1002/bse.2641

Leszkiewicz, A., Hormann, T., & Krafft, M. (2022). Smart Business and the Social Value of AI. In Smart Industry – Better Management (Advanced Series in Management, Vol. 28). Emerald Publishing Limited. doi:10.1108/S1877-636120220000028004

Lieder, M., & Rashid, A. (2016). Towards circular economy implementation: A comprehensive review in context of manufacturing industry. *Journal of Cleaner Production*, *115*, 36–51. doi:10.1016/j.jclepro.2015.12.042

Miller, D. D., & Brown, E. W. (2018). Artificial intelligence in medical practice: The question to the answer? *The American Journal of Medicine*, *131*(2), 129–133. doi:10.1016/j.amjmed.2017.10.035 PMID:29126825

Musleh Al-Sartawi, A. M., Hussainey, K., & Razzaque, A. (2022). The role of artificial intelligence in sustainable finance. *Journal of Sustainable Finance & Investment*, 1–6. doi:10.1080/20430795.2022.2057405

Nabavi-Pelesaraei, A., Rafiee, S., Mohtasebi, S. S., Hosseinzadeh-Bandbafha, H., & Chau, K. W. (2018). Integration of artificial intelligence methods and life cycle assessment to predict energy output and environmental impacts of paddy production. *The Science of the Total Environment*, *631*, 1279–1294. doi:10.1016/j.scitotenv.2018.03.088 PMID:29727952

Niet, I., van Est, R., & Veraart, F. (2021). Governing AI in electricity systems: Reflections on the EU artificial intelligence bill. *Frontiers in Artificial Intelligence*, *4*, 1–7. doi:10.3389/frai.2021.690237 PMID:34396090

Nishant, R., Kennedy, M., & Corbett, J. (2020). Artificial intelligence for sustainability: Challenges, opportunities, and a research agenda. *International Journal of Information Management*, *53*, 102104. doi:10.1016/j.ijinfomgt.2020.102104

Nti, E. K., Cobbina, S. J., Attafuah, E. E., Opoku, E., & Gyan, M. A. (2022). Environmental Sustainability Technologies in Biodiversity, Energy, Transportation and Water Management using Artificial Intelligence: A Systematic Review. *Sustainable Futures : An Applied Journal of Technology, Environment and Society*, *100068*. Advance online publication. doi:10.1016/j.sftr.2022.100068

Okorie, O., Salonitis, K., Charnley, F., Moreno, M., Turner, C., & Tiwari, A. (2018). Digitisation and the circular economy: A review of current research and future trends. *Energies*, *11*(11), 3009. doi:10.3390/en11113009

Perucica, N., & Andjelkovic, K. (2022). Is the future of AI sustainable? A case study of the European Union. Transforming Government: People. *Transforming Government*, *16*(3), 347–358. doi:10.1108/TG-06-2021-0106

Salas, J., Patterson, G., & de Barros Vidal, F. (2022). A Systematic Mapping of Artificial Intelligence Solutions for Sustainability Challenges in Latin America and the Caribbean. *Revista IEEE América Latina*, *20*(11), 2312–2329. doi:10.1109/TLA.2022.9904756

Sanders, N. R., Boone, T., Ganeshan, R., & Wood, J. D. (2019). Sustainable supply chains in the age of AI and digitization: Research challenges and opportunities. *Journal of Business Logistics*, *40*(3), 229–240. doi:10.1111/jbl.12224

Sarkis, J., Zhu, Q., & Lai, K. H. (2011). An organizational theoretic review of green supply chain management literature. *International Journal of Production Economics*, *130*(1), 1–15. doi:10.1016/j.ijpe.2010.11.010

Sciarretta, E., Mancini, R., & Greco, E. (2022). Artificial Intelligence for Healthcare and Social Services: Optimizing Resources and Promoting Sustainability. *Sustainability (Basel)*, *14*(24), 16464. doi:10.3390/su142416464

Tsolakis, N., Schumacher, R., Dora, M., & Kumar, M. (2022). Artificial intelligence and blockchain implementation in supply chains: A pathway to sustainability and data monetisation? *Annals of Operations Research*, 1–54. PMID:35755830

Van Wynsberghe, A. (2021). Sustainable AI: AI for sustainability and the sustainability of AI. *AI and Ethics*, *1*(3), 213–218. doi:10.1007/s43681-021-00043-6

Yee, F. M., Shaharudin, M. R., Ma, G., Zailani, S. H. M., & Kanapathy, K. (2021). Green purchasing capabilities and practices towards Firm's triple bottom line in Malaysia. *Journal of Cleaner Production*, *307*, 127268. doi:10.1016/j.jclepro.2021.127268

Yildiz Çankaya, S., & Sezen, B. (2019). Effects of green supply chain management practices on sustainability performance. *Journal of Manufacturing Technology Management*, *30*(1), 98–121. doi:10.1108/JMTM-03-2018-0099

Yusuf, Y. Y., Gunasekaran, A., Musa, A., El-Berishy, N. M., Abubakar, T., & Ambursa, H. M. (2013). The UK oil and gas supply chains: An empirical analysis of adoption of sustainable measures and performance outcomes. *International Journal of Production Economics*, *146*(2), 501–514. doi:10.1016/j.ijpe.2012.09.021

Chapter 3
Can Fossil Fuels Be Replaced by Entirely Renewable Energy

Ravi Gorli
https://orcid.org/0000-0002-4381-3250
Gitam University, India

ABSTRACT

A renewable energy supply system must be established. This chapter outlines a complete decarbonization strategy. It explains how renewable energy is improving the long-term sustainability of energy sources. A new IEA analysis shows that progress has been made and that additional improvements in global energy efficiency are possible. By 2050, the estimated annual cost of energy shift will be $1.7 trillion. In terms of output, renewable energy sources have surpassed fossil fuels. Never before has cutting-edge tech been so accessible. In 2017, solar thermal electricity with thermal storage hit a record low price of under 10 cents per kWh. Renewable energy utilization is rising. Transportation, heating, and other energy needs account for 20% of overall consumption. Companies can benefit greatly from using green energies like solar thermal, geothermal, and biofuels.

1.0 INTRODUCTION

It is critical to alter the global energy distribution network. To replace the current dependency on fossil fuels, and energy supply system based on renewable energy sources must be developed. This article outlines a comprehensive approach for decarbonizing the energy system, which is based on high energy efficiency and the utilization of renewable energy sources, among other things. It details the mechanism by which the change is occurring, as well as how renewable energy deployment is enhancing the long-term sustainability of energy sources. Additionally, this research demonstrates that decarbonization is technically feasible, less costly, and more beneficial than continuing business as normal (BAU). This may result in a world that is both more prosperous and less vulnerable in the long run.

Global temperatures should rise by no more than 2°C over the next century, according to the study's initial target, which has a 66 percent chance of being met. According to the majority of recent estimates, energy-related CO_2 emissions increased at a steady rate between 2014 and 2016 but jumped by 1.4 percent

DOI: 10.4018/978-1-6684-9130-0.ch003

in 2017 to a new record high of 32.5 Gt (IEA, 2018a). Even if we achieve the aspirational target of 1.5 degrees Celsius global warming, we will be a long way from attaining the global warming objective of well below 2 degrees Celsius. While advances were made in a variety of categories in 2017, the electrical industry made considerable strides in 2017. 2018 was a record year for renewable energy deployment, both in terms of new capacity built and total electricity generated (IRENA, 2018a). As a result of these advancements, the use of modern bioenergy, solar thermal, and geothermal energy, as well as other forms of end-use electrification (such as heat pumps), has increased significantly. Predictions of an increase in renewable energy's percentage of total final energy consumption (TFEC) were accurate to within 0.25 percent, a new high and record. On the other hand, renewable energy production demands a significant acceleration of advancement. Renewable energy must be increased from 19 percent of total final energy consumption (TFEC) in 2017 to two-thirds by 2050 to minimize global greenhouse gas emissions. The world economy's energy consumption will need to decline by an average of 1.8 percent per year until 2050 if it is to meet the goal of reducing global energy intensity by an average of 2.8 percent per year until 2050. Despite massive population and economic growth, it is projected that the world's energy consumption will have reduced marginally by 2050. Carbon dioxide emissions have increased in recent years, which can be linked to a stop in the expansion of energy efficiency. Despite this, a new study from the International Energy Agency (IEA) indicates that progress has been made and that there are various opportunities to further improve energy efficiency globally in the future, particularly in the developing world (IEA, 2018b).

This document includes a list of steps that can be taken to accelerate the global energy transition. The amount of expenditure required to change both supply and demand-side technology is explored in this study. The energy transition's benefits and costs are also considered. According to the estimate, energy change will cost an additional $1.7 trillion each year by 2050. (on average USD 6.3 trillion in the same year). More comprehensive measures of total well-being can have a major impact on results. According to the Reference Case, global GDP is predicted to increase by 1% by 2050. Growth is anticipated in light of current and future policies, including Nationally Determined Contributions (NDCs). As a result, millions of new employment would be created globally. Is there anything impeding our transition to a more sustainable energy system?

It is critical to rebuilding the global energy system. Apart from lowering carbon emissions, the energy revolution provides several other advantages. By utilizing this technology, we can lower the cost of energy access for all people, enhance health, and boost the security of our energy supply. IRENA is currently compiling a report on the geopolitics of energy transition (IRENA, 2018b). As a result, the energy business cannot fix the world's issues alone. Energy transitions must consider all facets of the economy and society. All individuals should have access to energy, and those harmed by the transition should receive assistance. Carbon emissions from energy consumption can be reduced in a variety of ways, but efficiency and renewable energy sources are critical for climate change mitigation. The handling of the switch is extensively addressed. The analysis makes use of IRENA's current technological capabilities to envision a future energy system transformation. Almost all of these ideas are immediately implementable, based on the research findings. However, in other instances, new technology is required. Numerous emerging technologies require support and advancement. Additionally, this category encompasses electric vehicles and novel energy storage systems. Renewable hydrogen, enhanced aviation biofuels, and renewable hydrogen derived from renewable sources are all included in this category. There are various benefits to a global energy transition, including avoiding global temperatures rising by more than 2°C, preventing the accumulation of energy assets that will become obsolete before their technological life spans, and minimizing the collateral damage caused by fossil fuel consumption.

Figure 1. Wind turbines and electric vehicle

2.0 STATUS OF THE ENERGY TRANSITION

Renewable energy is becoming more affordable, and more individuals are making the transition. Production of fossil fuels slows as renewable power generation grows. This year, renewable energy capacity was built at an annual growth rate of 8 percent to 9 percent since 2010. Over the past few years, renewable energy sources have outperformed traditional energy sources in terms of overall output. Renewable energy capacity, including solar PV and wind, increased by 94 GW in 2017. Offshore wind power is included in this figure (IRENA, 2018a). For the first time, more than a quarter of the world's total electricity generation came from renewable sources in 2013. In the recent decade, the cost of solar panels and wind turbines has declined dramatically. When renewable energy costs drop, more people will switch to using clean, renewable energy for all their needs. Cutting-edge technology has never before been more accessible. A record low price of fewer than 10 cents per kilowatt-hour (kWh) for solar thermal electricity with thermal storage was reached for the first time in 2017. The International Agency for the Promotion of Renewable Energy Resources (IRENA, 2018c). Technology and auction outcomes are expected to lower prices in the future. By 2020, the cost of solar photovoltaic systems is expected to fall by half. (in comparison to the period from 2015 to 2016). As of early 2017 and early 2018, onshore wind had an average cost per kWh of USD 6 cents, while the cost of solar PV was USD 10 cents. The United Nations (IRENA) (2018b). In the future, these averages could be dramatically reduced based on current auction results

There was a significant increase in the amount of renewable energy being used in electric systems in 2017. Most of the eastern region's electricity came from renewable sources like solar and wind. 50Hertz can run energy systems on a significant fraction of variable renewables, such as wind and solar, with high reliability (50Hertz, n.d.). Many countries around the world saw a rise in renewable energy use for days, weeks, or even months at a time. Many people today consider renewable energy systems to be an essential resource and a valuable financial investment.

Recent trends show that renewable energy is on the rise. Growth and decarbonization objectives for 2050 are not being met at this point in the process. If renewable electricity continues to grow at an ever-increasing rate, heating, transportation, and other energy-related activities will be possible. This is not a good sign for the rest of society. Transportation, heating, and other uses of electricity make up 20% of the total energy consumed (broadly defined as the end-use sectors of building, industry, and transport). It's estimated that coal and natural gas make up the remaining 20% of the energy consumed. End-use companies can gain significantly from the usage of green energies like solar thermal, geothermal, and biofuels. Due to the rising usage of renewable energy, energy consumption must rise. To reduce carbon emissions, electric vehicles can be used for transportation. A record-breaking 1.5 percent of this year's automotive sales were new electric vehicles, with 1.2 million of these vehicles sold (Spiegel, 2018). China has overtaken the United States to become the world's most populous country, surpassing the United States in population. Five years ago, the annual compound growth rate for electric vehicle sales was 52 percent... It is likely that by 2050, one billion electric vehicles will be on the road if the decarbonization approach stated in this article is adopted. Among all end-user industries, construction has the highest power consumption rates by far. The majority of the time, fossil fuels are used for heating and cooking. Hundreds of millions of people throughout the world rely on modern cookstoves to cook with conventional biomass. In 2017, heat pump technology reached new heights. The current construction laws in Japan, for example, call for energy-positive buildings to be the norm within the next few years. In the existing stock, low repair rates indicate that the industry confronts major hurdles. There must be three times as many improvements as there are now. In the industrial sector, this is the most demanding part of the job Lifecycle thinking and innovative solutions are required because of the high energy demands, the low carbon content of products, and large emissions from particular processes in some industries, amongst other factors. Renewable hydrogen feedstock (for example, for primary steel manufacturing) has continued to be used by the heavy industry since the beginning of 2017.

2.1 Energy-Related Carbon Dioxide Emissions

When fossil fuels are used, they emit large amounts of carbon dioxide into the atmosphere, considerably contributing to global warming. Numerous countries have boosted their efforts in the recent year to minimize greenhouse gas emissions. As the Reference Case indicates, new policies and actions, such as NDCs, are expected to result in CO2 emissions reductions over time. Carbon dioxide emissions from the Reference Case are predicted to be 11% lower in 125 years from their current level of 1,380 Gt in 2015. This development resulted in a 1.4 percent increase in CO2 emissions in 2017. The United Nations' nuclear safety and security agency (IEA, 2018a). Additionally, government-mandated emission reduction targets are woefully inadequate. According to the Reference Case, if current and anticipated policies are implemented, the world would exhaust its energy-related CO2 emission budget in less than 20 years. Emissions must be reduced by 470 Gt by 2050 to avert a more than 2°C increase in global temperatures (with a 66 percent likelihood). (in comparison to current and planned policies, as shown in Figure 2).

Figure 2. Cumulative energy-related CO2 emissions and emissions gap, 2015-2050 (Gt CO2)

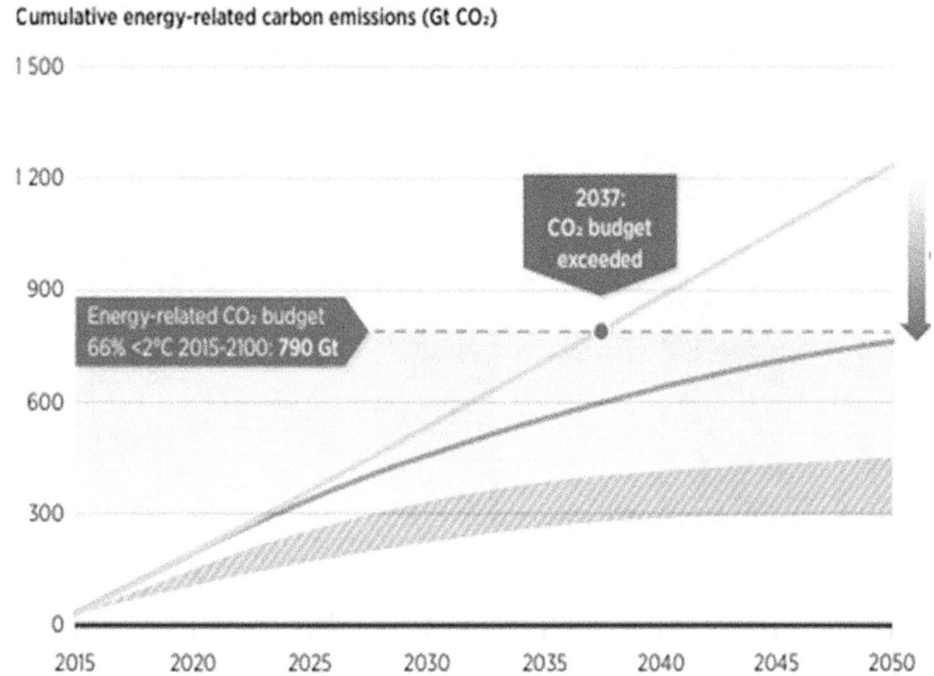

2.2 A Pathway for the Transformation

Renewable energy must increase its share of total primary energy consumption (TPES) from 15% in 2015 to 66% by 2050. This is illustrated in Figure 3. According to Reference Case, this number will only rise to 27% under existing and future legislative actions. Under the REmap scenario, the increase in 2050 is predicted to be from 64 to 222 EJ. As a result, solar and wind energy will account for more than half of the renewable energy mix, rather than bioenergy, which presently accounts for more than half. By 2050, bioenergy will account for approximately one-third of worldwide renewable energy demand. Due to the great synergies between renewable energy and energy efficiency, the REmap Case TPES would decline slightly below 2015 levels, despite significant population and economic expansion. The global economy's energy intensity must be reduced by an average of 2.8 percent each year until 2050 to achieve the necessary significant improvements in energy efficiency.

Figure 3. TPES and the share of renewable and non-renewable energy under the reference and REmap cases, 2015-2050 (EJ/yr)

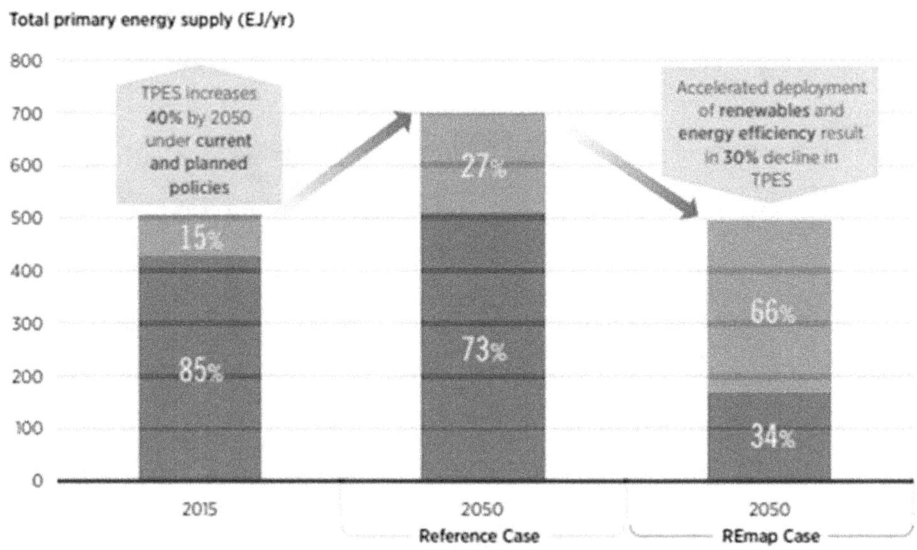

By 2050, the REmap Case would have resulted in a one-third reduction in current fossil fuel usage. Oil and coal prices would plunge the highest, by 70% and 85%, respectively. Although natural gas production is expected to decline by 30% from its current level by 2050, it will remain by far the most important fossil fuel.

Figure 4. The declining importance of fossil fuels

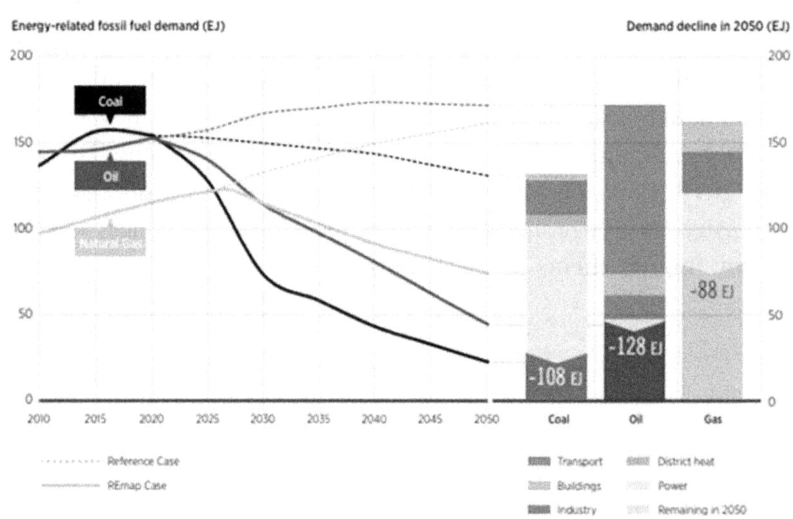

3.0 ANALYSIS AND INSIGHTS IN KEY SECTORS

3.1 Transport

In terms of energy use, transportation is lagging well behind the rest of the economy. The global renewable energy output of this industry was just 4% in 2015. Figure 7 illustrates this point. ethanol and biodiesel are the primary renewable energy sources in some countries. Even though electricity can assist lower the industry's carbon footprint, just 1% of it has been implemented in this company. Shipping and aviation, on the other hand, haven't made much progress in recent years. In the Reference Case, which is backed by the REmap Case, reducing transportation emissions to just 3 Gt of CO2 per year by 2050 represents a 70% reduction from current policy. Thirty percent of the reduction in greenhouse gas emissions can be attributed just to transportation.

3.2 Buildings

Currently, there are 150 billion square meters (m2) of residential and commercial floor space, and that amount is expected to grow to 270 billion square meters (m2) by 2050. Building emissions account for a large portion of total emissions, which can be decreased in a variety of methods. The energy transition is widely acknowledged, but nothing has been done to encourage it by the sector. In 2015, renewable energy sources provided around 36% of the energy used in buildings around the world (including traditional biomass). A 70 percent increase in the building sector's power demand is expected by 2050 because of an increase in energy demand (especially among rising economies) and the increasing electrification of heating. utilizing heat pumps and seasonal storage). For example, in the REmap scenario, the replacement of gas boilers with heat pumps and other energy-efficient forms of heating systems is included in this category.

3.3 Industry

In terms of CO2 emissions, the industrial sector is second only to transportation. It is responsible for a third of the world's greenhouse gas emissions. Even if IRENA's REmap Case is implemented, the industrial sector is estimated to release 5.1 gigatons of CO2 in the year 2050. (a little under half of which would be process-related emissions). There will be a massive increase in industrial emissions from 29 percent to 46 percent as a result of the REmap Case. Thermal and energy-intensive processes that are difficult to decarbonize are some of the most polluting in the petrochemical, chemical, and steel industries. To decarbonize the economy, REmap recommends that industry investment in low-carbon energy technology treble. USD 5 trillion in investments by 2050 would necessitate an extra USD 2.8 trillion in investments.

3.4 Power

To meet the demands of the energy transition, the power industry must be nearly carbon-free by 2050. Increase the amount of renewable energy and improve the system's scalability to meet this goal. More than 42 000 TWh will be consumed by Remap's end-use sectors by 2050, while the power sector will reduce its carbon emissions by 85%. (Figure 4). By 2050, renewables are expected to produce 85 percent of the world's electricity, up from a predicted 25 percent in 2017. From today's 800 GW to 13000

GW in 2050, solar and wind power capacity will be the driving force. During this period, the output of geothermal, biofuels, and hydropower would all rise by 800 GW. With solar and wind accounting for around 80 percent of that growth, the installed capacity of renewable electricity is predicted to grow by 400 GW each year. Increasing the current 2 percent of decentralized renewable energy generation to 20 percent by the year 2050. All existing coal-fired power facilities must be shut down by 2025. A $500 billion annual investment in renewable power generation is expected by 2050. Infrastructure and energy flexibility must be invested an additional USD 500 billion a year to achieve an 85 percent renewable energy power system (the Reference Case). Electricity will need to be decarbonized at a cost of almost $1 trillion per year for the next 30 years beginning in 2014.

Figure 5. The rising importance of solar and wind energy in the power sector

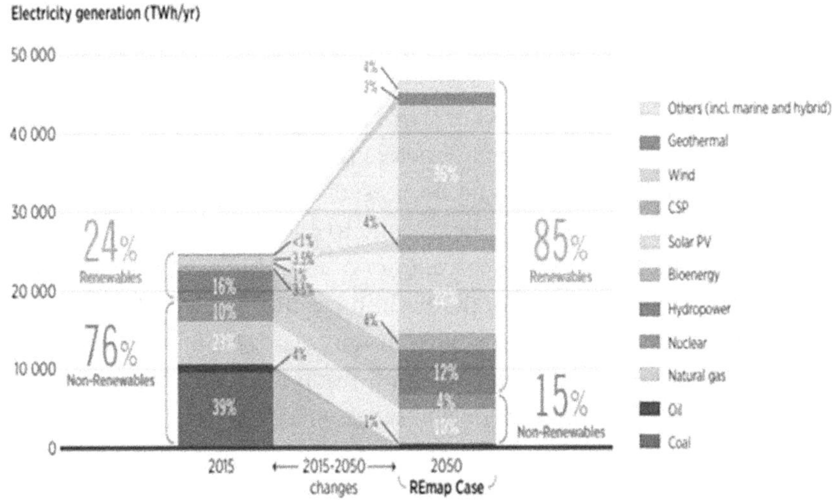

4.0 SOCIO-ECONOMIC BENEFITS OF THE ENERGY TRANSITION

The energy system is being reformed to satisfy both economic and environmental objectives in the context of a global economy that is reliant on fossil fuels for its primary energy source. The development of more cost-effective renewable energy and energy efficiency technology has allowed for more precise control over power, heat, and fuel consumption. Renewable energy and energy efficiency technologies are becoming more cost-effective. When handling energy transitions, it is necessary to take into account socioeconomic structures. The reforms to the energy sector described in the REmap transformation roadmap will have far-reaching consequences for the rest of the economy as a whole. A slew of benefits, including increased GDP, job creation, and improved overall well-being, can be derived from energy-economic integration. It is now possible to improve the situation by gaining an understanding of the elements that contributed to the transformation. Amid an economic revolution, different areas and countries have distinct economic structures, capabilities, and dynamics, and these differences must be recognized. Government officials can utilize economic diversification initiatives, support for adaptive local supply chains, and social protection measures for people who rely on declining industries to aid in the transition process and ensure that it is equitable for all involved (including fossil fuels).

5.0 HOW FINANCE AFFECTS THE ENERGY TRANSITION

Financing is critical to the development of the energy sector. Under present and future policies, investment is expected to fall short of the amount needed to complete the transition to a more sustainable economy. Between 2004 and 2017, investments in renewable energy surged eightfold, totaling USD 280 billion (Frankfurt School-UNEP Centre and BNEF, 2018). However, there is still a substantial amount of room for growth. There was a disproportionate amount of global renewable energy investment in 2016 coming from the OECD countries of China, Europe, and Western Europe; nearly 90% of this investment came from private sources in those countries (IRENA & CPI, 2018). More than 40 percent of direct renewable energy investment was funded by public funds in Sub-Saharan Africa, Latin America, and the Caribbean, and South Asia, compared to less than 10 percent in Europe. Increased investment rates will put a further burden on the financial system as a result of further delays in capital deployment within a set climate aim. Two factors have been linked to possible investment limits: capital costs and crowding out. In this study, the quantitative implications of these factors on the economic footprint of transition were assessed. As a starting point, we need to examine the cost of capital. Many factors influence the cost of financing, but the most important one is the perceived risk of the project by lenders. Those that embark on an energy transition run the danger of becoming more indebted as a result of their increased financial responsibilities. Higher debt ratios have a detrimental impact on the cost of capital as well as final energy costs, but the REmap Case still offers more benefits than the Reference Case despite these consequences. However, the cost of capital has a negative influence on both GDP and employment growth according to the following study. Higher capital expenditures harm the economy that increases with time; they are low until 2030 but become significant by 2050. If the problem isn't addressed, rising capital costs could have a substantial impact on the transition. As a result of country-specific risk concerns, countries and regions that are already burdened by high borrowing rates may face further difficulties. Because policymakers don't know exactly how much it will cost to make the switch, they must keep a close eye on the situation to see any potential roadblocks.

Secondly, there is a flight of capital as a result of the greater investment in infrastructure that must be made. Money from other businesses might be used to pay for the energy shift if that is what is needed. Net new investment from other sectors is assumed to contribute to half of the total new investment, according to the report's economic analysis. There are many unknowns in this transformation, therefore this study sought to uncover elements that lead to crowding out and evaluate how they can be disseminated throughout the global economy. Commercial banks, rather than the central bank, produce money when they issue new loans and establish corresponding deposits, according to the post-Keynesian theory of money supply. This point of view is backed up by facts and is widely accepted (Campiglio, 2016; Pollitt and Mercure, 2018). Ones in one region will not be displaced by larger investments in another because of this strategy A country's ability to create money depends on how policymakers use several tools and apply them in varying degrees to promote financial and economic stability. By setting a base rate, many developed countries limit the mandate of central banks to preserving the value of the currency (the interest rate at which the central bank lends money to commercial banks). As opposed to this, financial authorities in developing nations, such as central banks, use a wider range of policy instruments to manage the dynamics of credit expansion, such as reserve requirements and macroprudential measures (based on the review of Akinci and Olmstead-Rumsey, 2018; Campiglio, 2016; Cerutti et al., 2017; Zhang and Zoli, 2014). For their part, commercial banks have the power to produce new money, which gives them an advantage over private investors when it comes to the economy's overall health (Campiglio, 2016).

As a result, regions that rely more heavily on government support to finance the energy transition may be more susceptible to being forced out of the market. A greater share of public funds is needed to support the development of renewable energy sources in developing countries than more industrialized ones. This may be true, but there will be substantial regional disparities in the amount of crowding out as a result of the energy transition. This will exacerbate inequality as we move toward a more sustainable energy future. With 0% crowding out and 100 percent, we looked at the macroeconomic effects to see how crowding out might affect the economy. A large disparity in GDP and employment exists between the REmap Situation and the Reference Case when everyone is crammed into a small space (the employment impacts are shown in Figure 29). GDP and employment changes that are twice as great as those in the intermediate scenario (without crowding out) occur on average in most years (50 percent crowding out). However, in comparison to the 50% scenario, GDP and employment shrink less than in the case of a 100% scenario, but the decline is less pronounced than in the zero percent case.

6.0 CONCLUSION

We need to construct a green energy infrastructure. A comprehensive strategy to minimize GHG emissions is described here. Renewable energy helps ensure our energy infrastructure's long-term viability. Global energy efficiency will improve, according to a new IEA research (IEA). Energy change will cost $1.7 trillion per year by 2050. Renewable energy now outpaces fossil fuels. Technology has only recently gotten more inexpensive. $10/kWh Solar Thermal Energy With Thermal Storage In 2017. Renewable energy is popularizing. Transportation, heating, and other necessities absorb 20% of total energy. Solar thermal, geothermal, and biofuels could be economical in the future. Example CO2 emissions will be 11% lower in 125 years than in 2015. To stay below 2°C by 2050, we must reduce emissions by 470 Gt. A third of current fossil fuel usage could be cut using REmap by 2050. The oil and coal markets are expected to drop by 70% and 85%. IRENA's REmap Case calls for tripling industrial investment in low-carbon energy sources. Global renewable energy output should hit 85% by 2050. Decarbonizing the power supply will cost a trillion dollars in 30 years. Ethical and environmental changes in energy Energy, heat, and fuel regulation are now more precise. Between 2004 and 2017, $280 billion was invested in renewable energy. Capital costs boost jobs and economic growth. Leaner energy threatens debt-ridden nations. The authors found that REmap outperforms the Reference Case. The country's energy change will exacerbate crowding. Green energy will exacerbate existing inequalities. Renewable energy development demands more public financing in developing nations.

REFERENCES

50Hertz. (n.d.). *50Hertz - Reliable power supply to more than 18 million people*. https://www.50hertz.com/en/

Akinci, O., & Olmstead-Rumsey, J. (2018). How effective are macroprudential policies? An empirical investigation. *Journal of Financial Intermediation*, *33*, 33–57. doi:10.1016/j.jfi.2017.04.001

Campiglio, E. (2016). Beyond carbon pricing: The role of banking and monetary policy in financing the transition to a low-carbon economy. Ecological Economics, 121, 220–230. http://dx.doi.org/.ecolecon.2015.03.020 doi:10.1016/j

Cerutti, E., Claessens, S., & Laeven, L. (2017). The use and effectiveness of macroprudential policies: New evidence. *Journal of Financial Stability*, *28*, 203–224. doi:10.1016/j.jfs.2015.10.004

Coady, D., Parry, I. W. H., Sears, L., & Shang, B. (2015). *How Large Are Global Energy Subsidies?* IMF Working Paper.

Frankfurt School-UNEP Centre. (2018). Global trends in renewable energy investment 2018. Frankfurt School of Finance & Management.

IEA. (2015). *World Energy Balances 2015 edition (Database)*. IEA.

IEA. (2017). *World Energy Outlook 2017*. IEA.

IEA. (2018a). *Global Energy & CO2 Status Report 2017*. IEA.

IEA. (2018b). *Perspectives for the Energy Transition: The role of Energy Efficiency*. IEA.

IEA and IRENA. (2017). *Perspectives for the energy transition: Investment needs for a low-carbon energy system*. IRENA.

IRENA. (2016). *Renewable Energy Benefits: Measuring the Economics*. IRENA.

IRENA. (2017a). *Accelerating the Energy Transition through Innovation*. IRENA.

IRENA. (2017b). *Stranded Assets and Renewables: How the energy transition affects the value of energy reserves, buildings and capital stock*. IRENA.

IRENA. (2017d). *Renewable Energy and Jobs- Annual Review 2017*. IRENA.

IRENA. (2017e). *Untapped potential for climate action: Renewable energy in Nationally Determined Contributions*. IRENA.

IRENA. (2017f). *Renewable Energy Benefits: Leveraging Local Capacity for Solar PV*. LBNL.

IRENA. (2017g). *Renewable Energy Benefits: Leveraging Local Capacity for Onshore Wind*. IRENA.

IRENA. (2017h). *Renewable Energy Benefits: Leveraging Local Industries, International Renewable Energy Agency*. IRENA.

IRENA. (2018a). *Renewable Capacity Statistics 2018*. IRENA.

IRENA. (2018b). *New Global Commission to Examine Geopolitics of Energy Transformation*. http://www.irena. org/newsroom/pressreleases/2018/Jan/New-GlobalCommission-to-Examine-Geopolitics-of-EnergyTransformation

IRENA. (2018c). *Renewable Power Generation Costs in 2017*. IRENA.

IRENA. (2018d). *Renewable Energy Benefits: Leveraging Local Capacity for Offshore Wind*. IRENA.

IRENA & CPI. (2018). *Global landscape of renewable energy finance 2018*. CPI.

Lange, G.-M., Wodon, Q., & Carey, K. (2018). *The Changing Wealth of Nations 2018: Building a Sustainable Future*. World Bank. doi:10.1596/978-1-4648-1046-6

Pollitt, H., & Mercure, J.-F. (2018). The role of money and the financial sector in energy-economy models used for assessing climate and energy policy. *Climate Policy*, *18*(2), 184–197. doi:10.1080/14693062.2016.1277685

SE4ALL. (2016). *2015 SEforALL Global Tracking Framework, Sustainable Energy for All (SE4ALL)*. http://gtf.esmap.org/data/files/downloaddocuments/gtf-2105-full-report.pdf

Spiegel. (2018). *Neuzulassungen: Erstmals mehr als eine Million E-Autos (New Registrations: First time more than one Million Electric Vehicles)*. https://www.spiegel.de/auto/aktuell/emobilitaet-erstmals-ueber-eine-million-e-autoszugelassen-a-1197636.html

Zhang, L., & Zoli, E. (2014). *Leaning Against the Wind: Macroprudential Policy in Asia* (IMF Working Papers). International Monetary Fund (IMF).

Chapter 4
Data Envelopment Analysis and Its Application in Energy and Environment

Saransh Tiwari
https://orcid.org/0000-0002-3145-1432
Decision Sciences Area, Indian Institute of Management, Lucknow, India

Sanjeet Singh
https://orcid.org/0000-0001-5776-224X
Decision Sciences Area, Indian Institute of Management, Lucknow, India

ABSTRACT

The expansion of the industrial and manufacturing sectors is crucial to the economic well-being of all nations. Problems arise because this development is inevitably accompanied by air, water, and other types of pollution that are commonly linked to human health issues and climatic shifts. Thus, we need to think about how to strike a balance between economic growth and pollution reduction in order to keep global society as stable as possible. It is generally agreed that data envelopment analysis (DEA) is one approach of assessing sustainability. Therefore, this chapter is aimed at providing a comprehensive knowledge to the researchers and practitioners on the use of DEA in the field of energy and environment, including a discussion of the key principles and methodology in DEA used in environmental evaluation. This chapter, together with the explanation of conceptual and methodological advances, will serve as a road map for researchers' future investigations into the effects of DEA on energy and environmental problems.

INTRODUCTION

In recent years, environmental issues have become more widespread owing to fast industrialization and a growing human population. These issues pose a significant risk to the health and prosperity of nations worldwide. Loss of biodiversity, soil erosion, deforestation, water scarcity, and pollution of air and

DOI: 10.4018/978-1-6684-9130-0.ch004

water are some of the predominant local environmental challenges that are becoming more prominent as the impacts of environmental change keep on getting more severe (World Bank, 2008). In order for countries to meet the mutually agreed upon environmental sustainability goals and targets imposed by the international treaties such as the European Union climate and energy package (European Union, 2008), the Kyoto Protocol (United Nations, 1998), and the 2021 Glasgow Climate Pact, countries have experienced significant difficulties in reducing their consumption of natural resources, environmental pollution, and waste production. These difficulties have posed a challenge for countries to meet the goals and targets ensuring environmental sustainability. There is no room for debate regarding the importance of environmentally responsible practises and environmentally sustainable growth.

Even the United Nations' Intergovernmental Panel on Climate Change (IPCC), which was founded within the framework of the United Nations Environment Program, has admitted that the world's emissions of greenhouse gases are increasing. The IPCC recommended that countries reduce their emissions of greenhouse gases by about 40-70% from their levels in 2010 until 2050, with the goal of reducing emissions to nearly zero by the end of the 21st century (Sueyoshi et al., 2017). If we do not act soon, the globe could be forced to deal with catastrophic droughts, heat waves, and other disruptive climate changes, all of which have the potential to cause harm to people as well as social and economic systems.

The concerns raised and policy suggestions provided by these world organisations are important to be addressed to ensure sustainable development, but these organisations do not imply any methodological guidance to keep a track of the development in environmental progress. Research that is founded on facts and is driven by data can assist policymakers in maintaining a sharper focus on the policy goals they have established, facilitating interactions with important stakeholders, and achieving the highest possible return on investment in environmental projects. As a result, to keep a check on the environmental progress and sustainability via quantitative approach, energy and environmental researchers supported the application of Data Envelopment Analysis (DEA). DEA has the capability to assess the energy and environmental efficiency and thus, has the potential to facilitate and guide the policymakers regarding the effectiveness of their proposed policies intended to combat undesirable environmental factors as pollution and global warming, among others.

DEA is a powerful Operations Research tool embodied with the capability to monitor the performance of entities. It is a non-parametric mathematical computational technique that evaluates the relative efficiency of a set of entities (technically termed as decision-making units (DMUs)), where each entity is responsible for transforming multiple inputs to multiple outputs (Charnes et al., 1978). From a geometrical point of view, DEA floats a piece-wise linear surface that envelops all the observed data points (Cooper et al., 2004), then compares each DMU with the best-practice DMU that rests on the surface (efficient frontier). The soundness and versatility of the DEA approach is evident from the following advantages that it has over other methods:

(i) In terms of performance assessment, it is flexible enough to deal with multiple inputs and multiple outputs.
(ii) It does not require any information regarding the functional form of the production technology, that is, functional forms representing relationships between inputs and outputs.
(iii) It even works with a small sample size dataset.
(iv) As a polarising strategy (extreme-point method) that compares and contrasts all DMUs to the ideal (best practises DMU), DEA provides an evaluation of relative performance, which is of greater importance to decision-makers.

Energy and Environmental performance evaluation is a complex task owing to the fact that it involves undesirable factors to be taken into account and disposability assumptions. DEA owing to its advantages serves as a method of choice for evaluating performance in the context of complex relations between input and output factors.

Thus, this book chapter aims to highlight the DEA methodology and its extensions to accommodate complex scenarios in Energy and Environmental performance evaluation. This chapter will serve as basis for the researchers and policymakers to identify suitable DEA model to be used in their respective analysis to assess the efficacy and fruitfulness of the existing proposed policies.

Following the Introduction section, we describe the basic DEA methodological framework in the immediate next section. Thereafter, we will introduce and explain the extended DEA models that have the capability to handle the undesirable outputs in the complex Energy and Environmental performance evaluation. Next, we will classify published studies in this field based on their methodological aspects and application theme. In the section that will follow, we will provide suggestions and guidelines on choosing the suitable DEA model based on the need of the researcher and the application in which it is to be utilized. This section will explain model selection in terms of reference technology, returns to scale (RTS), disposability assumptions, and other handling other complexities which are application specific. Finally, we will discuss some of the limitations of the existing DEA models and directions for future research followed by concluding remarks.

DEA METHODOLOGY

Basic DEA Framework

Built on the foundations of Farrell's (1957) work, DEA is well-established and popular performance evaluation methodology. DEA serves to evaluate the relative efficiencies of a set of homogeneous entities, termed as DMUs, all of which consume same set of inputs to produce same set of outputs via similar operations and functioning. The DMUs can be hospitals, organizations, individuals, and countries, among others. Figure 1 presents the general structure of a production process. DEA has potentially strong advantages which have made it such a popular tool amongst economists and Operations and Management Science researchers. To highlight a few, DEA can handle multiple inputs and outputs in a single model, it does not need information regarding the weights to be allocated to respective inputs and outputs, and it does not require any prior assumption regarding the functional form of the production technology.

Figure 1. A general production process.

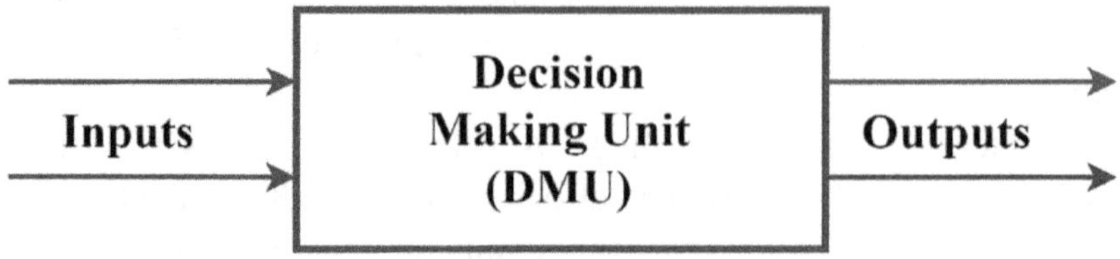

To get an insight of the basic DEA model, suppose that there are n DMUs whose performances are to be evaluated. Each DMU_j ($j = 1, \ldots, n$) uses m inputs x_{ij} ($i = 1, \ldots, m$) to produce s outputs y_{rj} ($r = 1, \ldots, s$). Utilizing the concept of efficiency in engineering, the efficiency of a DMU, say DMU_o ($o = 1, \ldots, n$), can be calculated by evaluating the ratio of its virtual output to its virtual input, where virtual output and virtual input represent weighted combination of outputs and weighted combination of inputs, respectively for the DMU under consideration. But the problem arises as to what are the suitable weights that should be allocated to outputs and inputs. To avoid this problem of arbitrariness in weights assignment to inputs and outputs, a DEA model, popularly known as Charnes, Cooper, and Rhodes (CCR) model, was proposed by Charnes et al. (1978). Using a mathematical programming paradigm, the model seeks appropriate optimal weights to be given to the inputs and outputs. Subject to the constraints that the ratio of virtual output to virtual input should be less than or equal to unity for all the DMUs, this model maximizes the same ratio for the DMU under evaluation. Ever since its inception, the mathematical modelling in DEA has grown significantly.

The CCR model is a linear-fractional programming model that can be transformed into an equivalent linear programming model by utilizing the Charnes-Cooper transformation (Charnes & Cooper, 1962). The linearized version of the CCR model is presented as follows:

$$\max \sum_{r=1}^{s} u_r y_{ro}$$

$$\text{s.t. } \sum_{i=1}^{m} v_i x_{io} = 1$$

$$\sum_{r=1}^{s} u_r y_{rj} - \sum_{i=1}^{m} v_i x_{ij} \leq 0, j = 1, \ldots, n \quad (1)$$

$$u_r, v_i \geq 0, r = 1, \ldots, s; i = 1, \ldots, m.$$

Model (1) is called the CCR model in the multiplier form. The efficiency for each of the DMU can be calculated by solving this model separately for each DMU. The dual of model (1) represents the CCR model in envelopment form, which is more popular among the researchers owing to its simpler structure and its ability to identify target input and output values. Here we present the CCR model in its envelopment form, as follows:

$$\min \theta$$

s.t. $\sum_{j=1}^{n} x_{ij}\lambda_j \leq \theta x_{io},, i=1,\ldots,m$

$$\sum_{j=1}^{n} y_{rj}\lambda_j \geq y_{ro},, r=1,\ldots,s \tag{2}$$

$\lambda j_{\geq}0$, j=1,…,n.

Model (2) is an input-oriented envelopment variant of the CCR model. It is worth noting that the CCR model has been developed with the constant RTS (CRS) assumption. In the next subsection, we will further discuss the characterization of RTS. This model works to reduce DMU's inputs in a proportionate manner without decreasing its outputs. The constraints of model (2) represent a piecewise linear production technology that has strong disposable inputs and outputs and exhibits CRS:

$\boldsymbol{T} = \{(\mathbf{x},\mathbf{y})$: $\sum_{j=1}^{n} \lambda_j x_{ij} \leq x_i,, i=1,\ldots,m,$

$\sum_{j=1}^{n} \lambda_j y_{rj} \geq y_r,, r=1,\ldots,s,$

$$\lambda j_{\geq}0, \text{ j=1,…,n}\}. \tag{3}$$

In equation (3), $\mathbf{x} = (x_1, x_2 \ldots, x_m)$ and $\mathbf{y} = (y_1, y_2 \ldots, y_s)$ represent the vectors of inputs and outputs, respectively. $\boldsymbol{T}$ contains all feasible combinations of inputs and outputs and is called the reference technology.

Figure 2 presents the general framework and structure of the DEA. As is evident from model (3) and equation (2), both the reference technology and efficiency measure need to be defined before a DEA model can be created. The reference technology is constructed based on the RTS assumption and operating characteristics and disposability consideration of inputs and outputs. The model's orientation and type define the characteristics of the efficiency measure. So, to construct a DEA model, we need a solid foundation of the efficiency measure and the underlying reference technology, both of which are the key constituents of any DEA model. In the sections that follow, we will investigate each of these constituents in detail.

Figure 2. General framework and structure of the DEA

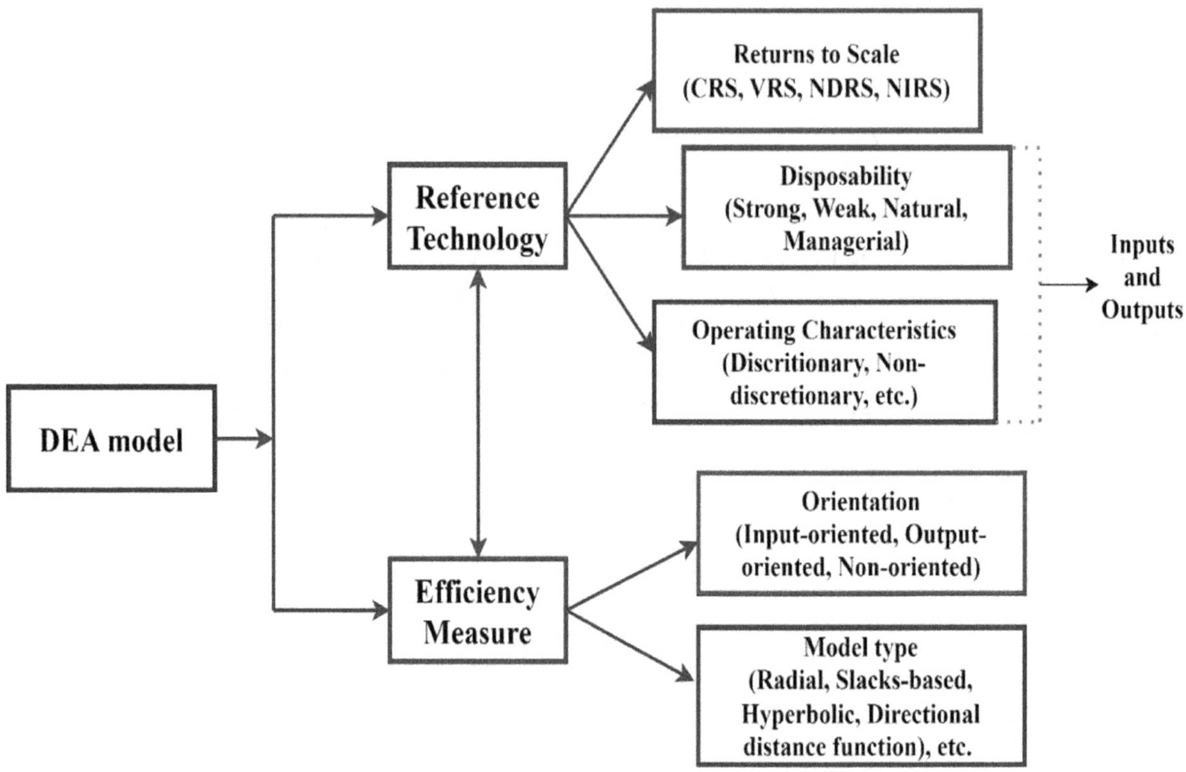

Returns to Scale (RTS) in DEA

RTS form an important subpart of the reference technology. It aids in delineating the efficient boundary against which DMUs are to be benchmarked. The RTS is ensured in the DEA model by means of the following constraint:

$$L \leq \sum_{j=1}^{n} \lambda_j \leq U,$$

where L and U refer to the lower and upper limits, respectively on the summation of the intensity variables λj. The suitable values of L and U determine the corresponding RTS that the DEA model possesses. Our DEA model can have any one of the following RTS:

(i) Constant RTS (CRS)

In CRS, we have $L = 0$ and $U = \infty$. The reference technology presented in equation (3) corresponds to the CRS with the consideration of strongly disposable inputs and outputs. Figure 3 shows the production feasible set and the efficient frontier for CRS.

Figure 3. CRS frontier

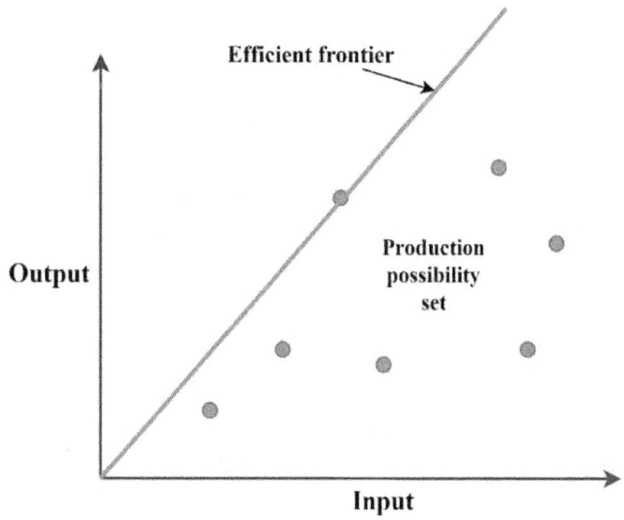

It is evident that in CRS, the proportional increase in input will produce equal proportional increase in output, that is, $\frac{\Delta y}{y} = \frac{\Delta x}{x}$. The condition that $L = 0$ and $U = \infty$ assumes that it is possible to reduce as well as expand the scale of the DMU.

(ii) Variable RTS (VRS)

In VRS, we have $L = 1$ and $U = 1$. The reference technology for the VRS with the consideration of strongly disposable inputs and outputs is defined as follows:

$\boldsymbol{T}_{VRS} = \{(\mathbf{x},\mathbf{y})\text{: } \sum_{j=1}^{n} \lambda_j x_{ij} \le x_i,\ i=1,\ldots,m,$

$\sum_{j=1}^{n} \lambda_j y_{rj} \ge y_r,\ r=1,\ldots,s,$

$\sum_{j=1}^{n} \lambda_j = 1$

$\lambda j_{\geq} 0,\ j=1,\ldots,n\}.$ (4)

Figure 4 shows the production feasible set and the efficient frontier for VRS.

Figure 4. VRS frontier

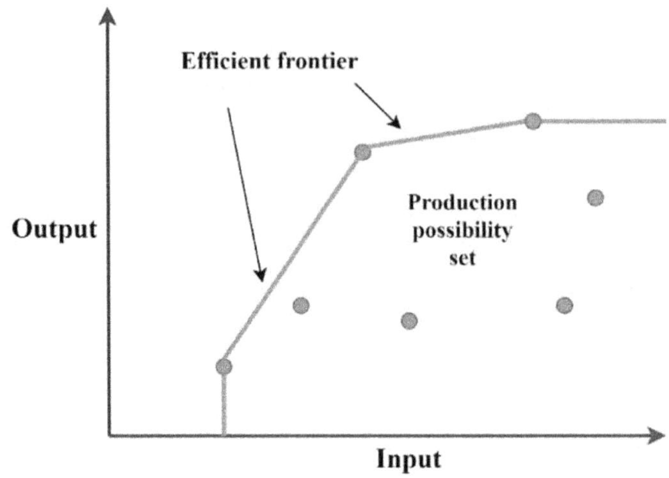

In VRS, for the initial part of the efficient frontier, we have $\frac{\Delta y}{y} \geq \frac{\Delta x}{x}$ which means that the proportional increment in output is greater than the proportional increment in input, for the second part of the efficient frontier, we have $\frac{\Delta y}{y} = \frac{\Delta x}{x}$ which means that the proportional rise in output is equal to the proportional rise in input, and for the last part of the efficient frontier, we have $\frac{\Delta y}{y} \leq \frac{\Delta x}{x}$ which means that the proportional increase in output is less than the proportional increase in input. Thus, the VRS ensures that it gives priority to smaller DMUs where the RTS is increasing as well as larger DMUs where the RTS is decreasing.

(iii) Non-decreasing RTS (NDRS)

In NDRS, we have $L = 1$ and $U = \infty$. NDRS is sometimes also known as the increasing RTS. The reference technology for the NDRS with the consideration of strongly disposable inputs and outputs can be presented as follows:

Figure 5. NDRS frontier

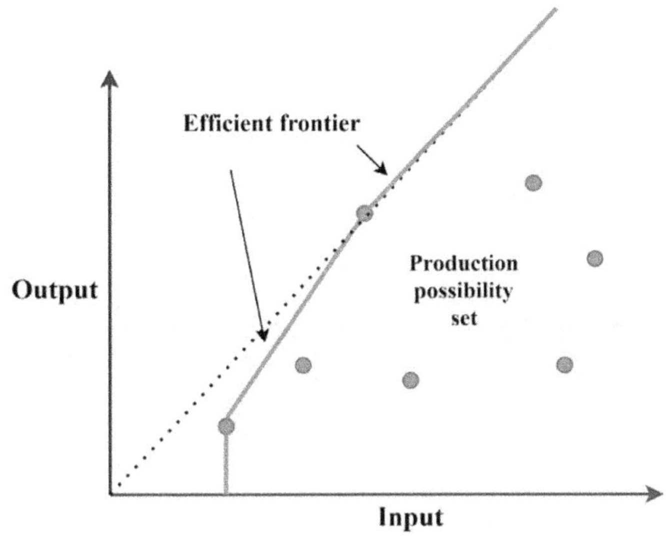

$$T_{NDRS} = \{(\mathbf{x},\mathbf{y}): \sum_{j=1}^{n} \lambda_j x_{ij} \leq x_i, i=1,\ldots,m,$$

$$\sum_{j=1}^{n} \lambda_j y_{rj} \geq y_r, r=1,\ldots,s,$$

$$1 \leq \sum_{j=1}^{n} \lambda_j$$

$$\lambda j_{\geq} 0, j=1,\ldots,n\}. \tag{5}$$

Figure 5 shows the production feasible set and the efficient frontier for NDRS.

In NDRS, by incorporating $1 \leq \sum_{j=1}^{n} \lambda_j$ it is assumed that although we can expand the scale of the DMU but it is not possible to reduce its scale. It is evident from the efficient frontier of NDRS that the proportional increase in output will always be larger than or equal to the proportional increase in input. Mathematically, we have $\frac{\Delta y}{y} \geq \frac{\Delta x}{x}$ in NDRS. Thus, NDRS puts emphasis on relatively smaller DMUs where RTS is increasing. In the first part of the efficient frontier $\frac{\Delta y}{y} > \frac{\Delta x}{x}$, while in the latter part of the efficient frontier $\frac{\Delta y}{y} = \frac{\Delta x}{x}$.

(iv) Non-Increasing RTS (NIRS)

In NIRS, we have $L = 0$ and $U = 1$. NIRS is sometimes also known as the decreasing RTS. The reference technology for the NIRS with the consideration of strongly disposable inputs and outputs is presented as follows:

$T_{NIRS} = \{(\mathbf{x},\mathbf{y})\text{: } \sum_{j=1}^{n} \lambda_j x_{ij} \leq x_i \text{, } i=1,\ldots,m,$

$\sum_{j=1}^{n} \lambda_j y_{rj} \geq y_r \text{, } r=1,\ldots,s,$

$0 \leq \sum_{j=1}^{n} \lambda_j \leq 1$

$\lambda_j \geq 0, j=1,\ldots,n\}.$ (6)

Figure 6 shows the production feasible set and the efficient frontier for NIRS.

Figure 6. NIRS frontier

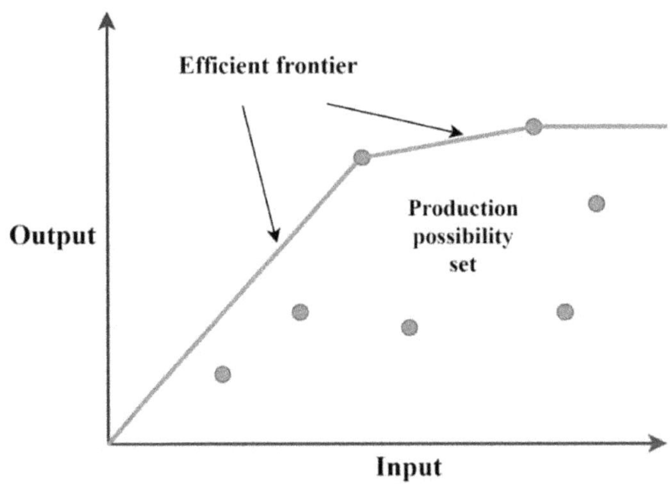

In NIRS, the condition $0 \leq \sum_{j=1}^{n} \lambda_j \leq 1$ ensures that the scale of the DMU cannot be expanded but it is possible to reduce its scale. It is evident from the efficient frontier of NIRS that the proportional increase in output will always be lesser than or equal to the proportional increase in input. Mathematically, we have $\frac{\Delta y}{y} \leq \frac{\Delta x}{x}$ in NIRS. Thus, NIRS puts emphasis on relatively larger DMUs where RTS is decreasing. In the first part of the efficient frontier $\frac{\Delta y}{y} = \frac{\Delta x}{x}$, while in the latter part of the efficient frontier $\frac{\Delta y}{y} < \frac{\Delta x}{x}$.

An essential point to take into consideration here is that although we have discussed four types of returns to scale (RTS), CRS and VRS have been popularly utilized in most of the existing research studies.

EXTENSIONS TO DEA MODELLING

In this section, we further elaborate the DEA modelling framework and discuss the extensions of DEA models by incorporating undesirable outputs.

Disposability Assumptions

In the conventional DEA setting, it is normally assumed that the inputs and outputs are freely (strongly) disposable, that is, if $(\mathbf{x}, \mathbf{y}) \in T$ and $\hat{x} \geq \mathbf{x}$ (or $\hat{y} \leq \mathbf{y}$), then $(\hat{x}, y) \in T$ (or $(x, \hat{y}) \in T$). However, this assumption usually does not work in line with the actual production process in real life. In a production process, desirable outputs are accompanied by undesirable outputs production. Consider for example an electricity generation power plant that utilizes fossil fuels to produce electricity. The electricity generated is the desirable output but it is also accompanied by the production of carbon dioxide, sulphur dioxide, and other harmful emissions, which are undesirable outputs. These undesirable outputs cannot be disposed-off costlessly (Zhou et al., 2008). There is always some cost associated with their reduction. It is therefore not wise to use the strong disposability reference technology in such cases. So, in an attempt to model the production of undesirable outputs, researchers have come up with extended disposability assumptions which are discussed as follows.

(i) Natural Disposability and Managerial Disposability

To illustrate the concept of natural and managerial disposability, let us now introduce undesirable outputs into our DEA framework. Suppose in the production process, accompanying the desirable outputs, q number of undesirable outputs are also produced. Let b_{tj} ($t = 1, \ldots, q$) denote the tth undesirable output of DMU$_j$. The reference technology for the natural disposability is similar to the strong disposability and can be expressed as follows:

$$T_{natural} = \{(\mathbf{x},\mathbf{y},\mathbf{b}){:}\ \textstyle\sum_{j=1}^{n}\lambda_j x_{ij} \leq x_i, i=1,\ldots,m,$$

$$\textstyle\sum_{j=1}^{n}\lambda_j y_{rj} \geq y_r, r=1,\ldots,s,$$

$$\textstyle\sum_{j=1}^{n}\lambda_j b_{tj} \leq b_t, t=1,\ldots,q,$$

$$L \leq \textstyle\sum_{j=1}^{n}\lambda_j \leq U$$

$$\lambda j_{\geq}0, \mathrm{j}=1,\ldots,\mathrm{n}\}. \quad (7)$$

The reference technology considering the managerial disposability assumption can be expressed as follows:

$T_{managerial} = \{(\mathbf{x,y,b})\text{: } \sum_{j=1}^{n}\lambda_j x_{ij} \geq x_i, i=1,\ldots,m,$

$\sum_{j=1}^{n}\lambda_j y_{rj} \geq y_r, r=1,\ldots,s,$

$\sum_{j=1}^{n}\lambda_j b_{tj} \leq b_t, t=1,\ldots,q,$

$L \leq \sum_{j=1}^{n}\lambda_j \leq U$

$\lambda j_{\geq} 0, j=1,\ldots,n\}.$ (8)

Here $\mathbf{b} = (b_1, \ldots, b_q)$ denotes the vector of undesirable outputs. In the constraint $L \leq \sum_{j=1}^{n}\lambda_j \leq U$, suitable values for L and U can be chosen depending on the RTS consideration.

The noticeable difference between natural and managerial disposability assumption is that in natural disposability $\sum_{j=1}^{n}\lambda_j x_{ij} \leq x_i$, $i=1,\ldots,m$, holds implying that a DMU can reach the efficient frontier by reducing its consumption of inputs. While in managerial disposability $\sum_{j=1}^{n}\lambda_j x_{ij} \geq x_i$, $i=1,\ldots,m$, holds implying that a DMU can attain efficient frontier by increasing a directional vector of inputs. Managerial disposability assumptions depict a production process more closely as it asserts that undesirable outputs can be decreased by employing more resources. For instance, in a production process the undesirable outputs may be caused by utilization of worn-out machines. So, in such cases, more resources can be employed to replace these machines and thus, reduce undesirable outputs.

Common trait shared by the two notions of disposability assumptions is that in both we have $\sum_{j=1}^{n}\lambda_j y_{rj} \geq y_r$, $r=1,\ldots,s$, and $\sum_{j=1}^{n}\lambda_j b_{tj} \leq b_t$, $t=1,\ldots,q$. These conditions signify that the efficient frontier for the desired outputs should be situated either above or on all of the data points, whilst the frontier for the undesirable outputs should be situated either below or on the data points.

(ii) Strong Disposability and Weak Disposability

The strong disposability is similar to the natural disposability and its reference technology can be referred to from equation (7). The reference technology for the weak disposability assumption is presented as follows:

$T_{weak} = \{(\mathbf{x,y,b})\text{: } \sum_{j=1}^{n}\lambda_j x_{ij} \leq x_i, i=1,\ldots,m,$

$\sum_{j=1}^{n}\lambda_j y_{rj} \geq y_r, r=1,\ldots,s,$

$\sum_{j=1}^{n}\lambda_j b_{tj} = b_t, t=1,\ldots,q,$

$L \leq \sum_{j=1}^{n}\lambda_j \leq U$

$\lambda j_{\geq}0, j=1,\ldots,n\}.$ (9)

The environmental DEA technology is another name for the weak disposability reference technology that is described in equation (9). The distinction between strong and weak disposability lies in the fact that weak disposability allows for the possibility of proportionate reduction of both desirable and undesirable outputs but does not allow for the possibility of reduction of solely undesirable outputs. The reference technology of weak disposability satisfies the property of null-jointness which states that if $(\mathbf{x}, \mathbf{y}, \mathbf{b}) \in T_{weak}$ and $\mathbf{b} = 0$, then $\mathbf{y} = 0$ (Färe et al., 2005). This property implies that it is possible to eliminate all the undesirable outputs only by ending the production process. Weak disposability asserts the fact that one of the ways to reduce unwanted outputs is to decrease the activity level of the production. Thus, it appears out to be more practical representation of the real-life production phenomenon. Energy and Environmental performance evaluations in existing studies have made extensive use of the idea of weak disposability. See, for example, Cui (2019), Piao et al. (2019), Sueyoshi and Goto (2012), and Yang and Pollitt (2010).

The operating characteristics of inputs and outputs also frame the reference technology. The operating characteristics refer to the discretionary and non-discretionary nature of variables, and categorical variables, among others. These are used depending on the specific nature of the application addressed by the research problem.

Models and Approaches to Handle Undesirable Outputs

In the usual case of evaluating performance of DMUs, only desirable outputs and inputs are considered. But when evaluating Energy and Environmental performance, undesirable outputs also come in the picture. The conventional DEA models work to evaluate performance based on the fact that greater the output, better is the performance. However, with undesirable outputs higher the better does not work, rather, lesser the undesirable output better the performance works. So, there arises a need for modifying the traditional DEA models to make them become more flexible and compatible to accommodate undesirable factors and their properties into their setting. Here, we introduce some of the most prominent techniques for addressing undesirable outputs across a range of efficiency measures.

(i) Ignoring Undesirable Outputs

Excluding undesirable outputs from the analysis is one strategy for dealing with them. Most of the researchers in the very beginning era of DEA development used this approach. But ignoring undesirable outputs from the production technology is not a very wise idea because it will lead to producing misleading results and will not give a comprehensive picture of DMUs performances (Yang & Pollitt, 2009). Excluding undesirable outputs from the study implies that they do not pose any value in the evaluation process. In point of fact, undesirable outputs are unavoidably created alongside desirable outputs. This cannot be avoided. Therefore, it is necessary to account them in the performance evaluation process. Some previous researches that have discarded undesirable outputs are presented in Table 1.

Table 1. Studies ignoring undesirable outputs from their analysis

Reference	Outcome
Hailu and Veeman (2001)	This study examined the increase of productivity in the pulp and paper sector in Canada. It showed that typical metrics that omit undesirable outputs greatly understate the true productivity figure.
Pathomsiri et al. (2008)	Through a comparison with results from models without unwanted outputs, this study assessed the efficiency of 56 American airports between 2000 and 2003.
He et al. (2013)	Based on an analysis of data from 50 Chinese iron and steel companies, this research concluded that ignoring unfavourable outcomes leads to skewed efficiency improvements and technological advancements.
Yang and Pollitt (2009)	The research analysed data from 582 China's coal-fired power stations to demonstrate the enormous impact that enforcing the technically proper disposability features on unwanted outputs has on the overall efficiency assessment.

Note: This table is adapted from Halkos and Petrou (2019).

(ii) Using Undesirable Outputs as Inputs in the Process

One of the popular ways of dealing with undesirable outputs is to treat them like normal inputs. In DEA, DMUs are evaluated such that lesser the input consumption better is the performance. Undesirable outputs also have the same tendency that lesser their amount better is the performance. So, treating the undesirable outputs as inputs is one of the potential ways of dealing with it. Some of the studies that have utilized this concept are as follows.

By modelling 24 European coal-fired power stations and treating emissions as inputs, Korhonen and Luptacik (2004) assessed the eco-efficiency of these facilities, with the understanding that reducing both inputs and unwanted outputs would improve efficiency. By conceptualizing bad loans as a byproduct of the loan production process, Fukuyama and Weber (2008) uncovered the full extent of inefficiencies in Japanese banking and the true cost of bad debt. The typical CCR model was expanded by Amirteimoori et al. (2006) into a DEA-like model that addressed relative efficiency by reducing undesirable outputs and raising undesirable inputs. Utilizing information on nitrogen excess, phosphate excess, and aggregate (direct and indirect) energy utilization of an imbalanced board of Dutch dairies, Reinhard et al. (2000) inferred thorough ecological efficiency measures for these foundations.

(iii) Data Transformation

Another possible way of dealing with undesirable outputs is by applying necessary transformation. It is well known that higher the value of undesirable outputs, poor is the performance while higher the value of desirable outputs better is the performance. So, by applying certain transformations on undesirable output data, if we can convert the undesirable outputs into desirable output form, then we can easily model them using the conventional DEA approach. Some of these transformations include the following:

(a) By multiplying the undesirable output data by (-1), that is $(b)*(-1) = -b$, where b is the undesirable output data. By doing so, we can convert undesirable outputs into desirable ones. But a problem may arise because the data will become negative and conventional DEA models require that all data are non-negative. So, this approach can be used only with extended DEA models which have the capability to deal negative data. See foe example, Kao (2020, 2022) and Tone et al. (2020).

(b) Another way to transform undesirable outputs into desirable form can be derived from the extension of the approach presented in (a). Approach states multiply the undesirable output data with (-1) and then add certain constant (α) to make all data points positive, that is, $(b)*(-1) + \alpha = (-b + \alpha)$. In this approach, the obtained data points can be evaluated by the conventional DEA methodology only and there will be no need to extend models to negative data. But an important point to note is that efficiency scores may depend on α.

(c) We can also convert the undesirable outputs into desirable form by taking the multiplicative inverse of the undesirable output data, that is, $(1/b)$. This approach can be best applied when all our undesirable output data are non-negative.

(iv) Hyperbolic Output Efficiency Measure

Hyperbolic output efficiency measure attempts to proportionally expand desirable outputs and contract undesirable outputs simultaneously in a single model. Efficiency index based on a hyperbolic function measure, proposed by Färe et al. (1989), asserts that the production process should be evaluated in accordance with the environmentally friendly standards. This implies the fact that the DMUs which produce lesser undesirable outputs and higher desirable outputs should be rated better than those with higher undesirable output production. Considering the strong disposability assumption, the hyperbolic output efficiency measure for a DMU, say DMU_o, is defined as follows:

$$\max \rho o$$

$$\text{s.t. } \sum_{j=1}^{n} x_{ij}\lambda_j \le x_{io},\ i=1,\ldots,m$$

$$\sum_{j=1}^{n} y_{rj}\lambda_j \ge \rho_o y_{ro},\ r=1,\ldots,s \tag{10}$$

$$\sum_{j=1}^{n} b_{tj}\lambda_j \le \frac{1}{\rho_o} b_{to},\ t=1,\ldots,q$$

$$L \le \sum_{j=1}^{n} \lambda_j \le U$$

$$\lambda j_{\ge} 0,\ j=1,\ldots,n\}.$$

Similarly, under the weak disposability assumption, we can define the hyperbolic output efficiency measure for a DMU, say DMU_o as follows:

$$\max \rho o$$

$$\text{s.t. } \sum_{j=1}^{n} x_{ij}\lambda_j \le x_{io},\ i=1,\ldots,m$$

$$\sum_{j=1}^{n} y_{rj}\lambda_j \geq \rho_o y_{ro},\ r=1,\ldots,s \tag{11}$$

$$\sum_{j=1}^{n} b_{tj}\lambda_j = \frac{1}{\rho_o} b_{to},\ t=1,\ldots,q$$

$$L \leq \sum_{j=1}^{n} \lambda_j \leq U$$

$\lambda j_{\geq}0$, j=1,…,n}.

Here in models (10) and (11), the constraint $L \leq \sum_{j=1}^{n} \lambda_j \leq U$ corresponds to RTS. A suitable value of L and U chosen will define the corresponding RTS. An important point to note is that models (10) and (11) are non-linear because of the constraints $\sum_{j=1}^{n} b_{tj}\lambda_j \leq \frac{1}{\rho_o} b_{to}$, $t=1,\ldots,q$, and $\sum_{j=1}^{n} b_{tj}\lambda_j = \frac{1}{\rho_o} b_{to}$, $t=1,\ldots,q$, respectively. However, these constraints can be linearized by using the Taylor series approximation around $\rho o = 1$ as follows:

$\Rightarrow \sum_{j=1}^{n} b_{tj}\lambda_j \leq \frac{1}{\rho_o} b_{to}$, $t=1,\ldots,q$, can be linearized into $\sum_{j=1}^{n} b_{tj}\lambda_j \leq 2b_{to} - \rho o_b t_{o,}$ t=1,…,q.

$\Rightarrow \sum_{j=1}^{n} b_{tj}\lambda_j = \frac{1}{\rho_o} b_{to}$, $t=1,\ldots,q$, can be linearized into $\sum_{j=1}^{n} b_{tj}\lambda_j = 2b_{to} - \rho o_b t_{o,}$ t=1,…,q.

(v) Directional Distance Function (DDF)

The directional distance function (DDF) method is another common strategy for dealing with undesirable outputs. This approach makes an effort to simultaneously expand the desirable outputs and concurrently reduce the undesirable outputs and inputs along a specified directed vector in a proportionate fashion. DDF approach also works on the same ideology as the hyperbolic measure that DMU should be evaluated as per environmental standards wherever required. Consider the scenario shown in Figure 7, which consists of a single input and a single output, in order to have a grasp on the DDF idea. Here the DMUs are labelled as A, B, and C. Given a directional vector $\mathbf{g} = (g^-, g^+)$ along which the DMUs are to be evaluated, it can be observed that DMU C is efficient as it lies on the efficient frontier, the directional distance score for DMU A is AD/O**g** and that for DMU B is BE/O**g**. The DDF score basically represents the extent of inefficiency in the DMU.

Figure 7. DDF conceptualization

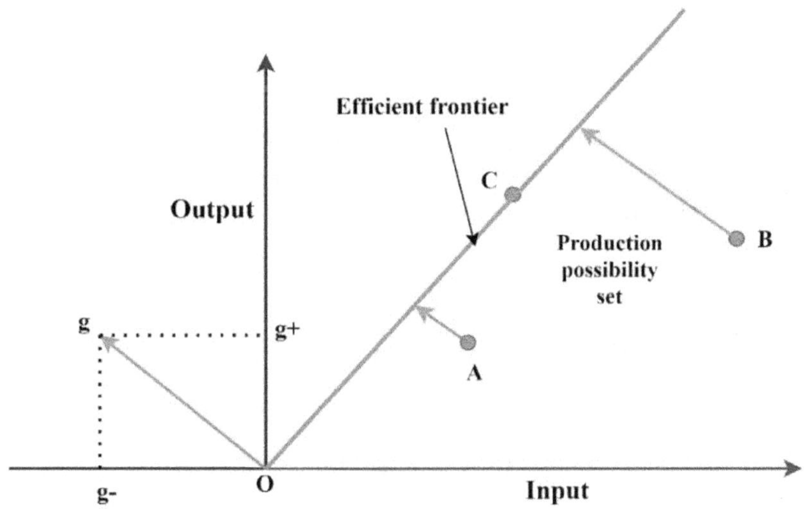

The DDF model for a DMU, say DMU$_o$, along a direction vector

$$\mathbf{g} = (g_1^-, \ldots, g_m^-, g_1^+, \ldots, g_s^+, g_1^{b-}, \ldots, g_q^{b-}),$$

under the strong disposability assumption can be defined as follows:

$$\max \beta o$$

$$\text{s.t. } \sum_{j=1}^{n} x_{ij}\lambda_j \leq x_{io} - \beta_o g_i^-, i=1,\ldots,m$$

$$\sum_{j=1}^{n} y_{rj}\lambda_j \geq y_{ro} + \beta_o g_r^+, r=1,\ldots,s \tag{12}$$

$$\sum_{j=1}^{n} b_{tj}\lambda_j \leq b_{to} - \beta_o g_t^{b-}, t=1,\ldots,q$$

$$L \leq \sum_{j=1}^{n} \lambda_j \leq U$$

$\lambda j_{\geq} 0$, j=1,…,n}.

Similarly, the DDF model for a DMU, say DMU$_o$, along a direction vector

$$\mathbf{g} = (g_1^-, \ldots, g_m^-, g_1^+, \ldots, g_s^+, g_1^{b-}, \ldots, g_q^{b-}),$$

considering the assumption of weak disposability can be defined as follows:

max βo

s.t. $\sum_{j=1}^{n} x_{ij}\lambda_j \leq x_{io} - \beta_o g_i^-$, i=1,…,m

$\sum_{j=1}^{n} y_{rj}\lambda_j \geq y_{ro} + \beta_o g_r^+$, r=1,…,s (13)

$\sum_{j=1}^{n} b_{tj}\lambda_j = b_{to} - \beta_o g_t^{b-}$, t=1,…,q

$L \leq \sum_{j=1}^{n} \lambda_j \leq U$

λj$_\geq$0, j=1,…,n}.

Here in models (12) and (13), the optimal value of the objective function, that is, β_o^* is the DDF score for the DMU$_o$. The DMU is said to be efficient if β_o^* = 0, and inefficient if $\beta_o^* > 0$.

For more details on how to choose the values for the direction vector

$$\mathbf{g} = (g_1^-, \ldots, g_m^-, g_1^+, \ldots, g_s^+, g_1^{b-}, \ldots, g_q^{b-}),$$

we recommend the readers to refer Tone (2016).

(vi) Slacks-Based Measure

Unlike other models discussed so far in this chapter, slacks-based DEA models evaluate the DMUs based on non-radial contraction of inputs and undesirable outputs and non-radial expansion of desirable outputs. Other models attempt to proportionally contract and expand factors involved in production process while slacks-based measure work on the approach to non-radially attain the efficient frontier. Non-oriented slacks-based DEA model for gauging efficiency of a DMU, say DMUo, under strong disposability assumption can be presented as follows:

$$\min \theta = \left(1 - \frac{1}{m}\sum_{i=1}^{m}\frac{s_i^-}{x_{i0}}\right) / \left(1 + \frac{1}{s+q}\left(\sum_{r=1}^{s}\frac{s_r^+}{y_{r0}} + \sum_{t=1}^{q}\frac{s_t^-}{b_{t0}}\right)\right)$$

s.t. $\sum_{j=1}^{n} \lambda_j x_{ij} + s_i^- = x_{i0}$, i=1,…,m

$\sum_{j=1}^{n} \lambda_j y_{rj} - s_r^+ = y_{r0}$, r=1,…,s (14)

$$\sum_{j=1}^{n}\lambda_j b_{tj} + s_t^- = b_{t0},\ t=1,\ldots,q$$

$$L \le \sum_{j=1}^{n}\lambda_j \le U$$

$$\lambda_j, s_i^-, s_r^+, s_t^- \ge 0, j=1,\ldots,n;\ i=1,\ldots,m;\ r=1,\ldots s;\ t=1,\ldots,q$$

Similarly, the slacks-based DEA model under the weak disposability assumption for assessing the efficiency of a DMU, say DMU$_o$, in the non-oriented form can be defined as follows:

$$\min\theta = \left(1-\frac{1}{m}\sum_{i=1}^{m}\frac{s_i^-}{x_{i0}}\right) / \left(1+\frac{1}{s+q}\left(\sum_{r=1}^{s}\frac{s_r^+}{y_{r0}}+\sum_{t=1}^{q}\frac{s_t^-}{b_{t0}}\right)\right)$$

$$\text{s.t.}\ \sum_{j=1}^{n}\lambda_j x_{ij} + s_i^- = x_{i0},\ i=1,\ldots,m$$

$$\sum_{j=1}^{n}\lambda_j y_{rj} - s_r^+ = y_{r0},\ r=1,\ldots,s \tag{15}$$

$$\sum_{j=1}^{n}\lambda_j b_{tj} + s_t^- = \left(1+\frac{1}{s}\sum_{r=1}^{s}\frac{s_r^+}{y_{r0}}\right) b_{t0},\ t=1,\ldots,q$$

$$L \le \sum_{j=1}^{n}\lambda_j \le U$$

$$\lambda_j, s_i^-, s_r^+, s_t^- \ge 0, j=1,\ldots,n;\ i=1,\ldots,m;\ r=1,\ldots s;\ t=1,\ldots,q$$

In models (14) and (15), the constraint $L \le \sum_{j=1}^{n}\lambda_j \le U$ corresponds to the RTS based on the suitable values of L and U chosen. It is to be noted that models (14) and (15) are non-linear but they are permissible for linearization by the use of the popular Charnes-Cooper transformation. To see the linearization process, one can refer to Chang et al. (2014).

Productivity Change Over Time: Malmquist Productivity Index (MPI)

Investigation of how levels of productivity fluctuate over time is a topic that generates a lot of interest among scholars and practitioners working in the fields of Energy and Environmental studies. However, the DEA models discussed so far in this chapter operate on cross-sectional data. One of the potential approaches to study productivity changes over time is the Malmquist productivity index (MPI).

MPI is defined as the product of catch-up (CU) effect and frontier-shift (FU) effect terms. The degree to which the efficiency of a DMU shifts from one time period to the next is quantified using the CU effect. On the other hand, examining how the technological periphery surrounding the DMU has changed over time is the focus of the FU effect. MPI analyses a DMU's productivity growth over time by combining the results of these two metrics. The MPI is representative of the total factor productivity growth of a DMU since it represents both the efficiency growth or decline of a DMU and the growth or decline of the reference technology over time.

Figure 8. MPI evaluation

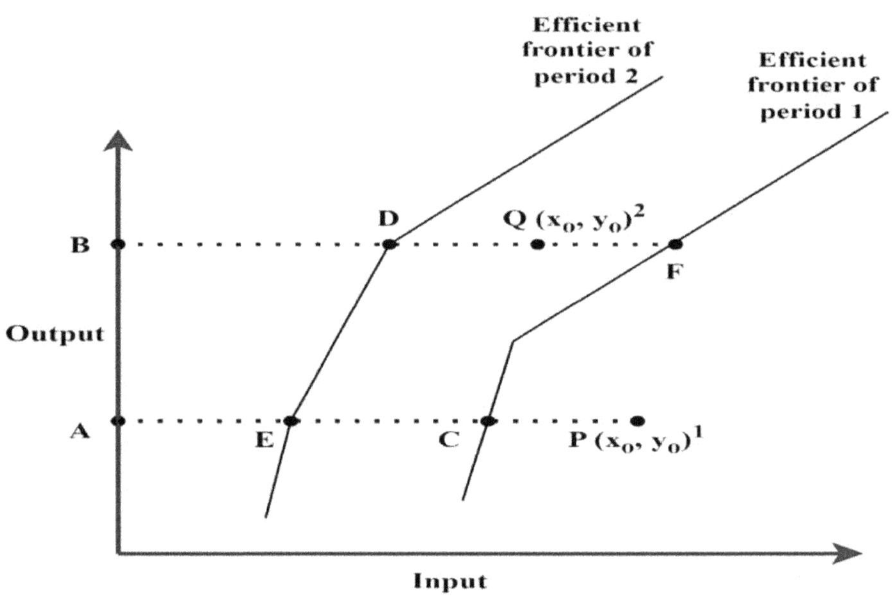

See Figure 8 for a simplified example of a system with a single input and a single output that exemplifies the CU effect, the FU effect, and the MPI. Let $(x_o, y_o)^1$ and $(x_o, y_o)^2$ denote DMU$_o$ $(o = 1, \ldots, n)$ in period 1 and 2, respectively. Let $\delta t^{2(}(xo_{,}\ yo_{)}t^{1)}$ denote the efficiency score of DMU $(xo_{,}\ yo_{)}t^1$ measured with respect to the frontier technology t2.

(i) Catch-Up (CU) Effect

The following formula encapsulates the CU effect and its definition:

CU effect = [Efficiency of $(x_o, y_o)^2$ with respect to period 2 frontier] / [Efficiency of $(x_o, y_o)^1$ with respect to period 1 frontier]

This can be evaluated using any suitable DEA model. The CU effect (based on input-orientation) can be estimated as follows for our straightforward case example shown in Figure 8:

$$\text{CU effect} = \frac{BD}{BQ} / \frac{AC}{AP}$$

Based on the definition, it is observed that CU > 1 indicates enhancement in efficiency of a DMU from period 1 to 2, CU = 1 indicates no change, and CU < 1 indicates fall in the efficiency.

(ii) Frontier-Shift (FU) Effect

FU effect accounts for the change in frontier technology (innovation) around a DMU by measuring the shift in the frontier from period 1 to 2. The frontier- shift at $(x_o, y_o)^1$ is given as:

$\Phi_1 = \frac{AC}{AE} = \frac{AC}{AP} / \frac{AE}{AP}$ = [Efficiency of $(x_o, y_o)^1$ with respect to period 1 frontier] / [Efficiency of $(x_o, y_o)^1$ with respect to period 2 frontier]

Similarly, the FU at $(x_o, y_o)^2$ is given as:

$\Phi_2 = \frac{BF}{BD} = \frac{BF}{BQ} / \frac{BD}{BQ}$ = [Efficiency of $(x_o, y_o)^2$ with respect to period 1 frontier] / [Efficiency of $(x_o, y_o)^2$ with respect to period 2 frontier]

Thus, the FU effect (Φ) is defined as the geometric mean of Φ_1 and Φ_2 as follows:

$$\text{FU effect } (\Phi) = \sqrt{\Phi_1 \Phi_2}$$

It is evident from the definition of FU effect that FU > 1 indicates progress in frontier technology around DMU_o from period 1 to 2, FU = 1 indicates no change in frontier technology, and FU < 1 indicates regress in the frontier technology.

(iii) Malmquist Productivity Index (MPI)

As defined MPI is calculated as the product of CU effect and FU effect.

$$\text{MPI} = (\text{CU effect}) \times (\text{FU effect})$$

For the case illustrated

$$\text{MPI} = \left(\frac{BD}{BQ} / \frac{AC}{AP}\right) \times \sqrt{\frac{AC}{AE} \times \frac{BF}{BD}} = \frac{AP}{BQ} \times \sqrt{\frac{BF}{AC} \times \frac{BD}{AE}}$$

In standard symbolic terms, MPI can be represented as:

$$\text{CU effect} = \frac{\delta^2\left((x_o, y_o)^2\right)}{\delta^1\left((x_o, y_o)^1\right)}$$

$$\text{FU effect} = \left[\frac{\delta^1\left((x_o, y_o)^1\right)}{\delta^2\left((x_o, y_o)^1\right)} \times \frac{\delta^1\left((x_o, y_o)^2\right)}{\delta^2\left((x_o, y_o)^2\right)}\right]^{1/2}$$

$$\text{MPI} = (\text{CU effect}) \times (\text{FU effect}) = \left[\frac{\delta^1\left((x_o, y_o)^2\right)}{\delta^1\left((x_o, y_o)^1\right)} \times \frac{\delta^2\left((x_o, y_o)^2\right)}{\delta^2\left((x_o, y_o)^1\right)}\right]^{1/2}$$

MPI > 1 indicates progress in the total factor productivity of DMU_o from period 1 to 2, MPI =1 indicates no change in total factor productivity, and MPI < 1 indicates regress in the total factor productivity of DMU_o.

APPLICATION OF DEA IN ENERGY AND ENVIRONMENT

In this section, we shall discuss about the various Energy and Environmental applications where DEA methodology has been used. We classify the existing studies based on their respective applications in the field of Energy as electricity, oil, gas, etc. and Environment.

(i) Application Areas

As a result of the inherent benefits that it has, DEA has been effectively used for a broad array of applications within the fields of Energy and Environment. Talking about the DEA application in the field of energy, electricity sector is most explored area. In the initial stages of its development, researchers mostly focused on assessing the efficiency of power facilities that generate electricity which was gradually extended to distribution and transmission divisions as well. Since the electricity sector was initially heavily regulated by regional and state governments, there has been a plethora of research into the topic. So, the researchers were motivated to study the inefficiency that may arise owing to the lack of discipline from market competition. It served as a tool that could assist the policymakers to decide on as to how to frame effective regulation schemes. Because it may happen that inefficient firms survived receiving the benefit from regulations. So, to keep a check on such issues regular monitoring of operational performance of the units was necessary. Other reason for the wide applicability of DEA in electricity sector can be attributed to the easily available data in public domains. Current research in electricity application area is more focussed towards integrating environmental factors as carbon dioxide emissions, etc. with the usual production process. Table 2 reports some of the relevant studies of the application of DEA in electricity sector.

Another applied area of DEA in energy sector is the oil, coal, and gas. The number of studies in these areas have been comparatively lesser than the electricity sector. One of the possible reasons for this can be lack of availability of data. Upstream and downstream processes, as well as the sector's overall business operations, are all investigated in the literature on the oil industry. Coal mining has typically been the main topic of writings about coal. Historically, the gas industry's transmission and distribution systems have been the primary focus of most of the research studies. The processing and use of gas have also been the subject of discussion among researchers. Table 2 reports some of the relevant studies of the application of DEA in oil, coal, and gas sector.

With the rising concern about climate changes and global warning, the nations have decided to progress ahead with using less of fossil fuels and enhancing dependence on renewable energy as solar, wind, and biomass. As the growing awareness on harnessing more of renewable energy has spread, researchers have also paid close attention to the topic of assessing the industry's performance. It is probably because of the wide availability of a diverse range of forms of biomass fuels, ranging from woody biomass to agricultural biomass, that biomass is the form of renewable energy technology that is the subject of the most research in DEA applications. Following the application of biomass is wind and solar photovoltaic in renewable energy, both of which have seen significant technological advancements and price reductions in recent years. Table 2 reports some of the relevant studies of the application of DEA in renewable energy sector.

Table 2. Studies on electricity, oil, gas, coal, and renewable energy applications

Energy Application Areas		
Electricity	**Oil, Gas, Coal**	**Renewable Energy**
An et al. (2015), Arabi et al. (2014), Azadeh and Mousavi Ahranjani (2014), Bagdadioglu et al. (1996), Chang and Yang (2011), Cook and Green (2005), Fallahi et al. (2011), Hampf and Rødseth (2015), Khalili-Damghani and Shahmir (2015), Mekaroonreung and Johnson (2012), Reza Salehizadeh et al. (2015), Sarkis and Cordeiro (2012), Sueyoshi and Goto (2015), Tone and Tsutsui (2007), Yadav et al. (2013)	Agrell and Bogetoft (2005), Azadeh et al. (2014, 2015), Fang et al. (2009), Halkos and Tzeremes (2011), Ismail et al. (2013), Ju et al. (2015), Liu et al. (2016), Luo et al. (2012), Munksgaard et al. (2005), Song et al. (2015), Sueyoshi and Goto (2012, 2015a), Toloo and Babaee (2015), Zhang et al. (2013)	Ali Azadeh et al. (2014), Babazadeh et al. (2015), Blokhuis et al. (2012), Chien and Hu (2007), Iglesias et al. (2010), Kagawa et al. (2013), Longo et al. (2015), Salazar-Ordóñez et al. (2013), Stallard et al. (2008), Sueyoshi and Goto (2014), Zografidou et al. (2016)

Following the application of DEA in energy sector is the environmental sector. The key feature of the studies in the environmental performance evaluation has been the incorporation of undesirable outputs with inputs and desirable outputs. As a result, DEA models with environmental reference technology have been widely explored in this area. The focus of the researchers has been towards studying environmental efficiency together with production efficiency. Table 3 reports some of the relevant studies of the application of DEA in environmental studies.

(ii) Key Findings

Based on the existing studies, some of the important observations that need to be discussed are as follows. Most of the studies that used only desirable outputs and inputs used the strong disposability assumption in the reference technology. This assumption works well for cases without undesirable outputs. But as the undesirable outputs are incorporated into the setting together with inputs and desirable outputs, weak disposability assumption in the reference technology has been widely employed. Since it more accurately depicts the production process, the weak disposability assumption in the presence of undesirable outputs seems more reasonable.

Table 3. Studies on environmental applications

Environmental Studies
Baležentis et al. (2016), Beltrán-Esteve and Picazo-Tadeo (2015), Blancard and Hoarau (2013), Callens and Tyteca (1999), Camioto et al. (2014), Y. T. Chang et al. (2013), C. Chen et al. (2012), J. Chen et al. (2015), L. Chen et al. (2016), Egilmez et al. (2014), Feng et al. (2015), Gómez-Calvet et al. (2016), Goto et al. (2014), Hathroubi et al. (2014), Huang et al. (2014), Khodakarami et al. (2015), Khoshnevisan et al. (2015), Lozano et al. (2009), Pandey and Singh (2021), Shortall and Barnes (2013), Zhu et al. (2014)

For the RTS assumption, CRS is the most popularly used, even though VRS seems to be a more reasonable. This can be attributed to the fact that in CRS the input-oriented efficiency scores are same as the output-oriented efficiency scores. So, the researchers can easily extrapolate their findings just by solving model in any orientation without any conflict regarding the selection of RTS assumption. Other potential reason can be the computation of MPI for a multi-period system. It may happen that MPI based on VRS may give infeasible solution in certain cases. Some studies have used both CRS and VRS with a purpose to report scale efficiencies of DMUs.

As for the efficiency measure, radial DEA has been a popular choice. But with the involvement of undesirable outputs in the setting, DDF and hyperbolic efficiency measures have been more common. Slacks-based efficiency measures have also been used in environmental settings owing to their ability of simultaneous non-radial contraction of inputs and undesirable outputs and expansion of desirable outputs and because of its higher discriminating power (Hu & Wang, 2006; Zhou et al., 2008). For evaluating productivity change over time, MPI has been the most popularly used index.

MODEL SELECTION

Owing to the availability of a large number of DEA models with wide variations in their properties and characteristics, it is quite obvious for the researchers working in Energy and Environmental studies to face difficulties in selecting the appropriate model. In this section, we focus on identifying suitable DEA model and approach based on the need of the researcher and application in which it is to be utilized. The idea is to briefly explain model selection in terms of reference technology, RTS, disposability assumptions, and handling other complexities which are application specific.

As is observed from the findings of the literature in Energy and Environmental performance evaluation, the DEA models in input orientation have been widely used. This can be attributed to the fact that in Energy and Environmental studies, higher priority is always given to meeting the demand. So, input conservation to produce the desired output seems to a reasonable logic. Thus, input oriented DEA models are comparatively more popular in Energy and Environmental performance evaluation.

When only desirable outputs and inputs are to be considered in the performance evaluation without including undesirable outputs, radial DEA models appear to be appropriate with the CRS and VRS and strong disposability assumptions in the reference technology. Information on both the technical and the scale efficiency of the DMUs may be gleaned from such a setup. Moreover, by the extension of these models, the researchers can even investigate the zone of operation of the DMUs, that is, whether DMUs are operating at sub-optimal scale size, optimal scale size, or supra-optimal scale size. If determining how DMU productivity has changed over time is a need, MPI may be utilized to do the investigation.

However, when undesirable outputs come into the picture of performance evaluation, weak disposability assumption appears to be a more suitable choice for the reference technology compared to strong disposability. This can be attributed to the fact that undesirable outputs are inevitably produced with the desirable outputs and they cannot be disposed-off costlessly. Weak disposability asserts the fact that undesirable outputs cannot be reduced in solo, their reduction has to be accompanied by the relevant decrease in desirable outputs. Moreover, the suitable choice for the efficiency measure can be hyperbolic efficiency measure, directional distance efficiency measure, and the slacks-based efficiency measure. During the process of performance evaluation, each of these efficiency measures has the potential to simultaneously reduce inputs and outputs that are undesirable while simultaneously increasing outputs that are desirable. Similarly, it is possible to link the MPI with environmental DEA technology to evaluate productivity changes of the DMUs over time using suitable model.

Choosing relevant inputs and outputs is a further difficulty when assessing performance. It is obvious that for a given application, there can be many inputs and outputs available. But in DEA, it is observed that as the number of inputs and outputs increase in comparison to the number of DMUs, the DEA models lose their discriminatory power (Charles et al., 2019). Therefore, it is necessary to pick the inputs and outputs for the analysis carefully. First step towards this endeavour can be listing the set of all possible inputs and outputs available. Then, using some preliminary judgement and statistical analysis, relevant and the most appropriate inputs can outputs can be filtered and selected. One of the other challenges can be availability of data, especially in the multi-period analysis, and the number of DMUs. Some popular rules for selecting number of DMUs, inputs, and outputs are as follows:

(i) $n \geq 2\times(m + s)$ as given by Golany and Roll (1989)
(ii) $n \geq 3\times(m + s)$ as suggested by Bowlin (1998) and Raab and Lichty (2002)
(iii) $n \geq 2\times m\times s$ as given by Dyson et al. (2001)
(iv) $n \geq \max\{m\times s, 3\times(\mathrm{m} + \mathrm{s})\}$ as suggested by Cooper et al. (2007),

where m = number of inputs, s = number of outputs, and n = number of DMUs. An important point to note is that these rules are based on empirical observations and are hard and fast to be followed. However, they provide initial basis for preparing the data.

Thus, once the data has been finalized, suitable DEA efficiency measure has been chosen together with suitable reference technology, the formulated DEA model can be solved using any optimization solvers such as CPLEX, pulp, Gurobi, DEAP, etc. The best part of DEA as a popular performance evaluation tool is that most of the DEA models can be linearized, so this provides the advantage of obtaining the optimal solutions. The efficiency scores obtained by solving DEA models can be further analyzed using suitable regression analysis techniques to obtain more insights and information based on the area of application.

LIMITATIONS OF DEA AND FUTURE OUTLOOK

Although DEA has numerous advantages which makes it a popular tool among researchers for performance evaluation, but like any other mathematical tool, DEA too has certain limitations. DEA is an approximation methodology for performance evaluation, so one has to be careful in using appropriate DEA model for guiding large policies and issues as climate change and global warming. Understanding limitations of the tool opens door for future research directions which the aim to eliminating the limitations.

As already highlighted, DEA loses its discriminatory power as the number of inputs and outputs increase in comparison to the number of DMUs. This may cause several inefficient DMUs to be reported as efficient with an efficiency score of unity. To overcome this issue, there is a need for researchers to work on more ranking based DEA models namely, cross-efficiency DEA, super-efficiency DEA, etc. There is a need to evolve the extensions of these models to incorporate undesirable outputs and their settings into the framework.

The DEA models discussed in this chapter are typically the 'black-box' models that do not consider the internal structure of the production process. As highlighted by Kao (2014) ignoring the internal divisions/structure of DMUs may give misleading results. Therefore, there is need for researchers to explore the internal divisions/structure of DMUs wherever possible using the Network DEA framework. Network DEA needs to be extended to accommodate undesirable output settings as weak disposability reference technology. This will help to explore and identify more precise sources of inefficiency in the DMU.

Also, most of the research involving efficiency changes over time is based on single period optimization in which the efficiency scores are computed for each period separately. However, this does not give a comprehensive picture of the overall phenomenon based on long term optimization. One of the potential approaches which take care of this issue is the dynamic DEA. Researchers can look forward to work on dynamic DEA incorporating undesirable outputs. Also, dynamic DEA has been recently incorporated with network DEA which provides an even more comprehensive performance evaluation tool. Investigating dynamic-network DEA and its application in Energy and Environmental performance evaluation can be yet another avenue worth exploring in the future.

The DEA models discussed in this chapter are deterministic in nature. They cannot accommodate imprecise data. One of the possible ways to handle this issue is integrating fuzzy modelling with DEA framework. Developing fuzzy DEA models with environmental technology can be potential area for future research.

CONCLUSION

With the growing awareness for the need of sustainable development and progress, there has been a great concern among researchers to quantify the extent of environmental deterioration caused by activities aimed at economic growth and development. There is a need to account for the extent of inefficiency and improvement required in order to meet sustainability goals and head towards an environmentally friendly future. We emphasis in this chapter the growing popularity among researchers for the utilization of DEA as a mathematical technique for performance evaluation and gauge extent of inefficiency.

The use of DEA in assessing performance has been capitalized in several applications as hospitals, education, industry, etc. In the recent times, the DEA method has been adapted for use in assessing environmental efficiency of entities as countries, regions, and organizations. There has been a rapid development in this field which is still growing and requires several issues still to be resolved. In this chapter, we attempt to bring together all the elements of DEA framework required for Energy and Environmental performance evaluation modelling.

We begin by explaining the basic DEA modelling framework followed by suitable RTS and disposability assumptions which constitute the reference technology for the DEA model. This chapter further explained the assumptions of weak and managerial disposability for constructing the reference technology in the presence of undesirable outputs to depict the real-life production process phenomenon. Further, we introduced the different efficiency measures incorporating undesirable outputs in the performance evaluation technique. The purpose is to provide the readers the knowledge to accommodate and model undesirable outputs in DEA modelling.

In addition, this chapter offers a summary of how DEA may be used in Energy and Environmental performance evaluation studies such as electricity, oil, gas, etc. Moving further, we outlined guidelines for selecting the suitable DEA model based on research challenges and fields of interest. Next, in this chapter, we brought forward to the researchers some of the limitations of the existing DEA approaches and suggested several avenues for further study.

Finally, we envision that our contribution in the domain of Energy and Environmental studies through this book chapter will advance the ever-growing research in this vast area.

REFERENCES

Agrell, P. J., & Bogetoft, P. (2005). Economic and environmental efficiency of district heating plants. *Energy Policy*, *33*(10), 1351–1362. doi:10.1016/j.enpol.2003.12.011

Amirteimoori, A., Kordrostami, S., & Sarparast, M. (2006). Modeling undesirable factors in data envelopment analysis. *Applied Mathematics and Computation*, *180*(2), 444–452. doi:10.1016/j.amc.2005.12.029

An, Q., Pang, Z., Chen, H., & Liang, L. (2015). Closest targets in environmental efficiency evaluation based on enhanced Russell measure. *Ecological Indicators*, *51*, 59–66. doi:10.1016/j.ecolind.2014.09.008

Arabi, B., Munisamy, S., Emrouznejad, A., & Shadman, F. (2014). Power industry restructuring and eco-efficiency changes: A new slacks-based model in Malmquist–Luenberger Index measurement. *Energy Policy*, *68*, 132–145. doi:10.1016/j.enpol.2014.01.016

Azadeh, A., Gaeini, Z., & Moradi, B. (2014). Optimization of HSE in maintenance activities by integration of continuous improvement cycle and fuzzy multivariate approach: A gas refinery. *Journal of Loss Prevention in the Process Industries*, *32*, 415–427. doi:10.1016/j.jlp.2014.10.006

Azadeh, A., Mokhtari, Z., Sharahi, Z. J., & Zarrin, M. (2015). An integrated experiment for identification of best decision styles and teamworks with respect to HSE and ergonomics program: The case of a large oil refinery. *Accident; Analysis and Prevention*, *85*, 30–44. doi:10.1016/j.aap.2015.08.016 PMID:26397195

Azadeh, A., & Mousavi Ahranjani, P. (2014). The impact of job security, satisfaction and stress on performance assessment and optimization of generation companies. *Journal of Loss Prevention in the Process Industries*, *32*, 343–348. doi:10.1016/j.jlp.2014.10.002

Azadeh, A., Rahimi-Golkhandan, A., & Moghaddam, M. (2014). Location optimization of wind power generation–transmission systems under uncertainty using hierarchical fuzzy DEA: A case study. *Renewable & Sustainable Energy Reviews*, *30*, 877–885. doi:10.1016/j.rser.2013.10.020

Babazadeh, R., Razmi, J., Pishvaee, M. S., & Rabbani, M. (2015). A non-radial DEA model for location optimization of Jatropha curcas L. cultivation. *Industrial Crops and Products*, *69*, 197–203. doi:10.1016/j.indcrop.2015.02.006

Bagdadioglu, N., Waddams Price, C. M., & Weyman-Jones, T. G. (1996). Efficiency and ownership in electricity distribution: A non-parametric model of the Turkish experience. *Energy Economics*, *18*(1), 1–23. doi:10.1016/0140-9883(95)00042-9

Baležentis, T., Li, T., Streimikiene, D., & Baležentis, A. (2016). Is the Lithuanian economy approaching the goals of sustainable energy and climate change mitigation? Evidence from DEA-based environmental performance index. *Journal of Cleaner Production*, *116*, 23–31. doi:10.1016/j.jclepro.2015.12.088

Beltrán-Esteve, M., & Picazo-Tadeo, A. J. (2015). Assessing environmental performance trends in the transport industry: Eco-innovation or catching-up? *Energy Economics*, *51*, 570–580. doi:10.1016/j.eneco.2015.08.018

Blancard, S., & Hoarau, J.-F. (2013). A new sustainable human development indicator for small island developing states: A reappraisal from data envelopment analysis. *Economic Modelling*, *30*, 623–635. doi:10.1016/j.econmod.2012.10.016

Blokhuis, E., Advokaat, B., & Schaefer, W. (2012). Assessing the performance of Dutch local energy companies. *Energy Policy*, *45*, 680–690. doi:10.1016/j.enpol.2012.03.021

Bowlin, W. F. (1998). Measuring Performance: An Introduction to Data Envelopment Analysis (DEA). *The Journal of Cost Analysis*, *15*(2), 3–27. doi:10.1080/08823871.1998.10462318

Callens, I., & Tyteca, D. (1999). Towards indicators of sustainable development for firms: A productive efficiency perspective. *Ecological Economics*, *28*(1), 41–53. doi:10.1016/S0921-8009(98)00035-4

Camioto, F. de C., Mariano, E. B., & Rebelatto, D. A. (2014). Efficiency in Brazil's industrial sectors in terms of energy and sustainable development. *Environmental Science & Policy*, *37*, 50–60. doi:10.1016/j.envsci.2013.08.007

Chang, D.-S., & Yang, F.-C. (2011). Assessing the power generation, pollution control, and overall efficiencies of municipal solid waste incinerators in Taiwan. *Energy Policy*, *39*(2), 651–663. doi:10.1016/j.enpol.2010.10.038

Chang, Y.-T., Park, H., Jeong, J., & Lee, J. (2014). Evaluating economic and environmental efficiency of global airlines: A SBM-DEA approach. *Transportation Research Part D, Transport and Environment*, *27*, 46–50. doi:10.1016/j.trd.2013.12.013

Chang, Y. T., Zhang, N., Danao, D., & Zhang, N. (2013). Environmental efficiency analysis of transportation system in China: A non-radial DEA approach. *Energy Policy*, *58*, 277–283. doi:10.1016/j.enpol.2013.03.011

Charles, V., Aparicio, J., & Zhu, J. (2019). The curse of dimensionality of decision-making units: A simple approach to increase the discriminatory power of data envelopment analysis. *European Journal of Operational Research*, *279*(3), 929–940. doi:10.1016/j.ejor.2019.06.025

Charnes, A., & Cooper, W. W. (1962). Programming with linear fractional functionals. *Naval Research Logistics Quarterly*, *9*(3–4), 181–186. doi:10.1002/nav.3800090303

Charnes, A., Cooper, W. W., & Rhodes, E. (1978). Measuring the efficiency of decision making units. *European Journal of Operational Research*, *2*(6), 429–444. doi:10.1016/0377-2217(78)90138-8

Chen, C., Zhu, J., Yu, J.-Y., & Noori, H. (2012). A new methodology for evaluating sustainable product design performance with two-stage network data envelopment analysis. *European Journal of Operational Research*, *221*(2), 348–359. doi:10.1016/j.ejor.2012.03.043

Chen, J., Song, M., & Xu, L. (2015). Evaluation of environmental efficiency in China using data envelopment analysis. *Ecological Indicators*, *52*, 577–583. doi:10.1016/j.ecolind.2014.05.008

Chen, L., Wang, Y.-M., & Wang, L. (2016). Congestion measurement under different policy objectives: An analysis of Chinese industry. *Journal of Cleaner Production*, *112*, 2943–2952. doi:10.1016/j.jclepro.2015.09.064

Chien, T., & Hu, J.-L. (2007). Renewable energy and macroeconomic efficiency of OECD and non-OECD economies. *Energy Policy*, *35*(7), 3606–3615. doi:10.1016/j.enpol.2006.12.033

Cook, W. D., & Green, R. H. (2005). Evaluating power plant efficiency: A hierarchical model. *Computers & Operations Research*, *32*(4), 813–823. doi:10.1016/j.cor.2003.08.019

Cooper, W. W., Seiford, L. M., & Tone, K. (2007). Data envelopment analysis: A comprehensive text with models, applications, references and DEA-solver software. In *Data Envelopment Analysis: A Comprehensive Text with Models* (2nd ed.). Applications, References and DEA-Solver Software. doi:10.1007/978-0-387-45283-8

Cooper, W. W., Seiford, L. M., & Zhu, J. (2004). Data envelopment analysis: History, models and interpretations. In L. M. Cooper, L. M. Seiford, & J. Zhu (Eds.), *Handbook on Data Envelopment Analysis* (pp. 1–39). Kluwer Academic Publishers. doi:10.1007/1-4020-7798-X_1

Cui, Q. (2019). Investigating the airlines emission reduction through carbon trading under CNG2020 strategy via a Network Weak Disposability DEA. *Energy*, *180*, 763–771. doi:10.1016/j.energy.2019.05.159

Dyson, R. G., Allen, R., Camanho, A. S., Podinovski, V. V., Sarrico, C. S., & Shale, E. A. (2001). Pitfalls and protocols in DEA. *European Journal of Operational Research*, *132*(2), 245–259. doi:10.1016/S0377-2217(00)00149-1

Egilmez, G., Kucukvar, M., Tatari, O., & Bhutta, M. K. S. (2014). Supply chain sustainability assessment of the U.S. food manufacturing sectors: A life cycle-based frontier approach. *Resources, Conservation and Recycling*, *82*, 8–20. doi:10.1016/j.resconrec.2013.10.008

European Union. (2008). *Energy and climate change elements of the final compromise*. Technical report, European Union.

Faere, R., & Grosskopf, S. (1989). *Multilateral Productivity Comparisons When Some Outputs are Undesirable : A Nonparametric Approach.* The MIT Press. https://www.jstor.org/stable/192

Fallahi, A., Ebrahimi, R., & Ghaderi, S. F. (2011). Measuring efficiency and productivity change in power electric generation management companies by using data envelopment analysis: A case study. *Energy*, *36*(11), 6398–6405. doi:10.1016/j.energy.2011.09.034

Fang, H., Wu, J., & Zeng, C. (2009). Comparative study on efficiency performance of listed coal mining companies in China and the US. *Energy Policy*, *37*(12), 5140–5148. doi:10.1016/j.enpol.2009.07.027

Färe, R., Grosskopf, S., Noh, D.-W., & Weber, W. (2005). Characteristics of a polluting technology: Theory and practice. *Journal of Econometrics*, *126*(2), 469–492. doi:10.1016/j.jeconom.2004.05.010

Farrell, M. J. (1957). The Measurement of Productive Efficiency. *Journal of the Royal Statistical Society. Series A (General)*, *120*(3), 253–281. doi:10.2307/2343100

Feng, C., Chu, F., Ding, J., Bi, G., & Liang, L. (2015). Carbon Emissions Abatement (CEA) allocation and compensation schemes based on DEA. *Omega*, *53*, 78–89. doi:10.1016/j.omega.2014.12.005

Fukuyama, H., & Weber, W. L. (2008). Japanese banking inefficiency and shadow pricing. *Mathematical and Computer Modelling*, *48*(11), 1854–1867. doi:10.1016/j.mcm.2008.03.004

Golany, B., & Roll, Y. (1989). An application procedure for DEA. *Omega*, *17*(3), 237–250. doi:10.1016/0305-0483(89)90029-7

Gómez-Calvet, R., Conesa, D., Gómez-Calvet, A. R., & Tortosa-Ausina, E. (2016). On the dynamics of eco-efficiency performance in the European Union. *Computers & Operations Research*, *66*, 336–350. doi:10.1016/j.cor.2015.07.018

Goto, M., Otsuka, A., & Sueyoshi, T. (2014). DEA (Data Envelopment Analysis) assessment of operational and environmental efficiencies on Japanese regional industries. *Energy*, *66*, 535–549. doi:10.1016/j.energy.2013.12.020

Hailu, A., & Veeman, T. S. (2001). Non-parametric Productivity Analysis with Undesirable Outputs: An Application to the Canadian Pulp and Paper Industry. *American Journal of Agricultural Economics*, *83*(3), 605–616. doi:10.1111/0002-9092.00181

Halkos, G., & Petrou, K. N. (2019). Treating undesirable outputs in DEA: A critical review. *Economic Analysis and Policy*, *62*, 97–104. doi:10.1016/j.eap.2019.01.005

Halkos, G. E., & Tzeremes, N. G. (2011). Oil consumption and economic efficiency: A comparative analysis of advanced, developing and emerging economies. *Ecological Economics*, *70*(7), 1354–1362. doi:10.1016/j.ecolecon.2011.02.010

Hampf, B., & Rødseth, K. L. (2015). Carbon dioxide emission standards for U.S. power plants: An efficiency analysis perspective. *Energy Economics*, *50*, 140–153. doi:10.1016/j.eneco.2015.04.001

Hathroubi, S., Peypoch, N., & Robinot, E. (2014). Technical efficiency and environmental management: The Tunisian case. *Journal of Hospitality and Tourism Management*, *21*, 27–33. doi:10.1016/j.jhtm.2014.03.002

He, F., Zhang, Q., Lei, J., Fu, W., & Xu, X. (2013). Energy efficiency and productivity change of China's iron and steel industry: Accounting for undesirable outputs. *Energy Policy*, *54*, 204–213. doi:10.1016/j.enpol.2012.11.020

Hu, J.-L., & Wang, S.-C. (2006). Total-factor energy efficiency of regions in China. *Energy Policy*, *34*(17), 3206–3217. doi:10.1016/j.enpol.2005.06.015

Huang, J., Yang, X., Cheng, G., & Wang, S. (2014). A comprehensive eco-efficiency model and dynamics of regional eco-efficiency in China. *Journal of Cleaner Production*, *67*, 228–238. doi:10.1016/j.jclepro.2013.12.003

Iglesias, G., Castellanos, P., & Seijas, A. (2010). Measurement of productive efficiency with frontier methods: A case study for wind farms. *Energy Economics*, *32*(5), 1199–1208. doi:10.1016/j.eneco.2010.03.004

Ismail, Z., Tai, J. C., Kong, K. K., Law, K. H., Shirazi, S. M., & Karim, R. (2013). Using data envelopment analysis in comparing the environmental performance and technical efficiency of selected companies in their global petroleum operations. *Measurement*, *46*(9), 3401–3413. doi:10.1016/j.measurement.2013.04.076

Ju, K., Su, B., Zhou, D., Zhou, P., & Zhang, Y. (2015). Oil price crisis response: Capability assessment and key indicator identification. *Energy*, *93*, 1353–1360. doi:10.1016/j.energy.2015.09.124

Kagawa, S., Takezono, K., Suh, S., & Kudoh, Y. (2013). Production possibility frontier analysis of biodiesel from waste cooking oil. *Energy Policy*, *55*, 362–368. doi:10.1016/j.enpol.2012.12.016

Kao, C. (2014). Network data envelopment analysis: A review. *European Journal of Operational Research*, *239*(1), 1–16. doi:10.1016/j.ejor.2014.02.039

Kao, C. (2020). Measuring efficiency in a general production possibility set allowing for negative data. *European Journal of Operational Research*, *282*(3), 980–988. doi:10.1016/j.ejor.2019.10.027

Kao, C. (2022). Measuring efficiency in a general production possibility set allowing for negative data: An extension and a focus on returns to scale. *European Journal of Operational Research*, *296*(1), 267–276. doi:10.1016/j.ejor.2021.03.061

Khalili-Damghani, K., & Shahmir, Z. (2015). Uncertain network data envelopment analysis with undesirable outputs to evaluate the efficiency of electricity power production and distribution processes. *Computers & Industrial Engineering*, *88*, 131–150. doi:10.1016/j.cie.2015.06.013

Khodakarami, M., Shabani, A., Farzipoor Saen, R., & Azadi, M. (2015). Developing distinctive two-stage data envelopment analysis models: An application in evaluating the sustainability of supply chain management. *Measurement*, *70*, 62–74. doi:10.1016/j.measurement.2015.03.024

Khoshnevisan, B., Bolandnazar, E., Shamshirband, S., Shariati, H. M., Anuar, N. B., & Mat Kiah, M. L. (2015). Decreasing environmental impacts of cropping systems using life cycle assessment (LCA) and multi-objective genetic algorithm. *Journal of Cleaner Production*, *86*, 67–77. doi:10.1016/j.jclepro.2014.08.062

Korhonen, P. J., & Luptacik, M. (2004). Eco-efficiency analysis of power plants: An extension of data envelopment analysis. *European Journal of Operational Research*, *154*(2), 437–446. doi:10.1016/S0377-2217(03)00180-2

Liu, J., Liu, H., Yao, X.-L., & Liu, Y. (2016). Evaluating the sustainability impact of consolidation policy in China's coal mining industry: A data envelopment analysis. *Journal of Cleaner Production*, *112*, 2969–2976. doi:10.1016/j.jclepro.2015.08.011

Longo, L., Colantoni, A., Castellucci, S., Carlini, M., Vecchione, L., Savuto, E., Pallozzi, V., Di Carlo, A., Bocci, E., Moneti, M., Cocchi, S., & Boubaker, K. (2015). DEA (data envelopment analysis)-assisted supporting measures for ground coupled heat pumps implementing in Italy: A case study. *Energy*, *90*, 1967–1972. doi:10.1016/j.energy.2015.07.024

Lozano, S., Villa, G., & Brännlund, R. (2009). Centralised reallocation of emission permits using DEA. *European Journal of Operational Research*, *193*(3), 752–760. doi:10.1016/j.ejor.2007.07.029

Luo, J., Yang, Y., & Chen, Y. (2012). Optimizing the drilled well patterns for CBM recovery via numerical simulations and data envelopment analysis. *International Journal of Mining Science and Technology*, *22*(4), 503–507. doi:10.1016/j.ijmst.2012.01.011

Mekaroonreung, M., & Johnson, A. L. (2012). Estimating the shadow prices of SO2 and NOx for U.S. coal power plants: A convex nonparametric least squares approach. *Energy Economics*, *34*(3), 723–732. doi:10.1016/j.eneco.2012.01.002

Munksgaard, J., Pade, L.-L., & Fristrup, P. (2005). Efficiency gains in Danish district heating. Is there anything to learn from benchmarking? *Energy Policy*, *33*(15), 1986–1997. doi:10.1016/j.enpol.2004.03.019

Pandey, U., & Singh, S. (2021). Environmental performance evaluation of European farms by assessing polluting factors in joint production. *Journal of Cleaner Production*, *328*, 129457. doi:10.1016/j.jclepro.2021.129457

Pathomsiri, S., Haghani, A., Dresner, M., & Windle, R. J. (2008). Impact of undesirable outputs on the productivity of US airports. *Transportation Research Part E, Logistics and Transportation Review*, *44*(2), 235–259. doi:10.1016/j.tre.2007.07.002

Piao, S.-R., Li, J., & Ting, C.-J. (2019). Assessing regional environmental efficiency in China with distinguishing weak and strong disposability of undesirable outputs. *Journal of Cleaner Production*, *227*, 748–759. doi:10.1016/j.jclepro.2019.04.207

Raab, R. L., & Lichty, R. W. (2002). Identifying Subareas That Comprise A Greater Metropolitan Area: The Criterion of County Relative Efficiency. *Journal of Regional Science*, *42*(3), 579–594. doi:10.1111/1467-9787.00273

Reinhard, S., Knox Lovell, C. A., & Thijssen, G. J. (2000). Environmental efficiency with multiple environmentally detrimental variables; estimated with SFA and DEA. *European Journal of Operational Research*, *121*(2), 287–303. doi:10.1016/S0377-2217(99)00218-0

Reza Salehizadeh, M., Rahimi-Kian, A., & Oloomi-Buygi, M. (2015). Security-based multi-objective congestion management for emission reduction in power system. *International Journal of Electrical Power & Energy Systems*, *65*, 124–135. doi:10.1016/j.ijepes.2014.09.031

Salazar-Ordóñez, M., Pérez-Hernández, P. P., & Martín-Lozano, J. M. (2013). Sugar beet for bioethanol production: An approach based on environmental agricultural outputs. *Energy Policy*, *55*, 662–668. doi:10.1016/j.enpol.2012.12.063

Sarkis, J., & Cordeiro, J. J. (2012). Ecological modernization in the electrical utility industry: An application of a bads–goods DEA model of ecological and technical efficiency. *European Journal of Operational Research*, *219*(2), 386–395. doi:10.1016/j.ejor.2011.09.033

Shortall, O. K., & Barnes, A. P. (2013). Greenhouse gas emissions and the technical efficiency of dairy farmers. *Ecological Indicators*, *29*, 478–488. doi:10.1016/j.ecolind.2013.01.022

Song, M., Zhang, J., & Wang, S. (2015). Review of the network environmental efficiencies of listed petroleum enterprises in China. *Renewable & Sustainable Energy Reviews*, *43*, 65–71. doi:10.1016/j.rser.2014.11.050

Stallard, T., Rothschild, R., & Aggidis, G. A. (2008). A comparative approach to the economic modelling of a large-scale wave power scheme. *European Journal of Operational Research*, *185*(2), 884–898. doi:10.1016/j.ejor.2007.01.021

Sueyoshi, T., & Goto, M. (2012a). Data envelopment analysis for environmental assessment: Comparison between public and private ownership in petroleum industry. *European Journal of Operational Research*, *216*(3), 668–678. doi:10.1016/j.ejor.2011.07.046

Sueyoshi, T., & Goto, M. (2012b). Weak and strong disposability vs. natural and managerial disposability in DEA environmental assessment: Comparison between Japanese electric power industry and manufacturing industries. *Energy Economics*, *34*(3), 686–699. doi:10.1016/j.eneco.2011.10.018

Sueyoshi, T., & Goto, M. (2014). Photovoltaic power stations in Germany and the United States: A comparative study by data envelopment analysis. *Energy Economics*, *42*, 271–288. doi:10.1016/j.eneco.2014.01.004

Sueyoshi, T., & Goto, M. (2015a). DEA environmental assessment in time horizon: Radial approach for Malmquist index measurement on petroleum companies. *Energy Economics*, *51*, 329–345. doi:10.1016/j.eneco.2015.07.010

Sueyoshi, T., & Goto, M. (2015b). Environmental assessment on coal-fired power plants in U.S. north-east region by DEA non-radial measurement. *Energy Economics*, *50*, 125–139. doi:10.1016/j.eneco.2015.04.016

Sueyoshi, T., Yuan, Y., & Goto, M. (2017). A literature study for DEA applied to energy and environment. *Energy Economics*, *62*, 104–124. doi:10.1016/j.eneco.2016.11.006

Toloo, M., & Babaee, S. (2015). On variable reductions in data envelopment analysis with an illustrative application to a gas company. *Applied Mathematics and Computation*, *270*, 527–533. doi:10.1016/j.amc.2015.06.122

Tone, K. (2016). Advances in DEA Theory and Applications: With Examples in Forecasting Models. *Advances in DEA Theory and Applications: With Examples in Forecasting Models*. doi:10.1002/9781118946688

Tone, K., Chang, T.-S., & Wu, C.-H. (2020). Handling negative data in slacks-based measure data envelopment analysis models. *European Journal of Operational Research*, *282*(3), 926–935. doi:10.1016/j.ejor.2019.09.055

Tone, K., & Tsutsui, M. (2007). Decomposition of cost efficiency and its application to Japanese-US electric utility comparisons. *Socio-Economic Planning Sciences*, *41*(2), 91–106. doi:10.1016/j.seps.2005.10.007

United Nations. (1998). *Kyoto Protocol to the United Nations Framework Convention on Climate Chance. Technical report*. United Nations.

World Bank. (2008). *Environmental sustainability: An evaluation of World Bank group support. Technical report*. World Bank.

Yadav, V. K., Chauhan, Y. K., Padhy, N. P., & Gupta, H. O. (2013). A novel power sector restructuring model based on Data Envelopment Analysis (DEA). *International Journal of Electrical Power & Energy Systems*, *44*(1), 629–637. doi:10.1016/j.ijepes.2012.08.013

Yang, H., & Pollitt, M. (2009). Incorporating both undesirable outputs and uncontrollable variables into DEA: The performance of Chinese coal-fired power plants. *European Journal of Operational Research*, *197*(3), 1095–1105. doi:10.1016/j.ejor.2007.12.052

Yang, H., & Pollitt, M. (2010). The necessity of distinguishing weak and strong disposability among undesirable outputs in DEA: Environmental performance of Chinese coal-fired power plants. *Energy Policy*, *38*(8), 4440–4444. doi:10.1016/j.enpol.2010.03.075

Zhang, H.-Y., Ji, Q., & Fan, Y. (2013). An evaluation framework for oil import security based on the supply chain with a case study focused on China. *Energy Economics*, *38*, 87–95. doi:10.1016/j.eneco.2013.03.014

Zhou, P., Ang, B. W., & Poh, K. L. (2008). A survey of data envelopment analysis in energy and environmental studies. *European Journal of Operational Research*, *189*(1), 1–18. doi:10.1016/j.ejor.2007.04.042

Zhu, Z., Wang, K., & Zhang, B. (2014). Applying a network data envelopment analysis model to quantify the eco-efficiency of products: A case study of pesticides. *Journal of Cleaner Production*, *69*, 67–73. doi:10.1016/j.jclepro.2014.01.064

Zografidou, E., Petridis, K., Arabatzis, G., & Dey, P. K. (2016). Optimal design of the renewable energy map of Greece using weighted goal-programming and data envelopment analysis. *Computers & Operations Research*, *66*, 313–326. doi:10.1016/j.cor.2015.03.012

Chapter 5
Generation Expansion Planning Using Renewable Energy Resources

Suriya Ponnambalam
https://orcid.org/0000-0002-8775-0474
Bharath Institute of Higher Education and Research, India

M. K. Ilampoornan
Bharath Institute of Higher Education and Research, India

ABSTRACT

Generation expansion planning using renewable resources to greenhouse gas emission has been analyzed in this chapter. Hydro, biomass, solar, and wind power plants have been used as renewable energy resources for expansion planning. In the developed model, two scenarios have been generated. GEP scenario represents power system expansion planning without any renewable power plant. On the other hand, the generation expansion with renewable energy resources scenario represents the role of renewable power plant in the system. The role of renewable power plant has been analyzed in two aspects, which are contribution to the production of electricity and to the reduction of GHG emission.

INTRODUCTION

Electricity requires abiding to increase in line through economic and population development. Gratifying the demand for electricity amid a reliable and safe system with the least cost combination of electric generation is the major challenge for the Generation Company (GENCO). On the other hand, cleaner generator technology will effect in a more expensive cost. Therefore, a choice of Generation Expansion Planning (GEP) must consider the two opposite aspects.

Introducing emission reduction in GEP results in more capacity of cleaner generator technology must be built in Tanatvanit, S, *et.al.* (2004). Renewable energy resources contribute to emission reduction. On the other hand, the implementation of the renewable-based power generator, such as with generator,

DOI: 10.4018/978-1-6684-9130-0.ch005

also increases reliability of power system presented in Kamalinia, S., & Shahidehpour, M. (2010). The impact of electricity generation on the environment can be used as a constraint in optimization modeling of GEP in Careri, F., *et.al* (2011). This model results in power system expansion to support green economy development. Many types of renewable energy resources have been considered in GEP model such as wind generator in Hemmati, R, *et.al* (2013) solar panel technology, Rajesh, K *et.al* (2016a and 2016b) and hydropower presented in Gil, E., Aravena, I., & Cárdenas, R. (2014). Specifically, power system reliability can be influenced by wind turbine penetration in Kadhem, A *et.al* (2017). Renewable energy implementation as a distributed generation has been published in Al Hasibi, R. A *et.al* (2018).

A more comprehensive model of GEP was introduced in Aghaei, J *et.al* (2012). This model has multi-objective that consist of minimizing cost, maximizing reliability, and minimizing emission. Renewable technology is also included in this study. However, the intermittent nature of renewable energy resources must be considered in the planning process. Several studied include uncertainty variable in the model to produce a stochastic model presented in Alizadeh, B., & Jadid, S (2014) Park, H., & Baldick, R. (2015) and Dehghan, S *et al* (2013).

Emission reduction relates to electricity generation can be achieved by reducing the usage of electricity. Reduction of electricity demand can be earned by energy efficiency, demand-side management (DSM), and demand response programs. Integrating energy efficiency program into power system planning has been introduced in Motie, S *et al* (2016) . The model of this study has been implemented in a competitive market. The energy efficiency of the side of energy resources can be used as a part of GEP model Ghaderi, A *et al* (2014). The policy development of energy efficiency will have a significant impact on SO2 and NOx emission reduction in Li, M., Patiño-Echeverri, D., & Zhang, J. J. (2019).

Emission reduction can be achieved by the implementation of DSM and load management for each demand sector in Malik, A. S. (2007). The integration of DSM and renewable energy penetration can be analyzed simultaneously. Emission reduction can be achieved by dealing the capacity addition of renewable energy-based generator as the impact of DSM application in Pina, A *et al* (2012). Integrated demand side and supply side management can be introduced simultaneously into power system expansion planning Karunanithi, K *et al* (2017). Similar to DSM, demand response with the main goal to reduce the peak load of the electrical power system is an important part of power system planning. The comparison of power system expansion policy with or without demand response is explained in Choi, D. G., & Thomas, V. M. (2012). A stochastic model of GEP with demand response has been published in Koltsaklis, N. E *et.al* (2015). Economical and environment benefit can be increased by implementing demand response in GEP model Conchado, A *et.al* (2016). In a microgrid, the optimal design of renewable energy resources has been investigated in Ruiz-Alvarez, S *et.al* (2017). For smaller system, the implementation of solar energy for residential sector has been published in Sarin, A *et.al* (2018).

The contribution of this study can be pointed out as:

1. The impact of renewable energy technology on the environment in GEP is analyzed.
2. Cost and environment benefit analysis relate to renewable energy implementation is conducted.

GENERATION EXPANSION PLANNING

A forecast is just an expectation of power demand in future values, however the accuracy of load forecast is troublesome as an outcome of the time and attributes of power system extensions. Accurate forecasts direct to extensive saving in operating and maintenance costs also it improves the reliability of power supply. An important activity of electric power providers are forecast the load demand using regression analysis, which may result in unexpected cost as it depends only on past values.

The consequences of load forecasting are obtained by providing the data obtained from regression analysis as input for Power System Expansion Planning (PSEP),which consecutively used to meet up the long term power demand with minimization of cost. One of the major problems confronting electric utility system planners is the choice of most excellent expansion in alternate for their system. If the forecasted demand is too high, it consequently requires too much generation capacity, causing the investment cost to be higher. Also, if the load forecast is low, generation capacity can't take care of out the demand. Henceforth, the load forecast uncertainty is a crucial parameter to be dealt with expansion planning. It is ambitious to realize our life in present times without the appearance of electric power. Load forecast plans and modes are increasing day-to-day because it is the effortless and cheapest transportable form of energy.

But as the demand for electric power carries on rising every day, an acceptable strategic planning for the hike of the power system is critical in future. There is a prerequisite to fabricate new assets in the sort of generation plants and transmission lines while thinking about the impacts of these new assets on power system operations and long haul motion of the economy.

Therefore, in modern years, there has been a vast inflation in the consumption of electrical energy. Hence, the PSEP has been viewed as a vital task for a long time. Power System Planning (PSP) can be performed for short, medium and long-term periods. PSP can likewise be performed at the generation levels, transmission levels along with distribution levels. Block diagram for power system expansion planning is shown in Fig.1.

The decision power system planners make are where to locate new investment, the capacity of new investment, the timing of new investment and what type of technology should be installed. Load forecasting is one of the most important tasks before doing the planning. The electric load forecasts for the complete planning period; planning constraints assure that the supply of generation will be sufficient for the period of the planning horizon.

The major difference between the electricity and other commodities is that electricity cannot be gathered in a large quantity economically, therefore electricity is consumed at the time of its production. Almost each attempt, the power system operator has made is to meet system demand with generation on a minute to minute basis. Generally, demand has the deficient adaptability to react to an irregularity amongst generation and demand, so changing generation following the demand difference turns the major way to keep demand/supply balance.

CHALLENGES IN EXPANSION PLANNING

The economic production and the associated development in economic activities increase the demand for electricity. Electricity is used in all kinds of human actions, such as industrial production, domestic/residential affair, agricultural activities, transportation, lighting and heating. There is no appropriate method to store the electric energy in large quantities; hence it is mandatory to keep up a continuous and a nearly spontaneous balance between the production and the consumption of electricity.

Some standby generator will be kept as reserve to satisfy the deviations in demand. If the supply system is inadequate to meet up the demand, load shedding is unavoidable. The increased installed capacity can reduce the power shortage. The overinvestment and the high operating costs may lead to increase energy cost and may reflect in the bill paid by the consumer. On the other hand, underinvestment and small generation margins may escort to the unavailability of power and the poor reliability to consumers. The identification and the examination of energy improvement and issues of energy policy options such as consumption, distribution and planning are most important.

RETIREMENTS OF GENERATING UNITS

When a generation unit stretches to its lifetime, most forth times, it enhances in vogue and inefficient. The modelling of the retirement of specific old incompetent generating units is considered in this thesis. Then, once a generation unit attains its lifetime a new investment is admitted to be made at the similar place.

This evaluation is important because of old units are being operated at the higher cost and a diminished practical capacity, which makes unreliability of the framework to be more serious. In recent years, reduction in greenhouse gases and the consumer welfare have attained more importance where the solution that only decreases the total planning costs may not be suitable for real power frameworks.

GRASSHOPPER OPTIMIZATION ALGORITHM

The main problem in GOA are affliction from global convergence speed, consequently it is easily cornered in local optima and cause low accuracy. Therefore, chaotic-based methods are accelerating into GOA to certify and deal with this problem. The main parameters of CGOA are C1and C2 which are accountable for the convergence of grasshoppers towards the target over the route of iterations.

The parameter C1 poises exploitation and exploration of the population, while C2 reduces the repulsion zone, comfort zone and attraction zone among grasshoppers. The values of these parameters are of extremely importance in determining the convergence speed and how to renew the location of the grasshoppers in CGOA. Through the CGOA is very influential, the solutions are at rest conflicting from each other at the last part of the search.

MULTI OBJECTIVE GEP WITH UNIT RETIREMENT

In the multi objective GEP with unit retirement problem, the decision variables are the quantity of the arrangement of generating units of each kind that will be formulated at each and every period of a planning horizon. The problem is considered with five objectives, i.e., minimization of cost, emission, energy price risk, fuel consumption and reliability cost. In the following, a detailed explanation of each objective and constraints of the problem is provided.

Let $Obj.f_z(v)$ be the zth objective function (z=1, 2, 3, 4, 5), v be a decision vector and η be the solution space which is given by Equation (1).

$$Obj.f_z(v) = \min[f_1(v), f_2(v), f_3(v), f_4(v), f_5(v)] \tag{1}$$

S.t. $v \in \eta$

In the multi objective function, $f_1(v)$ is the minimization of total cost, $f_2(v)$ is minimization of total emission, $f_3(v)$ represents minimizing energy price risk, $f_4(v)$ and $f_5(v)$ represents minimization of fuel consumption and outage cost respectively.

$v = X, E, P, I, U$

X represents the total cost, *E* represents total emission, *P* represents energy price risk, *I* represents fuel consumption and *U* represents outage cost.

Minimize Total Cost

This objective function is illustrated as the total present value of the investment cost ofrecently added generating units, fixed and variable operating and maintenance costs for newly added as well as existing generating units, plus the salvage cost related to the capacity of newly added generating units along the lifetime of the plants. These cost functions $f_1(x)$are explained from Equations(2) –(6). usually positive appropriate to the value of retired facilities.

$$Inv(M_t) = (1+r)^{-2t} \sum_{i=1}^{N} (CI_i * M_{t,i}) \times r(1+r)^{LT_i} / \left((1+r)^{LT_i} - 1\right) \tag{2}$$

$$Sal(M_t) = (1+r)^{-2(T+1)} \sum_{i=1}^{N} [CI_i * \delta_i^{2(T-t+1)} * M_{t,i}] \times f_{t,i} \tag{3}$$

where

$$f_{t,i} = (w^e_{(t-1),i} - w^e_{t,i})$$

$$w^e_{(t+1),i} \le w^e_{t,i}$$

$$w^n_{(t+1),i} \ge w^n_{t,i}$$

$$OM(X_t) = \sum_{g'=0}^{1} \left[(1+r)^{-(2.5+t'+g')} \times \sum_{i=1}^{N} ((w^e_{t,i} * X_{t,i} * FC_i) + VC_i * EES_{t,i}) \right] \tag{4}$$

where $t'=2(t-1)$ and $T'=2*T-t'$

$$Out(X_t) = OUC \times \sum_{g'=0}^{1}\left((1+r)^{-(2.5+t'+g')}\right) \tag{5}$$

where $OUC = EENS_t \times EENS\ cost$

$$EENS = \sum_{\{j|U_j>R_j\}} (U_j - R_j)P(U_j)Ot_j \tag{6}$$

Minimizing Total Emission

This objective function minimizes the summation of total emission ($i.e. CO_2$, SO_2, NO_X) over a planning horizon which is given by Equation (7).

$$min\, f_2(E_t) = \sum_{t=1}^{T}[CO_2(E_t) + SO_2(E_t) + NO_X(E_t)] \tag{7}$$

where $E_t = Ec_i + Es_i + En_i$

$$CO_2(E_t) = \sum_{i=1}^{N} Ec_i \times EES_{t,i} \tag{8}$$

$$SO_2(E_t) = \sum_{i=1}^{N} Es_i \times EES_{t,i} \tag{9}$$

$$NO_X(E_t) = \sum_{i=1}^{N} En_i \times EES_{t,i} \tag{10}$$

Minimizing Energy Price Risk

The energy price risk seeks to minimize the sum of fuel price unpredictability in the planning horizon which is given by Equation (11)

$$min\, f_3(P_t) = \sum_{i=1}^{N} P_{t,i} \times X_{t,i} \tag{11}$$

Minimize Fuel Consumption

The fuel consumption objective minimizes the sum of imported fuel consumption by conventional plants over a planning period which is specified in Equation (12)

$$min\, f_4(I_t) = (1+\alpha)^t \sum_{i=1}^{N} If_{t,i} \times y_{t,i} \quad (12)$$

Minimize Outage Cost

This objective function seeks to minimize the outage cost in order to maximize the reliability which is given by Equation (13).

$$min\, f_5(U_t) = OUC \times \sum_{g'=0}^{1} \left((1+\alpha)^{-(2.5+t'+g')} \right) \quad (13)$$

INTERNALIZATION OF EXTERNAL COSTS

While models considering environmental impacts as constraints in the form of emission taxes and penalties have the advantage of simplicity as they use existing deterministic models, their main limitation is that they assume a constant average emission level for each generation technology. In doing so, they preclude the impact that operating conditions of thermal power plants have on pollutant emission levels Conchado, A *et.al.* (2016). Another alternative is to consider environmental and health impacts of energy generation, whose costs are generally not directly borne by consumers. For example, every stage in the generation of electricity from coal-fired power plants releases harmful emissions that contribute to environmental and health degradation in the long term.

However, the low price of coal generated electricity does not account for the real cost that society ultimately pays. Internalisation of external costs refers to the integration of the environmental and health adverse effects into the decision making process.

Several authors have applied existing energy models to study the internalization of external costs of power production in GEP. A robust and exhaustive quantification of external costs is required to evaluate, in monetary terms, the impact of emissions from power plants on the environment and the human health Ruiz-Alvarez, S *et.al.* (2017). Energy models are ideal tools for such analysis as they are based on strong economic foundations that use rigorous mathematical formulations to process quantitative data and provide numerical solutions related to economics and the environment Sarin, A *et al* (2018).

INTEGRATION OF INTERMITTENT RENEWABLE ENERGY

RE sources have been the predominant drivers of the green revolution during the last decade, spurred by a combination of technological developments, innovation, decline in costs and government policies. In addition to offering emission-free electricity, RE ensures long-term energy security to countries. For these reasons and in the face of stringent carbon emission policies, many countries are considering the adoption of a larger share of renewable in their electricity mix. A series of compelling governmental actions are commonly applied to foster the development of RE. These include carbon taxes applicable to GHG emissions, feed –in tariffs guaranteeing lucrative wholesale prices for RE and tax credits for renewable electricity generation. The International Energy Agency (IEA) forecasts that the worldwide shares of RE technologies will increase to 57% of the load served by 2050 presented in Gallagher, K. S *et al.* (2012). Intermittent RE sources, notably wind and solar, are expected to account for an overwhelming majority of this share. Nevertheless, operation and planning of existing power systems have traditionally centred on fossil fuel generation that can be adjusted as required by varying fuel inputs to match variability on the load side. Integrating intermittent RE generation in the power system brings variability on the supply side as well. These technologies are characterized by fluctuations in the power output that can neither be fully anticipated nor controlled by the operator. Fluctuations in the RE resource availability can be cyclical, where they are related to diurnal and annual cycles or stochastic, where fluctuations cannot be forecasted based on historical data. Consequently, the integration of RE in the generation mix introduces more uncertainty in the power expansion problem. The task of balancing the supply and the load becomes more challenging. Likewise, ensuring supply adequacy so that there is enough generating capacity installed to satisfy peak load requirements plus a reserve margin turns out to be more complex. In-depth reviews of studies that determine the feasibility of integrating large amounts of wind power in power systems and the resulting operational impacts have been performed in Smith *et al* (2007) and Ela, E. *et al* (2009).

RECENT LONG-TERM CAPACITY PLANNING MODELS

In the wake of the increased integration of intermittent RE sources in the electricity grid, it is essential for the new long-term GEP paradigm to capture the operational characteristics of the generation fleet and gain insight into the actual dispatch practices in order to integrate flexibility issues. As observed throughout this chapter, traditional GEP models used to simply ignore short-term system operational details or account for them by using highly simplified assumptions. For example, cycling and load following features of individual power plants are often overlooked and the chronology of demand is entirely absent in the load duration curve used to approximate the annual load profile. Recent literature has mostly been directed towards findings ways to incorporate short-term power system operational details in GEP while ensuring that the problem remains computationally tractable. Typically, methods developed for this purpose integrate unit commitment and dispatch processes within the existing GEP models. In Ma, J *et al* (2013) proposed a unit commitment (UCC) algorithm to find the optimal generation portfolio that satisfies the flexibility requirements at a given wind penetration level. The UCC algorithm represents one of the first efforts to capture flexibility within GEP by integrating short-term operational decisions in the long-term planning. Rather than considering a fixed set of existing generating units and their operational costs as traditionally done in unit commitment planning, the UCC has the possibility of adding new generation plants along with their associated investment costs. Limitations of this work include the use of a deterministic methodology to model stochastic wind energy output and inability to study intra-hour variations in intermittent generation and demand.

In Palmintier, B. S., & Webster, M. D. (2015), GEP with unit commitment planning in another study to capture the influence of flexibility on long-term capacity planning. Computational tractability is ensured by clustering generators that share the same technical characteristics in terms of ramp rates, heat rate as well as minimum and maximum operating levels. A one-year simulation conducted at an hourly resolution on a 205-unit power system indicates not only that flexibility significantly affects generation mixes but also that ignoring flexibility significantly affects generation mixes but also that are infeasible to operate. A major drawback of this methodology relates to its deterministic approach to reserves. Once reserve constraints are satisfied, the model does not investigate scenarios where reserves may be inadequate due to capacity outages and forecast errors. In Batlle, C., & Rodilla, P. (2013), the cycling features of thermal units in a traditional least-cost optimization GEP model. The unit scheduling process considered start-up times and minimum generating levels of units together with usual cost functions in determining the most appropriate technologies to provide power at various intervals of the planning horizon. The main problems of this method relate to its unique economic objective function and the absence of uncertainties in its formulation. In Jin, S *et al* (2014) the issue of uncertainties by proposing a stochastic GEP model where long-term uncertainty in the wind resource is introduced by using multiple scenarios consisting of weekly time series of hourly wind power output data.

CONCLUSION

The role of a renewable-energy based power plant has been analyzed through this study. Electricity demand has been aggregated into four sectors. There are, four types of renewable energy power plant have been implemented in the model and the result of the optimization process shows that hydropower and solar power plan the most appropriate technology to be implemented in the system. To enhance the developed model, the optimization method can be combined with the uncertainty method to analyze renewable energy power plant in power system expansion planning. Moreover, energy supply chains can be developed based on the optimization method to result in more complex model based on economical, technological, and social perspectives.

REFERENCES

Aghaei, J., Akbari, M. A., Roosta, A., Gitizadeh, M., & Niknam, T. (2012). Integrated renewable–conventional generation expansion planning using multiobjective framework. *IET Generation, Transmission & Distribution*, *6*(8), 773–784. doi:10.1049/iet-gtd.2011.0816

Al Hasibi, R. A., Hadi, S. P., & Sarjiya, S. (2018). Integrated and simultaneous model of power expansion planning with distributed generation. *International Review of Electrical Engineering*, *13*(2), 116–127. doi:10.15866/iree.v13i2.14748

Alizadeh, B., & Jadid, S. (2014). Uncertainty handling in power system expansion planning under a robust multi-objective framework. *IET Generation, Transmission & Distribution*, *8*(12), 2012–2026. doi:10.1049/iet-gtd.2013.0674

Batlle, C., & Rodilla, P. (2013). An enhanced screening curves method for considering thermal cycling operation costs in generation expansion planning. *IEEE Transactions on Power Systems*, *28*(4), 3683–3691. doi:10.1109/TPWRS.2013.2249540

Careri, F., Genesi, C., Marannino, P., Montagna, M., Rossi, S., & Siviero, I. (2011). Generation expansion planning in the age of green economy. *IEEE Transactions on Power Systems*, *26*(4), 2214–2223. doi:10.1109/TPWRS.2011.2107753

Choi, D. G., & Thomas, V. M. (2012). An electricity generation planning model incorporating demand response. *Energy Policy*, *42*, 429–441. doi:10.1016/j.enpol.2011.12.008

Conchado, A., Linares, P., Lago, O., & Santamaría, A. (2016). An estimation of the economic and environmental benefits of a demand-response electricity program for Spain. *Sustainable Production and Consumption*, *8*, 108–119. doi:10.1016/j.spc.2016.09.004

Dehghan, S., Amjady, N., & Kazemi, A. (2013). Two-stage robust generation expansion planning: A mixed integer linear programming model. *IEEE Transactions on Power Systems*, *29*(2), 584–597. doi:10.1109/TPWRS.2013.2287457

Ela, E., Milligan, M., Parsons, B., Lew, D., & Corbus, D. (2009, July). *The evolution of wind power integration studies: Past, present, and future. In 2009 IEEE Power & Energy Society General Meeting*. IEEE.

Gallagher, K. S., Grübler, A., Kuhl, L., Nemet, G., & Wilson, C. (2012). The energy technology innovation system. *Annual Review of Environment and Resources*, *37*(1), 137–162. doi:10.1146/annurev-environ-060311-133915

Ghaderi, A., Moghaddam, M. P., & Sheikh-El-Eslami, M. K. (2014). Energy efficiency resource modeling in generation expansion planning. *Energy*, *68*, 529–537. doi:10.1016/j.energy.2014.02.028

Gil, E., Aravena, I., & Cárdenas, R. (2014). Generation capacity expansion planning under hydro uncertainty using stochastic mixed integer programming and scenario reduction. *IEEE Transactions on Power Systems*, *30*(4), 1838–1847. doi:10.1109/TPWRS.2014.2351374

Hemmati, R., Hooshmand, R. A., & Khodabakhshian, A. (2013). Reliability constrained generation expansion planning with consideration of wind farms uncertainties in deregulated electricity market. *Energy Conversion and Management*, *76*, 517–526. doi:10.1016/j.enconman.2013.08.002

Jin, S., Botterud, A., & Ryan, S. M. (2014). Temporal versus stochastic granularity in thermal generation capacity planning with wind power. *IEEE Transactions on Power Systems*, *29*(5), 2033–2041. doi:10.1109/TPWRS.2014.2299760

Kadhem, A. A., Wahab, N. A., Aris, I., Jasni, J., Abdalla, A. N., & Matsukawa, Y. (2017). Reliability assessment of generating systems with wind power penetration via BPSO. *Journal of International Journal on Advanced Science. Engineering and Information Technology*, *7*(4), 1248–1254.

Kamalinia, S., & Shahidehpour, M. (2010). Generation expansion planning in wind-thermal power systems. *IET Generation, Transmission & Distribution*, *4*(8), 940–951. doi:10.1049/iet-gtd.2009.0695

Karunanithi, K., Saravanan, S., Prabakar, B. R., Kannan, S., & Thangaraj, C. (2017). Integration of demand and supply side management strategies in generation expansion planning. *Renewable & Sustainable Energy Reviews*, *73*, 966–982. doi:10.1016/j.rser.2017.01.017

Koltsaklis, N. E., Liu, P., & Georgiadis, M. C. (2015). An integrated stochastic multi-regional long-term energy planning model incorporating autonomous power systems and demand response. *Energy*, *82*, 865–888. doi:10.1016/j.energy.2015.01.097

Li, M., Patiño-Echeverri, D., & Zhang, J. J. (2019). Policies to promote energy efficiency and air emissions reductions in China's electric power generation sector during the 11th and 12th five-year plan periods: Achievements, remaining challenges, and opportunities. *Energy Policy*, *125*, 429–444. doi:10.1016/j.enpol.2018.10.008

Ma, J., Silva, V., Belhomme, R., Kirschen, D. S., & Ochoa, L. F. (2013, July). *Evaluating and planning flexibility in sustainable power systems. In 2013 IEEE power & energy society general meeting*. IEEE.

Malik, A. S. (2007). Impact on power planning due to demand-side management (DSM) in commercial and government sectors with rebound effect—A case study of central grid of Oman. *Energy*, *32*(11), 2157–2166. doi:10.1016/j.energy.2007.05.004

Motie, S., Keynia, F., Ranjbar, M. R., & Maleki, A. (2016). Generation expansion planning by considering energy-efficiency programs in a competitive environment. *International Journal of Electrical Power & Energy Systems*, *80*, 109–118. doi:10.1016/j.ijepes.2015.11.107

Palmintier, B. S., & Webster, M. D. (2015). Impact of operational flexibility on electricity generation planning with renewable and carbon targets. *IEEE Transactions on Sustainable Energy*, *7*(2), 672–684. doi:10.1109/TSTE.2015.2498640

Park, H., & Baldick, R. (2015). Stochastic generation capacity expansion planning reducing greenhouse gas emissions. *IEEE Transactions on Power Systems*, *30*(2), 1026–1034. doi:10.1109/TPWRS.2014.2386872

Pina, A., Silva, C., & Ferrão, P. (2012). The impact of demand side management strategies in the penetration of renewable electricity. *Energy*, *41*(1), 128–137. doi:10.1016/j.energy.2011.06.013

Rajesh, K., Bhuvanesh, A., Kannan, S., & Thangaraj, C. (2016). Least cost generation expansion planning with solar power plant using differential evolution algorithm. *Renewable Energy*, *85*, 677–686. doi:10.1016/j.renene.2015.07.026

Rajesh, K., Karthikeyan, K., Kannan, S., & Thangaraj, C. (2016). Generation expansion planning based on solar plants with storage. *Renewable & Sustainable Energy Reviews*, *57*, 953–964. doi:10.1016/j.rser.2015.12.126

Ruiz-Alvarez, S., Patiño, J., Marquez, A., & Espinosa, J. (2017). Optimal design for an electrical hybrid micro grid in Colombia under fuel price variation. *Int. J. Renew. Energy Res*, *7*(4), 1535–1545.

Sarin, A., Gupta, R., & Jituri, V. V. (2018). Solar Residential Rooftop Systems (SRRS) in South Delhi: A Strategic Study with Focus on Potential Consumers' Awareness. *International Journal of Renewable Energy Research*, *8*(2), 954–963.

Smith, J. C., Milligan, M. R., DeMeo, E. A., & Parsons, B. (2007). Utility wind integration and operating impact state of the art. *IEEE Transactions on Power Systems*, *22*(3), 900–908. doi:10.1109/TPWRS.2007.901598

Tanatvanit, S., Limmeechokchai, B., & Shrestha, R. M. (2004). CO2 mitigation and power generation implications of clean supply-side and demand-side technologies in Thailand. *Energy Policy*, *32*(1), 83–90. doi:10.1016/S0301-4215(02)00260-4

Chapter 6
Green Building Certificates in Building Projects

Vasco Barata
Universidade Lusófona, Portugal

Jorge Vareda Gomes
https://orcid.org/0000-0003-0656-9284
Universidade Lusófona, Portugal

Mario Romão
https://orcid.org/0000-0003-4564-1883
ISEG, Universidade de Lisboa, Portugal

ABSTRACT

The importance of the building sector as a key player in the path of innovation associated with sustainable transition has been gaining preponderance at different levels of society. This radical change has been increasingly driven by governmental and non-governmental pressures, encouraging changes not only at the organizational level, but also at the macroeconomic level. Based on this premise, the present work aimed to study the influence of the use of green building certificates (GBC), not only in the innovation process, but also on what it promotes in collaborative processes in building projects. A multiple case study was used, and a comparison was made with academic references. It confirmed the importance of the role of the green project champion who, through the leadership process, has a direct influence on the development of the project in its different phases.

INTRODUCTION

The climate changes that we observe today are increasingly extreme and catastrophic, and man's role in accelerating these changes is all too evident. The most recent report by the Intergovernmental Panel on Climate Change (IPCC, 2021) presents relevant scientific facts on these topics, providing a detailed summary of the current state of the climate, the human influence and some future predictions.

DOI: 10.4018/978-1-6684-9130-0.ch006

In addition to the environmental problem, the increase in competitiveness resulting from the globalization of markets has accelerated the need for change. Organizations are forced to introduce innovations of the most varied types, to obtain competitive advantages over other organizations, to meet sustainable development goals and to be able to satisfy not only their customers, but also their employees (Atkinson, 2013; Porter, 1990).

The building sector is no exception and, nowadays, it accompanies other fundamental sectors for socio-economic development, in this long path of sustainable transition. In recent decades, there has been a widespread concern at an international level to improve building practices to minimize their harmful effects on the environment (Cole, 2003). This evolutionary process brought the most varied topics to the discussion, from the environmental impact itself, to recycling, to the concept of green buildings and building certificates that quantify the measures implemented, with building performance being one of the main focuses of the professionals in the sector (Ding, 2008). Currently, it is common knowledge that the building sector is one of the most responsible for the environmental degradation, not only because of the excessive use of natural resources in the constructions process and maintenance, but also by direct or indirect pollution caused (Amado et al., 2013; Ding, 2008).

With the notion of the emerging importance of the themes, in this article we present the results of a comparative study that contributed to clarify our understanding on how the use of Green Building Certificates (GBCs) influences the processes of innovation and collaboration in the building sector.

SUSTAINABLE DEVELOPMENT

According to the Commission Brundtland (1987, pp.15), the concept of sustainable development is defined as the "...development that meets the needs of the present without compromising the ability of future generations to meet their own needs".

The concept has been widely used, not only academically, but also politically and corporately (Redclift, 2005). Many definitions have emerged over the years from the original definition (Emas, 2015). When analysing the definition, it becomes clear that the idea of preserving resources for future generations is the fundamental feature that differentiates sustainable development policies from traditional environmental policies.

We can define the general objective of the sustainable development as long-term stability, not only of the environment, but also of the economy. For this to happen, it is necessary to identify and consider the different economic, environmental and social concerns, also defined as the Triple Bottom Line (Emas, 2015). In organizational context we can consider the concept of sustainable development as the continuous efforts made by organizations to balance environmental and social concerns with their economic goals, to minimize damage and increase environmental and societal benefits (Dyllick & Hockerts, 2002).

The building sector and the environment are intrinsically linked for the most diverse reasons, highlighting the use of natural resources, pollution or modification of ecosystems. In this way, the sector has a leading role when concerns about environmental impact and respective mitigation measures are pointed out (Ding, 2005). The transition to more sustainable practices is driven by increasing pressure caused by legal commitments, as well as the growing public concern with the building impact on the environment (Robichaud & Anantatmula, 2011).

Sustainable building is seen as a way for the sector to achieve the sustainable development, contributing to the protection of the environment, the preservation of natural resources and also to improving the quality of human life (Amado et al., 2013; Ding, 2005). Sustainable build seeks to incorporate, at all stages, from the definition of the project to the construction and buildings maintenance, the fundamental premises of the sustainable development at the economic, social and environmental levels, pillars of the Triple Bottom Line (Tan et al., 2011). However, Berardi (2012) argues that most approaches to the use of sustainable buildings and the respective assessment methods focus disproportionately on the environmental dimension, with energy consumption becoming the dominant measure.

Although there is a wide consensus on the importance of sustainable development in the building sector, the interests of the different actors in its adoption of a given project are still uncertain (Herazo & Lizarralde, 2015). These authors have summarized the main approaches on this topic and relate them to the GBCs, with the effects they have on collaboration and innovation, with the project planning and, consequently, on the own project.

Reed (2007) refers to the need to interact with the environment, using the sustainability of ecological systems as a basis for design, rather than just looking to cause less damage to the environment and proposes three levels of learning:

- Level I - correspond to the level of greening (efficiency).
- Level II - can be seen as alignment with sustainability (effectiveness).
- Level III - addresses an evolving macro understanding, where levels of reconciliation and regeneration question the project purpose.

These three levels of learning are further divided into five increasing and evolving stages: (1) conventional practices; (2) green buildings; (3) sustainable buildings; (4) sustainable designs and (5) restorative, reconciling and regenerative designs.

GREEN BUILDING CERTIFICATES (GBC)

The first GBCs emerged in the early 1990s (Pérez-Lombard et al., 2009) with the need to ensure the alignment of the different building processes with more sustainable practices (Bragança, 2005). For this, methodological structures capable of measuring and monitoring the environmental performance were created, not only of the different processes applied, but also for the building (Ofori-Boadu et al., 2012).

In the building sector, environmental certifications have as their main purposes the recognition and appreciation of buildings that contribute to a more sustainable future (Cole, 2005). They also help to create awareness among stakeholders and users, becoming increasingly a powerful tool to mirror the commitment to sustainable development (Ding, 2008).

From a chronological perspective, the BREEAM (Building Research Establishment Environmental Assessment Methodology) building certificate appear in Great Britain in the 1990s, and over the following decades, numerous building certificates emerged, originating in several countries or from different international institutions (Mateus & Bragança, 2011) (e.g., LEED – USA; CASBEE - Japan; NABERS – Australia; HQE – France; DGNB – Germany and SBTool in several countries).

According to Herazo & Lizarralde (2015), initially the certificates focused mostly on the so-called “Green Buildings”. Over the years and the consequent gain in experience, they also began to focus, among other strands, namely, on sociocultural, economic or more technical aspects, evolving a little towards the concept of “Sustainable Buildings”.

Whichever GBCs is used, obtaining it entails additional costs, which must be considered. However, experts often point to the fact that the long-term benefits associated with obtaining any certificate outweigh the short-term costs (Herazo & Lizarralde, 2015).

In addition to the GBCs, institutions such as the International Organization for Standardization (ISO), the European Committee for Standardization (ECS), the International Code Council (ICC) or even more recently the European Commission (EC), have made efforts to define minimum environmental and sustainability requirements for the construction sector. (Mateus & Bragança, 2011).

Although there is a great diversity of GBCs, in general, they all have the same purpose. However, LEED certification (Leadership in Energy and Environment Design Certification) is considered to be the most popular, with the widest dissemination worldwide and to which we will place greater emphasis.

LEADERSHIP IN ENERGY AND ENVIRONMENT DESIGN CERTIFICATION (LEED)

The LEED certification is an external program based on prerequisites and credits whose main purpose is to assist in the design, building and certification of green buildings. This certification is a voluntary and internationally recognized system that quantifies and classifies a building, or group of buildings, based on implemented strategies that aim to improve the performance of the same(s) in areas, such as: efficiency energy and hydraulics; reduction of CO2 emissions; improving the quality of the indoor environment; and application of different resources (Dobiás & Macek, 2014; Sebastien et al., 2007).

In the first versions, a less formalized approach was used, where points were assigned to each of the main categories based on an assessment made by an external team to the building process and responsible for the certification of the project. The LEED 2009 version added an innovation to the evaluation process, adding a weighting system that attributed points based on the ability to reduce the environmental impact produced (USGBC, 2018). LEED v4 incorporated the previously developed framework and complemented it with a set of criteria (impact categories). In addition to this change, an innovative digital tool was also added that allows not only to visualize the different combinations between the strategies to be implemented and the consequences of their use, but also allows for a statistical analysis of the results, generating greater analytical rigor and a better perception of the system (Owens et al., 2013). The latest version of the certification, LEED v4.1, presents itself as an improved version, having raised the bar even further about the applied building standards, presenting itself more and more as a management system for world-leading sustainable building projects.

LEED v4.1 promotes a structure that develops actions in six distinct fields, known as credit categories, namely: location and transportation; sustainable areas; water use efficiency; energy and atmosphere; materials and resources; and quality of the indoor environment. In each of the six credit categories, there are mandatory prerequisites and optional (credits) strategies, each with its own purpose and a set of requirements to be fulfilled. To obtain certification, projects are required to implement and document not only the prerequisites, but also enough credits to add a minimum of 40 points out of 100 available. The more points accumulated, the higher the certification awarded to the building (Owens et al., 2013). The existence of credits promotes flexibility and adaptability of the certification to the different particularities of the projects. The scoring system is defined based on the relevance of each credit to the overall objective of the certification, that is, the most valued credits are those that contribute most to the achievement of the goals defined in the six impact categories (Owens et al., 2013). To define the value of the different credits, an analysis is made of the impact they have on each of the impact categories. According to official information, LEED certification has numerous advantages in its use:

- Helps investors meet environmental, social and governance goals by providing a robust, globally recognized green building framework to measure and manage your real estate performance.
- Gives a competitive advantage, where more than half of the leaders of companies that applied the certification, see sustainability as a differentiating mark and that it contemplates a financial gain.
- Attracts occupants because, although rents increase on average by around 20%, buildings with LEED certification have, on average, higher occupancy rates than buildings without any certification.
- Promotes the happiness and health of the occupants of certified spaces.
- Saves energy and natural resources by reducing operating costs.

According to official data, it is estimated that between 2015 and 2018, buildings that have achieved LEED certification have achieved savings of around US$1.2 billion in energy consumption, US$149.5 million in water consumption, US$715.3 million in maintenance costs and US$54.2 million in waste.

LEED CERTIFICATION IN PORTUGAL

In Portugal, and similarly to what happens in the world, LEED certification is one of the most used. The official USGBC page provides an updated database with projects registered, under development or that have already obtained one of the possible certifications. Until today, 136 048 projects were registered on the site, spread across 169 different countries, of which 49 were in Portugal. Although at the national level there is still a small number of projects with some of the existing building environmental certifications, and in particular the LEED certification, the tendency for the emergence of new projects where they become essential is increasingly a reality. This growth trend, associated with the fact that many of the ongoing projects have not yet been registered in the international database, means that the real number of projects with LEED certification in Portugal may be a little higher than the number presented.

GREEN PROJECT CHAMPION

With the increase for this type of certificate, projects that intend to include measures in line with the sustainable development generally lack a project manager who is responsible for coordinating and integrating all project stakeholders (Herazo & Lizarralde, 2015). In the literature, this position is recognized as the "Green Project-Champion", where the person or group of people (representatives of the various stakeholders) have the function of influencing and mobilizing different stakeholders in decision-making processes (Gattiker & Carter, 2010), removing barriers to cooperation and enabling innovative ideas (Hartmann, 2008). They should show leadership skills with some propensity to risk and demonstrate strong innovative capacity (Howell & Higgins, 1990). When composed of more than one person, the team is commonly called a "Green Team", where the most experienced member becomes the leader (Bayraktar & Owens, 2010). For Bossink (2007a), these sustainable and innovative building project managers become leaders in knowledge management, as they are the focus for solving most problems related to the implementation process and obtaining certification.

INNOVATION

Sustainability-Oriented Innovation (SOI)

Oslo Manual (OECD, 2018) defined innovation as the implementation of a new or significantly improved product (good or service), or a process, or a new marketing method, or a new organizational method in business practice, in the organization of the workplace or in external relations.

One of the main ways to boost sustainable development over the last few decades has been through the introduction of innovations (Schaltegger & Wagner, 2011). According to Dias (2018), the connection between these concepts can be complex and a little in-depth analysis of these topics can lead to errors regarding the content inherent to the concept of SOI. SOIs at the organizational level, according to Adams et al. (2016), require intentional changes to be made in the company's philosophy and values, causing changes in products, processes or practices to serve the specific purpose of creating and realizing social and environmental value, in addition to economic returns. According to Klewitz & Hansen (2014), the concept of SOI is similarly described and is presented as the direction to be followed, where the deliberate management of economic, social and ecological aspects is necessary for them to become integrated into the environment design of new products, processes and organizational structures. According to the same authors, at the organizational level, the discussion about the focus of business towards sustainability through innovations initially focused on eco-innovations.

Eco-innovations can be represented in the form of processes, organizational structures, products or technologies (Beise & Rennings, 2005; OECD, 2018). Process eco-innovations are directly related to the production of goods and services, where the main objective is to promote eco-efficiency and eco-efficiency (Huber, 2008).

Organizational eco-innovations are related to the reorganization of company interactions and structures with new forms of management (Hansen et al., 2009). Eco-innovations of products and technologies are related to the improvement or development of new products or services, where, on the one hand, ecological design improves products using more ecologically acceptable products, with greater durability or with lower energy consumption. and, on the other hand, sustainable technological development presents entirely new products (Hart & Milstein, 2003).

Later, and through the inclusion of the social dimension in an environmental perspective, a more holistic view of sustainability within the organization began to be taken and, consequently, several authors developed numerous terms to contribute to the path of business sustainability through of innovations (Klewitz & Hansen, 2014; Rennings et al., 2006). The SOI concept comprises the improvements related to the developments obtained in the previous stages. It arises, as well as the long process on the path to sustainability, and this same path is supported by the deliberate management of different economic, social and environmental aspects in the same direction (Adams et al., 2016; Klewitz & Hansen, 2014).

INNOVATION AND COLLABORATION IN THE BUILDING SECTOR

In the last two decades, there has been a growing investigation and debate about buildings innovation. Toole et al., (2013) present a definition of innovation more directed to the sector, where they emphasize that innovation is the act of introducing a significant improvement in a process, product or system new to the organization, which can cause changes in the vision of individuals and which results in a competitive advantage, increase in value for the customer or benefit for the interested parties.

The improvements caused by the success of innovations represent an increase in value for a wide range of stakeholders, from investors to suppliers, and their path is guided by a sequence of decisions taken in an environment of uncertainty (Herazo & Lizarralde, 2015).

INNOVATION AND COLLABORATION ANALYSIS TOOL IN THE BUILDING SECTOR

Toole et al. (2013) developed a Maturity Assessment Model (MAM) that allows organizations to identify the main weaknesses and improvements in their innovation-related capabilities to be made through eight key areas of innovation influence. Bosink (2007b) develop an Inter-Organizational Innovation Model (IOIM) that focus on the process of cooperation between organizations to improve and implement innovations in sustainable building. The theoretical model is composed of eight stages of interorganizational innovation and 22 interaction patterns.

Herazo & Lizarralde (2015) developed an analytical tool based on theoretical models of Toole et al. (2013) and Bossink (2007b), which can be applied to different levels of strategic-tactical management of an organization, that is, in internal management, as well as in the relationships of a temporary multi-organization.

The study of Herazo & Lizarralde (2015) identified four tensions that occur during the process of applying the GBCs by temporary multi-organizations. These tensions resulted from the confrontation of the results obtained with the different "push-and-pull" factors identified in the bibliography, such as: (1) bottom-up or top-down participatory processes (Fraser et al., 2006); (2) long- or short-term efficiency (Eriksson et al., 2014); (3) the individual and collective approach (Fellows & Liu, 2013); and (4) competition and collaboration (Brandenburger & Nalebuff, 1998).

Each of the tensions brings two diametrically opposed concepts into confrontation, which necessarily leads to reflection and the search for a position of balance between them. The four tensions identified are as follows:

- Strategic/Tactics: The balance between the strategic plan and the tactical plan. Most decisions are typically taken from a strategic point of view in the early stages of the project. However, leadership and motivation to fulfil them in tactical terms in the operational stages is also highly relevant to the success of the project.
- Collaborative/Competitive: To face the high complexity and natural fragmentation of the building sector, organizations need to collaborate with each other. On the other hand, economic and social perspectives and technological developments force them to be increasingly competitive.
- Participatory/Effective: This type of tension between participation in decision-making processes and efficiency is spread between the different actors in the construction process and the users.

Sustainable building tools, in relation to this tension, allow the different actors to share experiences in terms of decision-making oriented towards sustainability.

- Individuals/Collectives: The building sector has, for the most part, an individualistic bent, but where cooperation and teamwork are also required. This balance is even more important when the decision processes of a given project consider its sustainability.

The way in which these tensions are managed causes benefits or losses both in the performance of organizations and in the project itself.

Herazo & Lizarralde (2015) identified and defined multiple factors that, when applied to GBCs in each project, influence innovation and collaboration in the building sector.

MATERIALS AND METHODS

This study aims to make a comparative assessment between the innovation process in the Portugal building sector associated to the application of GBCs and the study developed by Herazo & Lizarralde (2015).

We sought to identify the main converging and divergent points and, in this way, assess the degree of maturity of the main players in the constructive process for innovation and collaboration associated with sustainable development.

A qualitative methodology was used based on a multiple case study with a detailed approach to the influence of the use of GBC on innovation in the construction sector.

To obtain data comparable to the Herazo & Lizarralde (2015) study, a similar conceptual model was applied, that is, the projects were chosen according to the same criteria and the questions asked in the interviews were similar. The interviewees also occupied identical hierarchical positions, as much equivalent data were collected and observation campaigns were carried out identical to those described by the authors. The two chosen case studies met the following requirements: (1) Projects where LEED certification is being applied; (2) Projects that are being developed by large and complex organizations, with large private, public or mixed capital; (3) Projects that are being promoted by customers/owners with a view to hosting their own business activity; (4) Projects where there is a significant involvement with sustainable development, and (5) Projects in progress, or executed in the last three years, to facilitate the consultation of the necessary information and the carrying out of the interviews.

Ten interviews were conducted, organized into eight open-ended questions focusing on green certification processes of Herazo & Lizarralde (2015) and based on the 8 key areas presented in the model of Toole et al. (2013) (Table 1).

Table 1. Summary the case study respondents

Case Study	Client	Contractor	Architect / LEED Consultant	Project Manager / Inspection
A	1	2	1	2
B	1	1	1	1

During the data collection process, it was intended to explore the organizational aspects related to the green certification processes.

The information collected in the interviews was complemented with documentary information collected, visits to building places, photographs, project drawings, attendance at meetings and meeting minutes (Table 2).

Table 2. Documentary information collected

Document	Case Study A	Case Study B
Owner's strategic plans		×
Owner's annual reports		×
Official website information	×	×
Work meetings	×	
Press documents	×	×
Photographs	×	×

Table 3 presents the eight questions asked to each of the interviewees to collect as much information as possible and to recognize the intra-organizational relationships that influence the innovation processes in each case of study (Herazo & Lizarralde, 2015).

Table 3. Questions (adapted from Herazo & Lizarralde, 2015 and Toole et al., 2013)

Key Area	Question
Organizational Processes	How have Sustainable Development strategies been applied at the organizational and project decision-making levels?
Resource allocation	Was there a budget or a team specifically designated to obtain certification?
Risk Perspective	Did the customer find green certification risky?
Culture	Was the organizational environment open to new ideas, processes and internal and external feedback in a long-term perspective?
Customer Focus	How did the design teams (project and consultancy) focus on satisfying customer needs?
Learning	How was knowledge transmitted between projects?
Collaboration	What mechanisms were used to communicate, coordinate and collaborate with internal and external stakeholders?
Leadership	Has a Green Project Champion been designated?

After collection and transcription, all the information was processed in order not only to analyse the main areas of innovation suggested by Toole et al. (2013), but also to analyse the most common innovative practices at the inter-organizational level suggested in the model presented by Bossink (2007b).

This exercise allowed obtaining results equivalent to those presented by Herazo & Lizarralde (2015), to comparatively evaluate both results, based on the four tensions identified.

DISCUSSION

General Characteristics of Case Studies

Concerning the case study A (Table 4), the current owners bought land with a project and a closed GBC. After the review, they decided not only to change the certification system, but also to upgrade their level from a BREEAM GOOD certification to a LEED GOLD certification. These changes necessarily led to structural changes at the design level.

The owner of Case Study B (Table 4) is a relevant company of the Portuguese economy and intends to build an office building for its own use as a headquarters building. For different reasons, it is also a particularly interesting case, since it is the fourth building that this company has built with a LEED certification, which allows it to have a high degree of decision maturity and know-how in matters related to certifications.

Table 4. Summary table of the case studies presented

Characteristics	Case Study A	Case Study A
Client	Private	Private
Use	Multifunctional installations for any type of rent	Offices for the company's own use
Building area	About 6,500 m2	About 5.500 m2
Project budget	Approximately 60 million	Approximately 40 million
Certification	LEED GOLD	LEED GOLD
Completion estimates	2023	2022
Key green building strategies	Rainwater recovery systems, flow reducers, photovoltaic panels, facades with improved shading levels, parking for bicycles, electric vehicles	Rainwater recovery systems, photovoltaic panels, parking for bicycles, electric vehicles
Type of funds and investment	Private investment fund	Private company

COMPARISON WITH THEORETICAL MODELS

The collected data were confronted with the theoretical models of Toole et al., (2013) and Bosink (2007b). Table 5 presents the relationships between the eight key areas of Toole's et al. (2003) model and the certification process.

Table 5. Summary of key areas by case study (Toole et al., 2013).

Key Areas	Case Study A	Case Study B
Organizational Processes	Transfer of expectations related to sustainable development fundamentally from top to bottom (Top-down)	Top-down transfer of sustainable development-related expectations and needs
Allocation of resources	Budget and specific human resources to obtain certification; More horizontal structure	Budget and specific human resources to obtain certification; more hierarchical structure
Perspective of Risk	Certifications are seen as an opportunity to enhance the value of the building to be constructed; Viewed as a contractual obligation	Certifications are not seen as a risk, but as a tool to help financial savings; Viewed as a contractual obligation
Culture	Most of the stakeholders are receptive to the principles of sustainable development, which improve the project	All stakeholders are receptive to the principles of sustainable development, which improve the project
Customer Focus	Support from external consultants to carry out the project and obtain LEED certification	Support from external consultants to carry out the project and obtain LEED certification
Learning	Little evidence of formal learning, but commitment by all stakeholders to familiarize themselves with issues related to certification	Clear definition and clarification of communication strategies on certification, but with little clear evidence of formal learning
Collaboration	Collaboration between consultants and architects for strategic integration of the sustainable development premises in the initial phases of the project. Collaboration between project inspection/ management and contractor to define project execution tactics	Collaboration between consultants and architects for strategic integration of the sustainable development premises in the initial phases of the project. Collaboration between LEED consultants and contractor to define project execution tactics
Leadership	A more strategic and tactical Green Project Champion	A more strategic and tactical Green Project Champion

The Table 6 presents the stages of innovation and the patterns of inter-organizational interaction identified in each case study (Bosink, 2007b). The eight stages of innovation are identified, but the twenty-two interaction patterns were not detected in both cases.

Table 6. Stages of inter-organizational innovation (Bosink, 2007b)

Innovation Stages	Case Study A	Case Study B
Autonomous innovation	Autonomous innovation Innovation portfolio	Autonomous innovation Innovation portfolio
Networking	Choose to innovate or are forced to innovate. Prefer to work with known partners. Perceive an influential position in the innovative network	Choose to innovate or are forced to innovate. Prefer to work with known partners. Perceive an influential position in the innovative network;
Exploration (Execution)	Determine what experience is needed;	Determine what experience is needed. Develop a cooperative portfolio
Training	Enter contracts	Enter contracts
Organization	Establish a joint organization Establish control bodies Develop an architectural project	Establish control bodies Develop an architectural project
Planning	Allocate experiences Facilitate cooperation and communication	Allocate experiences Facilitate cooperation and communication
Co-innovation	Coordinate the realization of innovation	Coordinate the realization of innovation
Dismantling	Dismantle the joint organization	Dismantle the joint organization

ASSESSMENT OF TENSIONS INFLUENCING INNOVATION AND COLLABORATION

Analysing in detail all the testimonies, three of the four tensions already mentioned are identified. It should be noted that although identified, in most cases these tensions are presented in a more tenuous way, compared to previous studies.

Strategic/Tactical Tensions

For both cases, two common points are identified. Firstly, the commitment of the respective project owners, in a mostly strategic perspective and position, with the main premises of the Sustainable Development. Second, the concentration of most strategic/tactical decisions in the initial design phase.

In collaborative terms, in case study A, some communication and information transfer failures between the contractor and the LEED consultant are identified, which are compensated by the more horizontal organizational structure and the more active role of project management/supervision in matters related to the most operational part. Even so, it generated some tension between the parties, more evident in the contractor than in the LEED consultant.

In case study B, the well-hierarchical structure promotes a very effective and efficient collaboration between all the intervening parties. This is headed, in most cases, by the LEED consultant who, in addition to explaining in detail the entire project environment at the beginning of the project, is prepared in a very accessible way to help with any problem already in the construction and execution phase of the project.

Collaborative/Competitive Tensions

Collaborative/competitive tensions were not perceived in any of the interviews carried out in the two case studies. Through the information collected, it is possible to identify several reasons for this to have happened.

Firstly, the fact that the organizations that make up the temporary multi-organizations of the two case studies, although they are part of the same business sector, are normally not direct competitors. Another important factor in that these tensions were not clearly identified was the low level of knowledge that is still widespread on these issues related to GBC, on the part of most of the tactical actors in the process.

The data presented show that although the collaboration is transversal to the two case studies, they have different characteristics. Dissatisfaction with the way of working or the degree of perception of knowledge of the LEED consultants who provide consultancy to the projects, are two of the clearest disruptive points among the case studies under analysis.

Another relevant factor for these tensions not having been identified is related to the good relationship that unites all organizations during a certain period. In general, for large projects, which involve several organizations, and which lack the inclusion of innovative processes, there is a natural collaborative tendency inherent to a common desire for success and a contractual obligation that is transversal to all. This perception was clearly recorded in all interviews.

Participatory/Effective Tensions

The participatory/effective tensions were identified in a much more tenuous way than is generally referred to in the literature. This can be justified by the impossibility of carrying out interviews with future users of buildings, since it is generally between decision-makers and users that the greatest tensions between the participatory decision-making process and the short-term efficiency of the same.

In case study A, no type of participatory program was carried out with the community, but the need to incorporate the innovative increase in the final product that certification entails is evident, with a view to giving future companies using the infrastructure another key tool. to improve talent retention capacity. Regarding case study B, this same participatory work with employees was not possible to clearly clarify with the study carried out. However, during the interview with the owner, it was possible to identify some statements that show care for the work environment and the well-being of its employees, namely, the carrying out of satisfaction surveys and the inclusion of some recommendations made by the same. Some of these recommendations promoted some changes in the owner's existing buildings, such as the incorporation of a gym and support facilities in one of the headquarters buildings or the reduction of less healthy items in the vending machines scattered throughout the facilities and consequent replacement of them with healthier products.

Individual/Collective Tensions

This type of individual/collective tensions become even more visible when an innovative process is included, such as obtaining a GBC, in a very mechanized process, typical of the construction sector. According to Herazo & Lizarralde (2015), if there is a more collectivist organizational culture, the execution of a given project and obtaining a certification occur with less tensions between the work teams, with this type of tensions being more visible in the relationships between the intervening members. of the process from a more operational perspective, or green members, and the green project champion.

At this point, project owners can internalize the process, create a team to which they belong, or hire an external company that focuses on aligning the different actors with the Sustainable Development principles. Clarifying the position of the green project champion in each of the case studies helps to understand and identify this type of tensions more clearly.

In case study A, we have two types of green project champion, that is, a work team was created that distributes functions within the organizational process and that responds to the various green members individually. On the one hand, we have the LEED consultant who acts as a link from the main guidelines imposed by the certification to their implementation and subsequent project execution. On the other hand, we have the project management/inspection office, which assumes a more tactical role in helping green members in the best possible execution for the implementation of the processes imposed by the certification. With this, we point out, for example, aspects such as helping with the choice of materials incorporated in the work or the control and validation of the entire supply of these same materials, so that the process does not suffer individual delays.

In case study B, we also have a case of a company that was hired to assume the role of green project champion from the beginning of the whole process, which meant that all sustainability guidelines were defined in advance and delivered to the architects in a very early stage of the project, so that they were easily integrated. In this way, as the process was clarified from the beginning, the consultancy firm assumes a role of great preponderance and almost hegemony in the control and definition of decision-making in all phases of the project and thus, this organization takes the most part of the doubts in operational terms of how to proceed in the face of a given event.

Through this analysis, it is understood that the role of the green project champion is preponderant to reduce tensions between the more individualistic perspectives and the weight of collectivism in our temporary multi-organizations. However, it is also necessary to analyse these tensions from the perspective of the individual risk of each of the actors. At this point, a great homogeneity of opinions is identified, since both the more strategic/tactical and the more operational actors, with the appropriate differences of opinion, consider the use of GBCs as not very risky.

CONCLUSION

Similar to the study by Herazo & Lizarralde (2015), the results presented were initially compared with theoretical models by Toole et al., (2013), and Bosink (2007b). From this analysis, the different tensions affecting the temporary multi-organizations that influence the construction process when implementing a GBC were identified. This comparison aimed to evaluate the two scenarios to identify the converging and divergent points between them.

The study carried out provided a considerable amount of information regarding the two theoretical models, information that allowed the identification of three of the four tensions identified in the bibliography. The collaborative/competitive tension, based on all the documents consulted, the visits to the construction sites and all the interviews carried out, was not identifiable due to the poorly developed state of maturity in relation to issues related to the Sustainable Development, which the building sector at the national level still presents now.

Another conclusion in line with the literature is that leadership in situations where it is necessary to include innovative processes has a special influence on success and on the reduction of entropies created around them. In this perspective, the role of the green project champion, as a new member of the temporary multi-organizations, will positively influence the general understanding and will create the most consensus possible around these themes.

As presented in the academic literature, the identification of tensions in the relationships between different organizations are related to the stages of inter-organizational innovation proposed in Bossink's (2007b) model, making it therefore easy to identify many of them, in each case. Therefore, inter-organizational innovation processes arise because of the pressure imposed by stakeholders to implement the Sustainable Development principles. In the case studies, this pressure came essentially from the owners of the work and was consummated in a contractual obligation, implemented in the project, essentially by the LEED consultants, which all parties must comply with.

In addition, the same developers work as the centrepiece in enhancing these same phases of inter-organizational innovation in the projects, to reduce the inherent tensions caused between the more strategic and more tactical dimensions. In the case studies presented, this function of enhancing connections between the stakeholders is identified, although they place most of the technical responsibility on the LEED consultant, who assumes the role of green project champion, keeping to himself the final decision that must be in accordance with their previously stipulated objectives for the project.

One of the points that was not identified in the case studies, and which was quite explicit in the literature, was the different perceptions of risk for the implementation of GBC in projects. The addition of work for the more tactical actors, who manage the entire operational process, was not seen as something negative or hindering, it was seen as a learning opportunity.

In a similar way to that presented by Herazo & Lizarralde (2015), the tensions identified from the analyses carried out on the theoretical models for the case studies used, reveal the changes that the actors must carry out when a certain certification is imposed by the owner of the work. In the case studies, these changes are not always immediately perceived and identified by the stakeholders, due to the lack of experience of working with them.

Finally, it is concluded that, similarly to what is defined in the scientific literature, the role of the green project champion in the case studies used, also assumes an increasingly preponderant role, headed mostly by consulting companies designated for this purpose. Furthermore, the collaboration between the different actors of the temporary multi-organizations created is a fundamental prerequisite for the success of the process.

LIMITATIONS AND FUTURE WORK

As a final limitation to point out, there is a few numbers of projects available to use as case studies. Noting the existence of few projects that meet the selection criteria and the difficulty of directly contacting the stakeholders were the main conditioning factors. So, it becomes impossible to extrapolate the conclusions reached, as a rule, to the entire panorama of building at the national level. The present study is then positioned as an indicative study of the influence of the use of GBC on innovation and collaboration in construction projects, pointing out the key points to be considered and the main consequences that it causes in them.

As future investigations, it is suggested the possibility of repeating the same study in a time horizon of five to ten years with as many projects as possible, in order to assess the main developments associated with the theme and also the possibility of expanding the study to other certifications, or new equivalent programs, of governmental or private origin, in order to understand if the constraints, tensions, benefits and opportunities are similar or if they vary significantly.

REFERENCES

Adams, R., Jeanrenaud, S., Bessant, J., Denyer, D., & Overy, P. (2016). Sustainability-oriented Innovation: A Systematic Review. *International Journal of Management Reviews*, *18*(2), 180–205. doi:10.1111/ijmr.12068

Amado, M. P., Lucas, V., & Ribeiro, M. (2013) Sustainable Construction: Value of Certification. *Proceedings of the ACSEE - International Conference on Advances in Civil, Structural and Environmental Engineering, Institute of Research Engineers and Doctors*, 180-187. https://doi.org/10.3850/ 978-981-07-7965-8_83

Atkinson, R. D. (2013). *Competitiveness, Innovation and Productivity: Clearing up the Confusion*. The Information Technology & Innovation Foundation.

Bayraktar, M. E., & Owens, C. R. (2010). LEED Implementation Guide for Construction Practitioners. *Journal of Architectural Engineering*, *16*(3), 85–93. doi:10.1061/(ASCE)AE.1943-5568.0000013

Beise, M., & Rennings, K. (2005). Lead markets and regulation: A framework for analyzing the international diffusion of environmental innovations. *Ecological Economics*, *52*(1), 5–17. doi:10.1016/j.ecolecon.2004.06.007

Berardi, U. (2012). Sustainability Assessment in the Construction Sector: Rating Systems and Rated Buildings. *Sustainable Development (Bradford)*, *20*(6), 411–424. doi:10.1002/sd.532

Bossink, B. A. (2007a). Leadership for sustainable innovation. *International Journal of Technology Management & Sustainable Development*, *6*(2), 135–149. doi:10.1386/ijtm.6.2.135_1

Bossink, B. A. (2007b). The interorganizational innovation processes of sustainable building: A Dutch case of joint building innovation in sustainability. *Building and Environment*, *42*(12), 4086–4092. doi:10.1016/j.buildenv.2006.11.020

Bragança, L. (2005). *Princípios de Desenho e Metodologias de Avaliação da Sustentabilidade das Construções*. Revista Espaços. http://repositorium.sdum.uminho.pt/handle/1822/4943%0A

Brandenburger, A., & Nalebuff, B. (1998). Co-opetition: 1. A Revolutionary Mindset that Combines Competition and Co-operation. 2. In The Game Theory Strategy that's Changing the Game of Business. Currency Doubleday.

Cole, R. J. (2005). Building environmental assessment methods: Redefining intentions and roles. *Building Research and Information*, *35*(5), 455–467. doi:10.1080/09613210500219063

Commission Brundtland. (1987). Our Common Future. Oxford University Press. doi:10.1080/07488008808408783

Dias, A. S. (2018). *Inovação Orientada para a Sustentabilidade : Uma análise ao caso da Corticeira Amorim*. Universidade de Aveiro, Departamento de Economia, Gestão, Engenharia Industrial e Turismo.

Ding, G. (2005). Developing a multicriteria approach for the measurement of sustainable performance. *Building Research and Information*, *33*(1), 3–16. doi:10.1080/0961321042000322618

Ding, G. (2008). Sustainable construction-The role of environmental assessment tools. *Journal of Environmental Management*, *86*(3), 451–464. doi:10.1016/j.jenvman.2006.12.025 PMID:17289255

Dobiás, J., & Macek, D. (2014). Leadership in Energy and Environmental Design (LEED) and its impact on building operational expenditures. *Procedia Engineering*, *85*, 132–139. doi:10.1016/j.proeng.2014.10.537

Dyllick, T., & Hockerts, K. (2002). Beyond the business case for corporate sustainability. *Business Strategy and the Environment*, *11*(2), 130–141. doi:10.1002/bse.323

Emas, R. (2015). *Brief for GSDR 2015 The Concept of Sustainable Development: Definition and Defining Principles*. https://asset-pdf.scinapse.io/prod/2184349672/2184349672.pdf

Fellows, R., & Liu, A. (2013). Use and misuse of the concept of culture. *Construction Management and Economics*, *31*(5), 401–422. doi:10.1080/01446193.2013.794296

Fraser, E., Dougill, A., Mabee, W., Reed, M., & McAlpine, P. (2006). Bottom up and top down: Analysis of participatory processes for sustainability indicator identification as a pathway to community empowerment and sustainable environmental management. *Journal of Environmental Management, 78*(2), 114–127. doi:10.1016/j.jenvman.2005.04.009 PMID:16095806

Gattiker, T., & Carter, C. R. (2010). Understanding project champions' ability to gain intra-organizational commitment for environmental projects. *Journal of Operations Management, 28*(1), 72–85. doi:10.1016/j.jom.2009.09.001

Hansen, E. G., Grosse-Dunker, F., & Reichwald, R. (2009). Sustainability innovation cube - A framework to evaluate sustainability-oriented innovations. *International Journal of Innovation Management, 13*(4), 683–713. doi:10.1142/S1363919609002479

Hart, S. L., & Milstein, M. B. (2003). Creating sustainable value. *The Academy of Management Perspectives, 17*(2), 56–69. doi:10.5465/ame.2003.10025194

Hartmann, A. (2008). Overcoming resistance to innovation: The integration champion in construction. In P. S. Brandon & S.-L. Lu (Eds.), *Clients Driving Innovation* (pp. 157–167). Blackwell. doi:10.1002/9781444301342.ch16

Harty, C. (2005). Innovation in construction: A sociology of technology approach. *Building Research and Information, 33*(6), 512–522. doi:10.1080/09613210500288605

Herazo, B., & Lizarralde, G. (2015). The influence of green building certifications in collaboration and innovation processes. *Construction Management and Economics, 33*(4), 279–298. doi:10.1080/01446193.2015.1047879

Howell, J. M., & Higgins, C. A. (1990). Champions of technological innovation. *Administrative Science Quarterly, 35*(2), 317–341. doi:10.2307/2393393

Huber, J. (2008). Technological environmental innovations (TEIs) in a chain-analytical and life-cycle-analytical perspective. *Journal of Cleaner Production, 16*(18), 1980–1986. doi:10.1016/j.jclepro.2008.01.014

IPCC. (2021 in press). Summary for Policymakers. In *Climate Change 2021: The Physical Science Basis. Contribution of Working Group I to the Sixth Assessment Report of the Intergovernmental Panel on Climate Change*. Cambridge University Press.

Klewitz, J., & Hansen, E. G. (2014). Sustainability-oriented innovation of SMEs: A systematic review. *Journal of Cleaner Production, 65*, 57–75. doi:10.1016/j.jclepro.2013.07.017

Mateus, R., & Bragança, L. (2011). Sustainability assessment and rating of buildings: Developing the methodology SBToolPT-H. *Building and Environment, 46*(10), 1962–1971. doi:10.1016/j.buildenv.2011.04.023

OECD. (2018). Oslo Manual 2018: Guidelines for Collecting, Reporting and Using Data on Innovation, 4th Edition, The Measurement of Scientific, Technological and Innovation Activities. OECD Publishing. https://doi.org/ doi:10.1787/978

Ofori-Boadu, A., Owusu-Manu, D. G., Edwards, D., & Holt, G. (2012). Exploration of management practices for LEED projects: Lessons from successful green building contractors. *Structural Survey, 30*(2), 145–162. doi:10.1108/02630801211228743

Owens, B., Macken, C., Rohloff, A., & Roseberg, H. (2013). LEED v4 - Impact Category and Point Allocation Process. USGBC.

Pérez-Lombard, L., Ortiz, J., González, R., & Maestre, I. R. (2009). A review of benchmarking, rating and labelling concepts within the framework of building energy certification schemes. *Energy and Building*, *41*(3), 272–278. doi:10.1016/j.enbuild.2008.10.004

Porter, M. E. (1990). The Competitive Advantage of Nations. *Harvard Business Review*. Advance online publication. doi:10.1080/13602389400000006

Redclift, M. (2005). An Oxymoron Comes of Age. *Sustainable Development (Bradford)*, *13*(4), 212–227. doi:10.1002/sd.281

Reed, B. (2007). Shifting from 'sustainability' to regeneration. *Building Research and Information, 35*(6), 674–680. doi:10.1080/09613210701475753

Rennings, K., Ziegler, A., Ankele, K., & Hoffmann, E. (2006). The influence of different characteristics of the EU environmental management and auditing scheme on technical environmental innovations and economic performance. *Ecological Economics*, *57*(1), 45–59. doi:10.1016/j.ecolecon.2005.03.013

Robichaud, L. B., & Anantatmula, V. S. (2011). Greening Project Management Practices for Sustainable Construction. *Journal of Management Engineering*, *27*(1), 48–57. doi:10.1061/(ASCE)ME.1943-5479.0000030

Schaltegger, S., & Wagner, M. (2011). Sustainable entrepreneurship and sustainability innovation: Categories and interactions. *Business Strategy and the Environment*, *20*(4), 222–237. doi:10.1002/bse.682

Sebastien, H., Heike, A., Nishil, B., & Arpad, H. (2007). Leadership in Energy and Environmental Design (LEED) - A critical evaluation by LCA and recommendations for improvement. *International Journal of Life Cycle Assessment, 12*(Special Issue 1), 25. doi:10.11436/mssj.15.250

Tan, Y., Shen, L., & Yao, H. (2011). Sustainable construction practice and contractors' competitiveness: A preliminary study. *Habitat International*, *35*(2), 225–230. doi:10.1016/j.habitatint.2010.09.008

Toole, T. M., Hallowell, M., & Chinowsky, P. (2013). A tool for enhancing innovation in construction organizations. *Engineering Project Organization Journal*, *3*(1), 32–50. doi:10.1080/21573727.2012.717531

USGBC. (2018). *Leed and the International Green Construction Code*. https://www.usgbc.org/about/programs/green-codes

Chapter 7
Machine Learning-Based Load Forecast for Energy Markets

Mauparna Nandan
https://orcid.org/0000-0002-9797-3866
Techno Main Saltlake, Kolkata, India

K. S. Sastry Musti
https://orcid.org/0000-0003-4384-7933
Namibia University of Science and Technology, Nambia

ABSTRACT

Time series-based forecasting is one of the most popular machine learning methods due to its effectiveness in estimation of future values based on observations, interpolations, and interpretations of past data. Forecasting the load in power distribution networks is an essential step in energy trading and system operation. Machine learning, specifically time series-based forecasting, can be used in the precise prediction of energy consumption in power networks. Precise predictions have the potential to reduce operating and maintenance expenses, enhance the dependability of power supply and delivery systems, and enable informed decisions for future development endeavors. This chapter employs time series analysis to forecast energy usage in 10-minute intervals specifically for the city of Tétouan in Morocco by applying gradient boosting algorithm. Past and present data trends have been presented along with various accuracy parameters such as mean absolute error (MAE), mean squared error (MSE), and root mean squared error (RMSE).

INTRODUCTION

Forecasting electricity demand is crucial for strategic planning and facility expansion within the electricity sector. Traditionally, load forecasting is done using statistical projections. With the advent of soft computing and machine intelligence techniques, new approaches have been suggested for the estimation of load. Specifically, significant advancements have been made in time series forecasting techniques, leading to enhanced accuracy. These algorithms have a wide range of applications due to their improved

DOI: 10.4018/978-1-6684-9130-0.ch007

precision. Time series refers to a set of data points recorded at regular intervals of time. It encompasses different elements, including a trend that represents long-term patterns, cyclical patterns characterized by repeated up and down movements, seasonal variations that occur regularly within specific months or quarters, and irregular components which can't be explained by known factors (Popeangă & Lungu, 2014). Forecasting involves making predictions about future values based on historical and current data in a time series. It plays a crucial rule in planning and controlling future power system operations. Load forecasting, specifically, is important for anticipating the demand for electricity. Load forecasting can be categorized into three types: short-term load forecasting (STLF), medium-term load forecasting (MTLF), and long-term forecasting (LTLF). STLF focuses on predicting load demand up to one day or at most one week ahead, MTLF covers the forecasting range of one day to several months, while LTLF looks further into the future, forecasting more than a year in advance (Pedregal & Trapero, 2010). STLF is utilized for scheduling electricity generation and transmission, MTLF aids in planning fuel purchases, and LTLF aims to develop and optimize the power supply and delivery system, which includes generation units, transmission networks, and distribution systems (Almeshaiei & Soltan, 2011).

Based upon the 2014 Census, Tétouan is a Moroccan city situated in the northern part of the country, overlooking the Mediterranean Sea. It spans over an area of 11,570 km^2 and comprises of a population of approximately 380,000 individuals. The summer weather in Tétouan city is extremely hot and humid. According to a study as provided by the state-owned organization, known as the National Office of Electricity and Drinking Water (ONEE), Morocco's energy production in 2019 was predominantly sourced from coal (38%), followed by hydroelectricity (16%), fuel oil (8%), natural gas (18%), wind (11%), solar (7%), and other sources (2%) (Morocco-Energy, 2019). Morocco is one of the developing nations that is dedicated to the growth, progress and enhancement of its economy.

Energy is a significant catalyst for economic growth, and one of its key sources is electricity, which is considered a crucial commodity. The improvement of living standards, economic development, and industrial progress are all closely linked to access to electricity. In fact, the presence or absence of electricity has become an economic benchmark that determines the advancement or backwardness of a country. Presently, electricity consumption is one of the most significant indicators for governments seeking development, and accurate long-term forecasting of electricity usage is vital for enhancing energy efficiency and avoiding costly errors (Elsaraiti, Ali, Musbah, Merabet, & Little, 2021). Accurate power forecasting is essential for effective planning of electricity consumption and for the implementation of decision support systems that guide the decision-making process within the power system (Nichiforov et al., 2017). Precise forecasting of electricity consumption also plays a critical role for policymakers in formulating effective electricity supply policies. However, the availability of limited data and variables often hampers the acquisition of valuable information necessary for achieving satisfactory prediction accuracy (Li & Zhang, 2018). Considering the significant reliance on non-renewable sources, accounting for 64% of the energy mix, forecasting energy consumption can be instrumental in assisting stakeholders in effectively managing procurement and inventory. Furthermore, Morocco aims to decrease its dependence on energy imports by expanding the production capacity of renewable sources. However, it is widely acknowledged that renewable sources such as wind and solar may not be consistently available throughout the year. Therefore, gaining a comprehensive understanding of the energy requirements of the country, starting with an intermediate-sized city, could serve as a progressive measure in strategizing the utilization of these resources.

The literature offers a range of forecasting methods, including multiple linear regression, time series models, Auto regressive integrated moving average (ARIMA) process, artificial neural network, Fuzzy time series, Fuzzy neural network, and others. However, the current study solely focuses on reviewing time series models to forecast UTHM electricity consumption. Thus, the next section provides review of relevant literature in the key areas of power system load forecasting, time-series applications and machine learning in general.

BACKGROUND STUDY

Numerous models have been developed to enhance the precision of electricity consumption forecasts. This study focuses on evaluating the effectiveness of the ARIMA model in comparison to existing technologies by utilizing Mean Absolute Percentage Error (MAPE) as a metric for forecast accuracy (Jain et al., 2018). The ARIMA model is employed to determine the optimal time period for individual household electrical energy consumption (Chujai et al., 2013). Additionally, the ARIMA model is utilized for forecasting wind speed, with various error ratio measures employed to assess the accuracy of the predictions (Elsaraiti, Ali, Musbah, Merabet, & Little, 2021). In another study, six time series methods, namely simple moving average (SMA), weighted moving average (WMA), simple exponential smoothing (SES), Holt linear trend (HL), Holt-Winters (HW), and centered moving average (CMA), are applied and compared to evaluate their accuracy in forecasting monthly electricity consumption (Lee et al., 2018). A comparative study was conducted using the SARIMA and ARIMA models on the same dataset for electricity consumption forecasting. The results indicated that the SARIMA model outperformed the ARIMA model (Elsaraiti, Ali, Musbah, Merabet, & Little, 2021). For predicting electricity power demand in a wholesale power transmission company, a combination of classical 'ad hoc' statistical models, advanced ensemble techniques, and neural networks were employed (Kalimoldayev et al., 2020). Monthly electricity consumption forecasting was approached by integrating seasonal and trend decomposition using Loess (STL) with ARIMA models (Sun et al., 2019). In the context of electricity consumption prediction in the Philippines, the ARIMA model was utilized, and the values of (p, d, q) were determined by analyzing ACF and PACF plots as well as assessing data stationarity (Delima, 2019).

In another study by Nizami et al. (2018), the authors applied the ARIMA model to forecast electricity consumption in Pakistan. The model incorporated historical data, seasonal patterns, and trends to accurately predict future consumption levels. Zhang Zhang et al. (2019) conducted research on short-term power load forecasting using a hybrid model that combined ARIMA and Long Short-Term Memory (LSTM) neural networks. The hybrid model demonstrated improved accuracy compared to standalone ARIMA or LSTM models. Liu et al. (2019) explored electricity consumption forecasting in China using a seasonal hybrid ARIMA-Exponential Smoothing (ETS) model. The model effectively captured both long-term trends and short-term seasonality, resulting in accurate predictions. In a study by Dey et al. (2020), the authors employed an AutoRegressive Distributed Lag (ARDL) model to forecast electricity consumption in India. The ARDL model considered the dynamic relationship between electricity consumption and various influencing factors, such as temperature, GDP, and population. Tan et al. (2020) proposed a hybrid model that combined the ARIMA model with the Least Squares Support Vector Machines (LSSVM) algorithm for electricity consumption forecasting. The hybrid model enhanced the accuracy predictions by effectively capturing complex nonlinear relationships. El-Latif et al. (2021) investigated the application of SARIMA and artificial neural network models for short-term load forecasting in Egypt. The models were trained using historical load data and weather variables, resulting in reliable load forecasts for improved grid management.

These studies demonstrate the diverse approaches and techniques employed in utilizing time series analysis for power consumption forecasting, ranging from traditional ARIMA models to hybrid models incorporating machine learning algorithms. The integration of historical data, seasonal patterns, and external variables contributes to accurate and reliable electricity consumption predictions.

PROPOSED METHODOLOGY

Power consumption refers to the amount of electrical energy used by various devices, buildings, or systems over a specific period of time. Analyzing power consumption data is essential for understanding energy usage patterns, identifying trends, optimizing energy efficiency, and making informed decisions about resource allocation and demand forecasting.

Time series analysis is a statistical technique used to analyze data points collected over consecutive and equally spaced intervals of time. It involves studying the patterns, trends, and dependencies present in the data to gain insights and make predictions. Time series analysis is widely used in various fields, including finance, economics, weather forecasting, and energy management. When applied to power consumption data, time series analysis enables us to explore and understand the temporal patterns and factors affecting energy usage. It helps identify recurring patterns, seasonal variations, and trends in power consumption thereby, allowing us to optimize energy usage, predict future demand, and develop effective energy management strategies.

The key steps involved in analyzing power consumption using time series analysis are illustrated below:

1. **Data Collection:** The first step is to collect the power consumption data over a specific period. The data can be obtained from various sources, such as smart meters, sensors, or energy management systems. The data must include timestamps indicating when each measurement was taken.
2. **Data Preprocessing:** The collected data may contain missing values, outliers, or noise. Preprocessing steps, such as data cleaning, handling missing values, and removing outliers, are necessary to ensure the data's quality and reliability. Additionally, data may need to be resampled or aggregated to a desired time resolution (e.g. hourly, daily) for analysis.
3. **Exploratory Data Analysis:** This step involves visualizing and understanding the characteristics of the power consumption data. Exploratory data analysis techniques, such as line plots, histograms, and seasonal decomposition, help identify trends, seasonality, and other patterns present in the data.
4. **Time Series Modeling:** Time series modeling aims to capture and represent the underlying patterns and dynamics of power consumption. Common modeling techniques include autoregressive integrated moving average (ARIMA), seasonal ARIMA (SARIMA); exponential smoothing (ETS), and state space models. These models provide a mathematical framework for forecasting and understanding the behavior of power consumption over time.
5. **Model Evaluation and Selection:** Once the models are developed, they need to be evaluated and compared using appropriate performance metrics. The chosen model should have good predictive accuracy and capture the relevant features of the power consumption data. There exist various accuracy parameters like MSE, MAE, RMSE, R^2 etc. which provide an insight on the magnitude of errors from where the accuracy of the proposed system can be easily computed. It's important to consider these parameters collectively in order to gain a comprehensive understanding of a model's predictive performance.

6. **Forecasting and Analysis:** With a selected model, we can forecast future power consumption based on historical data and identified patterns. Forecasting helps in planning for future energy requirements, load balancing, and resource allocation. Various techniques, such as one-step ahead forecasting or multi-step forecasting, can be applied depending on the specific needs of the analysis.
7. **Optimization and Energy Management:** The insights gained from time series analysis can be used to optimize energy consumption, improve energy efficiency, and develop effective energy management strategies. For example, identifying peak load periods can help in implementing demand response programs or scheduling energy-intensive operations during off-peak hours.

Thus, time series analysis provides a powerful framework for understanding and analyzing power consumption patterns. By applying this technique, we can uncover valuable insights, make accurate predictions, and optimize energy usage to enhance sustainability and cost-efficiency in various domains, from individual households to large-scale industrial facilities.

RESULTS AND DISCUSSIONS

In this section, we now describe each step as mentioned in the previous section in a detailed manner with respect to the current problem under study.

Data Collection

The data for the current project has been collected by Amendis, a public service operator which further relies on the Supervisory Control and Data Acquisition System (SCADA). The distribution network is supported by three zone stations, namely Quads, Smir and Boussafou, which supply power to distinct areas within the city. As a result, there are three potential target variables on account of the differentiated power sources and zones. The dataset comprises of 52, 416 energy consumption data in a 10-minute window spanning over a time period from 1st January, 2017 to 30th December, 2017. Some of the major characteristics of the dataset are given below:

1. Date Time: A time frame of 10 minutes.
2. Temperature: The weather's temperature measured in degrees Celsius.
3. Humidity: The weather's humidity represented as a percentage.
4. Wind Speed: The speed of the wind measured in kilometers per hour.
5. Power Consumption of Zone 1: The amount of electricity consumed in Zone 1, measured in kilo-watts (KW).
6. Power Consumption of Zone 2: The amount of electricity consumed in Zone 2, measured in kilo-watts (KW).
7. Power Consumption of Zone 3: The amount of electricity consumed in Zone 3, measured in kilo-watts (KW).

Data Preprocessing

The necessary libraries for importing and analyzing the data in the project include Pandas and NumPy, as well as visualization libraries like Matplotlib and Seaborn. The libraries imported in the current study are depicted in Figure 1.

Figure 1. Python libraries imported

```
#Importing libraries
import pandas as pd
import numpy as np
import matplotlib.pyplot as plt
import seaborn as sns
```

A snapshot of the dataset employed in the study is depicted in Figure 2.

Figure 2.A snapshot of the dataset

	DateTime	Temp	Hum	Wind
0	1/1/2017 0:00	6.559	73.8	0.083
1	1/1/2017 0:10	6.414	74.5	0.083
2	1/1/2017 0:20	6.313	74.5	0.08
3	1/1/2017 0:30	6.121	75	0.083
4	1/1/2017 0:40	5.921	75.7	0.081

	Diff Fl	Zone 1 PC	Zone 2 PC	Zone 3 PC
0	0.119	34055.7	16128.9	20241
1	0.085	29814.7	19375.1	20131.1
2	0.1	29128.1	19006.7	19668.4
3	0.096	28228.9	18361.1	18899.3
4	0.085	27335.7	17872.3	18442.4

Upon observation, it becomes evident that each feature exclusively contains numerical values. As a result, it can be inferred that any potential prediction task would require regression techniques rather than classification. Also, since the dataset's index is represented by numerical values, therefore, the DateTime column can be utilized as the index instead. This will simplify the process of plotting charts. The process of transforming DateTime into index is depicted in Figure 3.

Figure 3. Transforming DateTime into index

```
#transforming DateTime column into index
df = df.set_index('DateTime')
df.index = pd.to_datetime(df.index)
```

The command "df.set_index()" enables us to replace the current index with the DateTime column, while the command "pd.to_datetime()" transforms the format of the new index into a more readable format.

Exploratory Data Analysis (EDA)

In this subsection, data exploration and visualization is carried out in a detailed manner. To begin with, in order to gain a greater insight into the dataset, the histogram corresponding to the different columns of the dataset are plotted. Histograms enable us to examine measures of central tendency as well as to find the range and standard deviation, to determine the skewness or symmetry of the dataset which in turn helps to identify patterns in the dataset and consequently, informed decisions. The histogram plot is illustrated in Figure 4.

Figure 4. Histogram plot

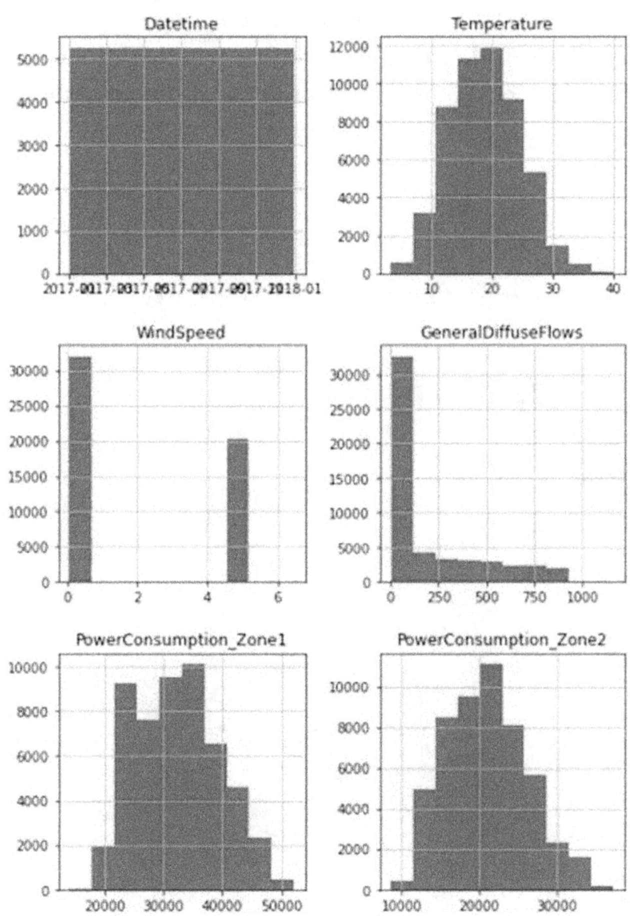

Next, a correlation matrix is generated using the seaborn library. Typically, the correlation can be categorized as follows:

- If the correlation is less than 0.3, it is considered to be weak.
- If the correlation falls between 0.3 and 0.6, it is categorized as moderate.
- If the correlation ranges from 0.6 to 0.9, it is regarded as strong.
- If the correlation is greater than 0.9, it is considered as extremely strong.

On the contrary, the sign of the correlation signifies whether the two factors exhibit growth in the same direction or in opposite directions. For example, a correlation of -0.9 suggests that as one variable increases, the other decreases. However, it's worth noting that the ranges for determining correlation strength can vary depending on the level of uncertainty in the environment. In quantitative finance, a correlation of 0.3 or 0.4 may be regarded as strong. The correlation plot corresponding to the current problem is depicted in Fig.5.

Now, from Fig.5., let's us examine the overall trends of the factors in relation to power consumption. Firstly, temperature exhibits a positive correlation with power consumption in all the zones. This could be attributed to a higher usage of air conditioning systems in response to warmer temperatures. On the other hand, humidity displays a slight negative correlation with power consumption in all the three areas. Additionally, there is a strong correlation observed among the power consumption in Zone 1, Zone 2 and Zone 3. This suggests that while there may be slight variations in consumption patterns among the different areas, they generally tend to increase or decrease together.

Figure 5. Feature correlation matrix

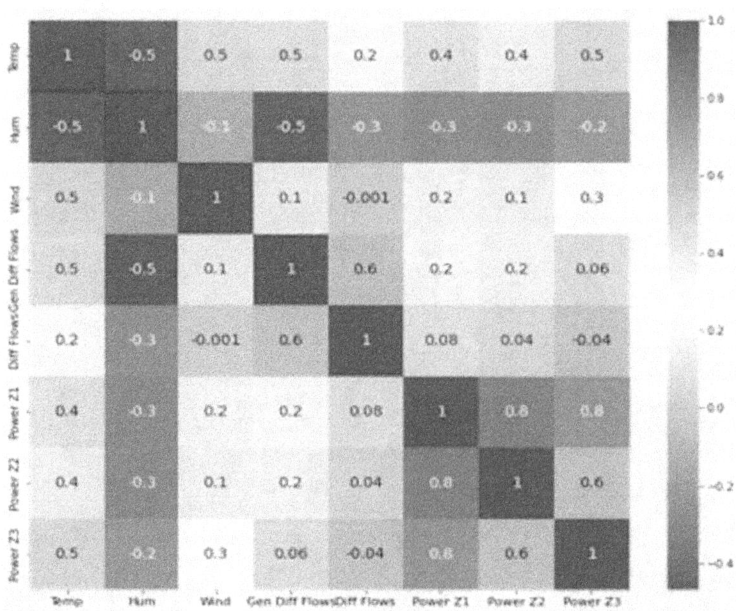

Time Series Modeling

The next stage involves the combination of feature creation and exploratory data analysis (EDA). This step, referred to as the third step, is particularly relevant in time series analysis and forecasting. In such cases, when the available feature options are limited and only the "time" aspect is accessible, it becomes crucial to extract as much valuable information from it as possible. To accomplish this, the provided function generates additional columns based on the DateTime index. By executing the create_features function on df, all the features specified within the function can be promptly generated. This allows for the instantaneous creation of the desired set of features for the dataset under study. The features created for the current problem are illustrated in Figure 6.

Figure 6. Create feature function

```
def create_features(df):
    """
    Create time series features based on time series index.
    """
    df = df.copy()
    df['hour'] = df.index.hour
    df['dayofweek'] = df.index.dayofweek
    df['quarter'] = df.index.quarter
    df['month'] = df.index.month
    df['year'] = df.index.year
    df['dayofyear'] = df.index.dayofyear
    df['dayofmonth'] = df.index.day
    df['weekofyear'] = df.index.isocalendar().week
    return df

df = create_features(df)
```

The time series analysis stage involves handling missing data, creating visualizations, analyzing correlation functions, examining stationarity, and performing decomposition which are illustrated below:

1. **Missing values:** The dataset has been checked for null values and it has been found that there are no null values which can be verified from Figure 7 given below.

Figure 7. Null values

	column	percent
0	Temperature	0.0
1	Humidity	0.0
2	WindSpeed	0.0
3	GeneralDiffuseFlows	0.0
4	DiffuseFlows	0.0
5	PowerConsumption_Zone1	0.0
6	PowerConsumption_Zone2	0.0
7	PowerConsumption_Zone3	0.0

2. **Visualization:** The following graphs in Figure 8 and Figure 9 display the entire time series as well as the last month's segment. These visuals reveal that there are two distinct seasonal patterns within the time series: a daily pattern and a weekly pattern respectively.

Figure 8. Zone 1 power consumption time series

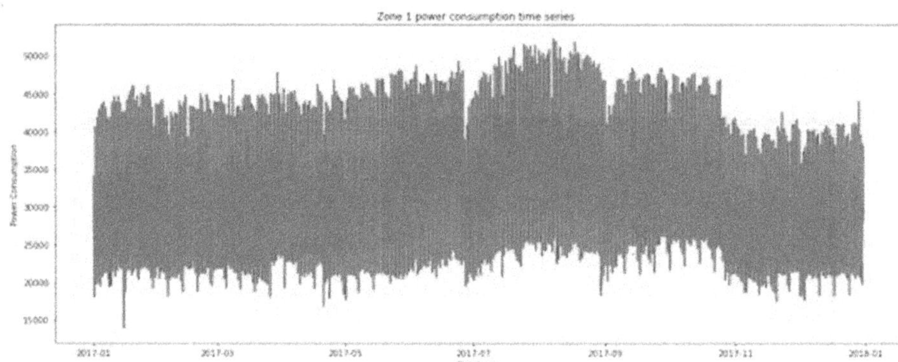

Figure 9. Zone 1 power consumption time series (December)

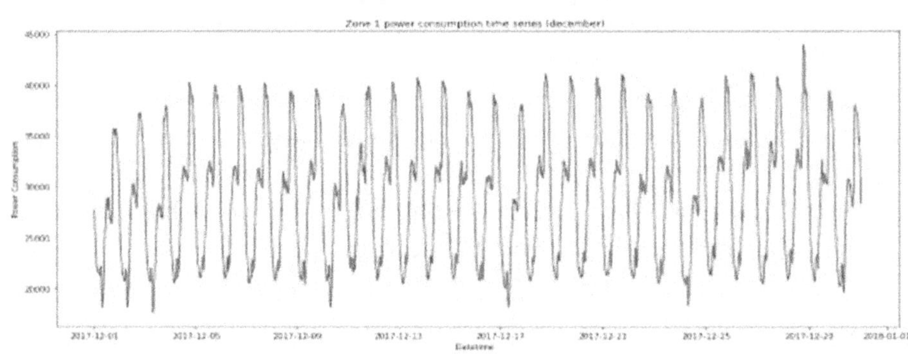

3. **Autocorrelation Analysis:** Autocorrelation analysis in time series involves examining the relationship between a variable and its lagged values. It helps to identify patterns, dependencies and seasonality within the data. This analysis is valuable for understanding the temporal dynamics and predicting future values in a time series. ACF (Autocorrelation Function) and PACF (Partial Autocorrelation Function) are two important tools employed in time series analysis to study and model the dependencies between observations.

    (i) **ACF** measures the correlation between a time series and its lagged values at different time intervals. It helps to identify the presence of any significant patterns or seasonality in the data.
    (ii) **PACF** on the other hand, measures the correlation between two observations in a time series while considering the intermediate lags. It helps to identify the direct influence of a specific lag on the current observation, while accounting for the indirect effects of other lags.

The plots of the 144 lags seasonality (corresponding to one day) and the non-stationarity of the time series are displayed in Figure 10 below.

Figure 10. ACF and PACF plots

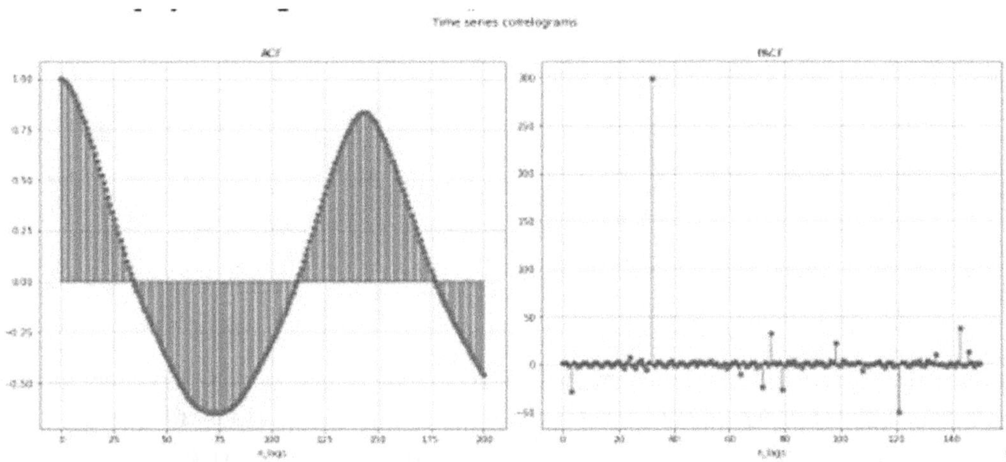

4. **Moving Average and Standard Deviation:** The Simple Moving Average (SMA) and Standard Deviation (SD) are two commonly used statistical measures in data analysis.

    (i) The simple moving average (**SMA**) is a vital tool that aids in smoothing the predictions. It calculates the average value of a variable over a specified number of consecutive periods. The moving average can be calculated through the following code snippet as highlighted in Figure 11.

Figure 11. Simple moving average (SMA) calculation

```
#Calculating moving average
df['SMA30'] = df['Zone 1 Power Consumption'].rolling(30).mean()
df['SMA15'] = df['Zone 1 Power Consumption'].rolling(15).mean()
```

To specify the number of days to be included in the simple moving average (SMA), the command “rolling(n_days).mean()” is utilized. This command calculates the mean of the specified number of days in the rolling window for the SMA.

(ii) The Standard Deviation (**SD**) is a measure of the dispersion or variability of the dataset. It quantifies how much the individual data points deviate from the mean value.

The plot of the SMA and the STD for the current dataset is depicted in Figure 12.

Figure 12. SMA and STD plots

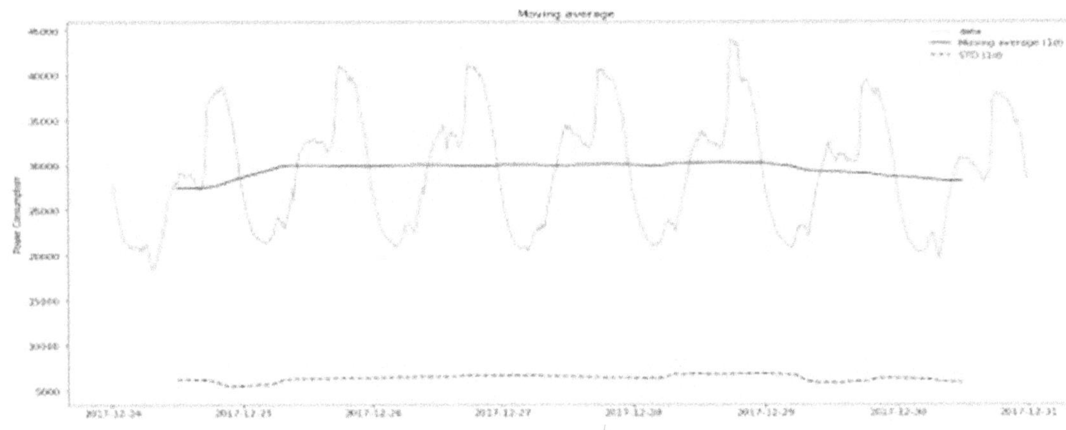

5. **Stationarity Tests:** In time series analysis, stationarity refers to a property where the statistical properties of a time series remain constant over time. There are various statistical tests available to determine the stationarity of a time series. The stationarity tests carried out for the current study are illustrated below.

   (i) **Augmented Dickey-Fuller (ADF) Test:** The ADF test is widely used to check for stationarity. It examines whether a unit root is present in the time series data. The null hypothesis of the test is that the series has a unit root (non-stationary), while the alternative hypothesis is that it is stationary. If the p-value obtained from the test is below a chosen significance level, the null hypothesis is rejected, indicating stationarity.

   (ii) **Kwiatkowski-Philips-Schmidt-Shin (KPSS) Test:** The KPSS test is another common test used to access stationarity. It examines the null hypothesis that the series is stationary against the alternative hypothesis of a unit-root (stationary). Unlike the ADF test, the KPSS test is more suitable for detecting trends and level shifts in the data. A significant p-value indicates non-stationarity.

The corresponding results obtained from the ADF and KPSS tests are displayed below in Figure 13 and Figure 14 respectively.

Figure 13. ADF test result

```
Results of Dickey-Fuller Test:
Test Statistic                     -4.508960
p-value                             0.000189
Lags Used                           5.000000
Number of Observations Used      1002.000000
Critical Value (1%)                -3.436893
Critical Value (5%)                -2.864429
Critical Value (10%)               -2.568308
dtype: float64

Differentiations needed according to ADF: 0
```

Figure 14. KPSS test result

```
Results of KPSS Test:
Test Statistic            0.128273
p-value                   0.100000
Lags Used                19.000000
Critical Value (10%)      0.347000
Critical Value (5%)       0.463000
Critical Value (2.5%)     0.574000
Critical Value (1%)       0.739000
dtype: float64

Differentiations needed according to KPSS: 0
```

Based on the results of the ADF and KPSS tests, as well as the analysis of the moving average plot, it can be concluded that the time series is stationary. However, there is contradictory evidence suggesting that it is not stationary. The presence of seasonality in the time series and the dissimilarity of the correlograms from that of a white noise correlogram provide indications against stationarity. Additionally, the fact that the PACF values exceed '1' further supports the argument that the time series is not stationary.

6. 6. **Decomposition:** Time series decomposition is a technique to break down a time series into its underlying components, typically consisting of trend, seasonality and residual (or error) components. The goal of decomposition is to gain a better understanding of the individual patterns and variations in the time series. This information is useful for forecasting, anamoly detection, and further analysis of time series data. The most commonly used method for time series decomposition is the

additive decomposition, which assumes that the components of the series add up to produce the observed values. The additive decomposition model can be represented as:

Y(t) = Trend(t) + Seasonality(t) + Residual(t)

The **'trend'** component represents the long-term pattern or the direction of the series. It captures the overall increasing or decreasing trend over time.

The **'seasonality'** component represents the repeated patterns that occur within a fixed time period. It reflects the regular, periodic fluctuations that happen within the time series. Seasonality can be daily, weekly, monthly, quarterly, or annual, depending on the data frequency.

The **'residual (or error)'** component represents the random or unpredictable fluctuations that cannot be explained by the trend and seasonality. It represents the noise or variability remaining after removing the trend and seasonal patterns. The decomposition of the time series corresponding to the current dataset into its components is displayed in Figure 15.

Figure 15. Time series decomposition

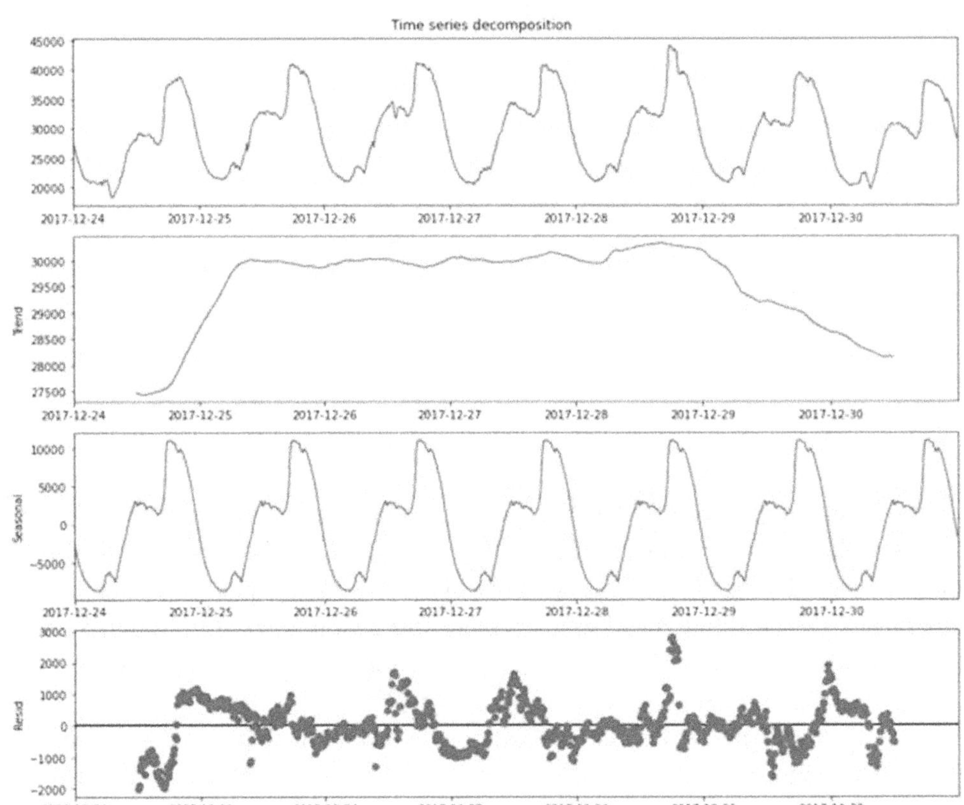

Once the features are generated, we can now visualize the distribution of each feature by using boxplots. These boxplots provide insights into the minimum, 25th percentile, average, 75th percentile, and maximum consumption within a specific timeframe. By plotting the data on an hourly basis, it is observed that Zone 1 exhibits a significantly higher average consumption compared to Zone 2 and Zone 3. Furthermore, for all the three areas, the average consumption tends to increase during the evening hours. The boxplot of hourly power consumption in KW is displayed in Figure 16.

Figure 16. Boxplot chart of hourly power consumption in KW

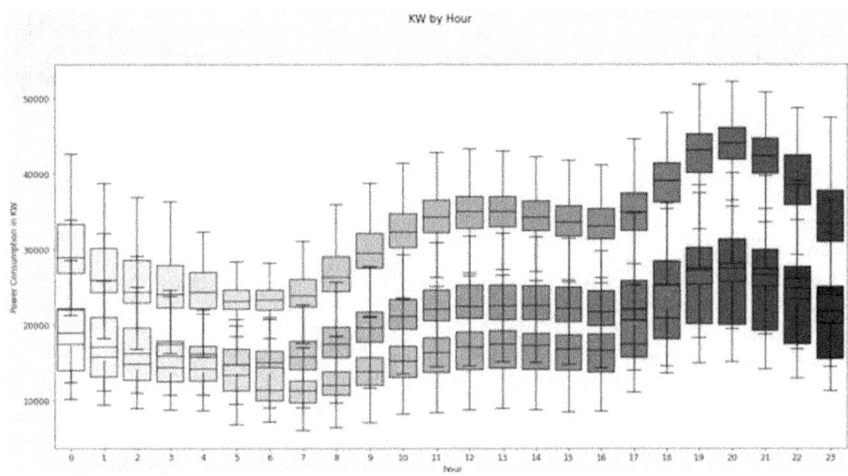

As anticipated, the temperature exhibits a rising trend during the central hours of the day, followed by a decline in the evening. Consequently, only 50% of the time, power consumption and temperature display a simultaneous movement. As the temperature begins to decrease in the early evening hours, there is a sudden surge in power consumption. The boxplot chart of hourly temperature in oC is displayed in Figure 17.

Figure 17. Boxplot chart of hourly temperature in °C

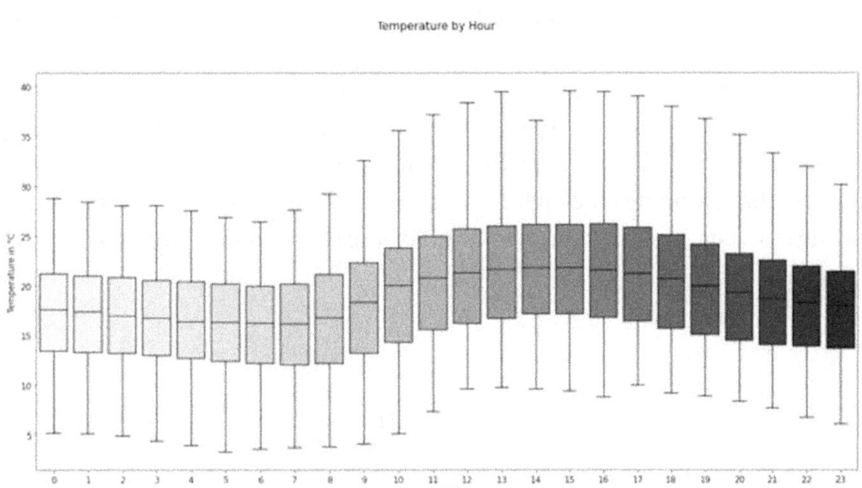

In contrast, humidity demonstrates an upward trend during the colder hours and a downward trend during the hottest hours throughout the day which is depicted in Figure 18.

Figure 18. Boxplot chart of hourly humidity in %

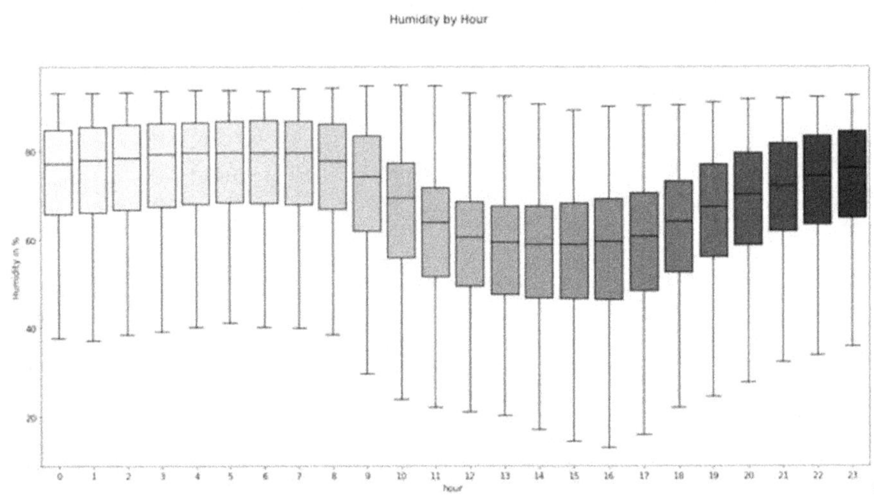

Train Test Split:

At this point, it is time to divide the dataset into training and testing sets. In a typical scenario, new records are randomly selected for training the algorithm and the remaining records are utilized for testing. However, in this case, it is crucial to maintain the time sequence between the training and testing data. In this case, 70% of the available data spans from 1st January to 10th October will serve as the training set. On the other hand, the time period spanning from 2nd October to 30th December will be reserved for the testing set.

A detailed examination of the dataset will reveal that the columns including Temperature, dayofyear, hour, dayofweek, SMA30 and SMA15 have been selected as model input. However, before proceeding with the final training phase, the calculation of the feature importance corresponding to each column is performed. Any columns that yielded a feature importance value of '0' were subsequently excluded from the model.

To begin with, the target variable is initially chosen to be Zone 1 Power Consumption. This selection aligns with the scenario's context, where analyzing each area's specific electricity requirements separately makes more sense. Another approach to address this issue would involve creation of a 4th column that combines that combines the energy demand of all the three areas, thereby, providing a more comprehensive solution. The input and target variables corresponding to the current study are defined as follows in Figure 19.

Figure 19. Input and target variables

```
#defining input and target variable
X_train = df.loc[:'10-01-2017',['Temperature','dayofyear', 'hour',
'dayofweek', 'SMA30', 'SMA15']]
y_train = df.loc[:'10-01-2017', ['Zone 1 Power Consumption']]
X_test = df.loc['10-01-2017':,['Temperature','dayofyear', 'hour',
'dayofweek', 'SMA30', 'SMA15']]
y_test = df.loc['10-01-2017':, ['Zone 1 Power Consumption']]
```

The corresponding training and testing dataset split is depicted in Figure 20.

Figure 20 Training and testing dataset split

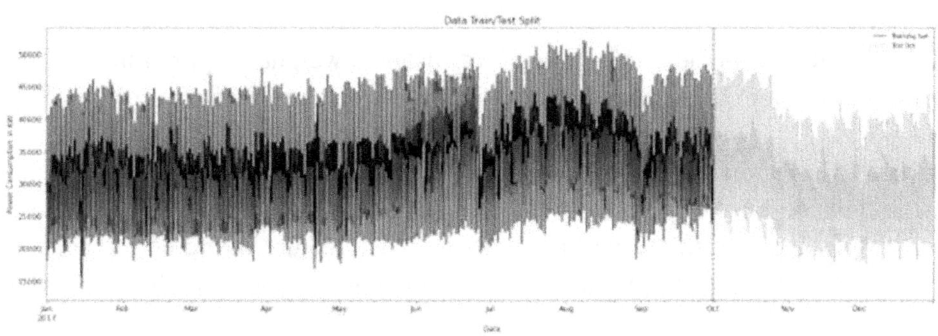

Based on the chart, it is evident that electricity consumption shows a peak towards the end of August and reaches its lowest point in January. This observation highlights the influence of temperature trends, which we have considered as one of the features in our model.

Predictions

To forecast values for the test dataset, the Gradient Boosting algorithm have been utilized. The gradient boosting algorithm is a machine learning technique that builds an ensemble of weak prediction models, typically decision trees in a sequential manner. It is a type of boosting algorithm that aims to create a strong predictive model by iteratively improving upon the weaknesses of the previous models. In each iteration, the algorithm calculates the gradient of the loss function with respect to the predicted values from the previous model. This gradient is used to adjust the parameters of the new model to minimize the loss. The process continues, with each subsequent model focusing on the residual errors of the ensemble's combined predictions.

Gradient boosting has several advantages, including its ability to handle complex datasets, handle different types of variables, and capture non-linear relationships between features and the target variable. Popular implementations of the gradient boosting algorithms include XGBoost, LightGBM and CatBoost.

For the current problem, XGBoost short for Extreme Gradient Boosting and a variant of the Gradient Boosting algorithm, is a powerful and popular machine learning algorithm known for its effectiveness in handling complex datasets and achieving high predictive accuracy is implemented. It combines the strengths of gradient boosting and regularization techniques to create a robust and optimized model. The most significant hyper parameters of XGBoost are highlighted below:

- **n_estimators:** It specifies the number of boosting rounds or decision trees to build. Increasing this parameter can potentially improve the model's performance, but it may also lead to overfitting.
- **max_depth:** This hyperparameter sets the maximum depth of each decision tree. Deeper trees can capture more complex relationships in the data, but they also increase the risk of overfitting. Care should be taken to balance the depth based on the dataset and problem at hand.
- **learning_rate:** It controls the step size at each boosting iteration. A lower learning rate makes the model more conservative, reducing the impact of each tree on the final prediction. However, a smaller learning rate requires more boosting rounds to converge.
- **reg_alpha and reg_lamda:** These are regularization parameters that control L1 and L2 regularization respectively. They add penalties to the model's loss function to prevent overfitting and encourage simpler models.
- **min_child_weight:** It sets the minimum sum of instance weights required in a child node. Higher values promote a more conservative tree structure by preventing nodes with low weights.

It is important to experiment and tune these parameters using techniques like cross-validation to find the optimal combination for the specific problem and dataset.

By creating a plot that illustrates the importance of each feature in predicting accurate results, we can easily observe significant patterns and trends. The feature importance plot of XGBoost algorithm is given below in Figure 21.

Figure 21. Feature importance XGBoost regressor

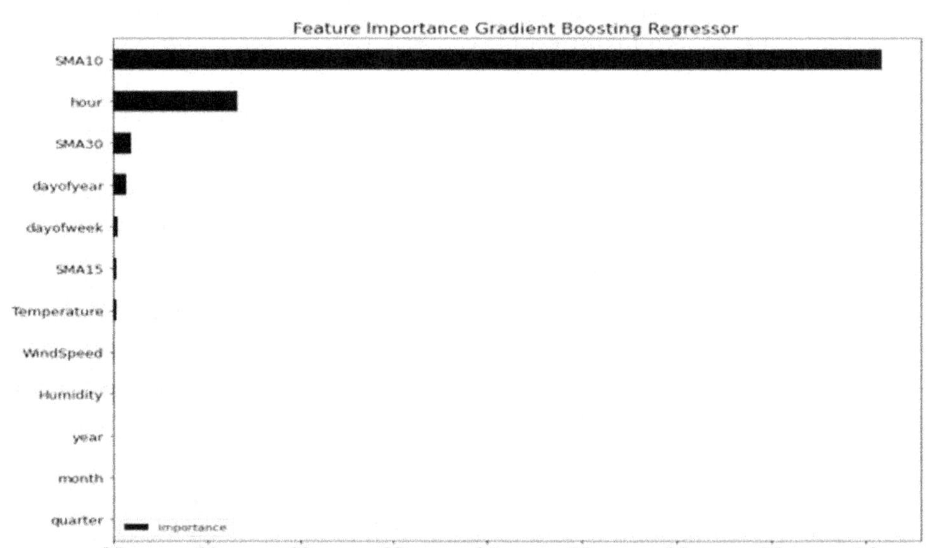

The 15-day Simple Moving Average (SMA) and the specific movement hour are the primary factors that hold the greatest significance. Out of all the external factors, temperature is the only one that has a noticeable impact on predicting energy consumption. As anticipated, the remaining factors in the original dataset have negligible influence on the forecasting aspect of the project, as their importance was essentially zero. Once the model is integrated back into the initial dataset, its overall performance can be finally accessed.

In general, the predictions align with the expected downward trend during the cold season, although the model encounters difficulties in accurately identifying the peaks in November and December as depicted in Figure 22.

Figure 22. Actual records vs prediction records for entire zone 1

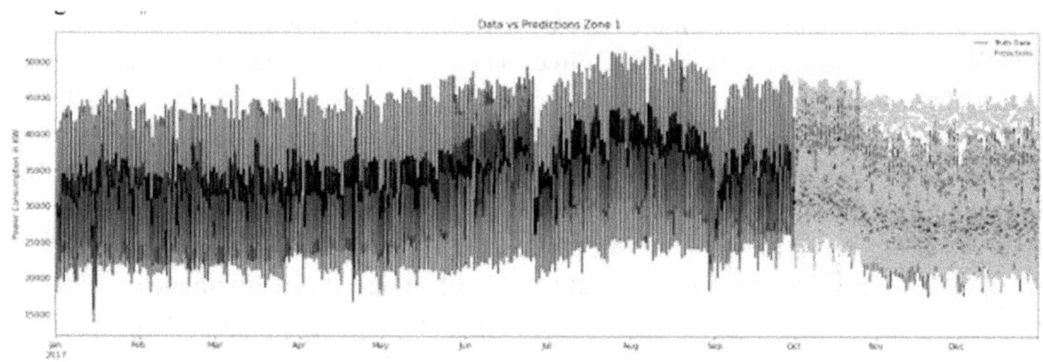

By focusing on the initial 15 days of October, it becomes evident that the predictions closely align with the test data for that particular month as shown in Figure23.

Figure 23. October data vs. prediction comparison

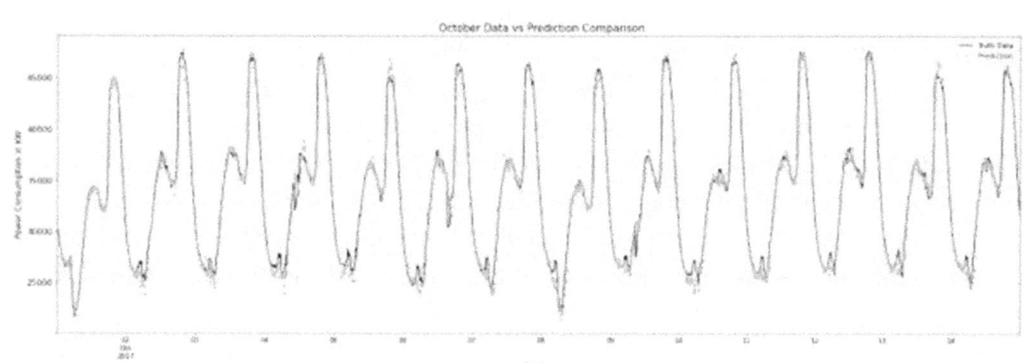

However, during the month of December, the predicted peaks significantly deviate from the actual records. To enhance future models, various options can be explored such as tuning hyper parameters, improving feature engineering, and considering other potential enhancements as displayed in Figure 24.

Accuracy Metrics

Accuracy metrics are quantitative measures used to evaluate the performance of a predictive model by comparing its predictions to the actual outcomes or observations. Some commonly used accuracy metrics employed in the current problem include:

1. **Mean Squared Error (MSE):** MSE provides a measure of the average magnitude of errors made by the model. It squares the errors, which amplifies large errors, thereby, making it sensitive to outliers. MSE is commonly used in optimization algorithms as it is differentiable and allows for easy computation of gradients. However, since it involves squared values, it may not always be directly interpretable in the same unit as the target variable.
2. **Mean Absolute Error (MAE):** MAE calculates the average of the absolute differences between the predicted and actual values. It provides a measure of the average magnitude of the errors made by the model, without considering the direction of the errors. MAE is less sensitive to outliers compared to MSE.
3. **Root Mean Squared Error (RMSE):** RMSE is the square root of the MSE. It represents the square root of the average squared differences between the predicted and actual values. RMSE is in the same unit as the target variable, making it more interpretable than MSE.
4. **Mean Absolute Percentage Error (MAPE):** MAPE calculates the average of the absolute percentage differences between the predicted and actual values. It provides a measure of the average relative error of the model's predictions. MAPE is useful in the case in understanding the magnitude of the errors relative to the actual values.
5. **R-squared (R^2):** R^2 is a statistical measure that represents the proportion of the variance in the dependent variable (target variable) that is predictable from the independent variables (features) in the model. It ranges from 0 to 1, with 1 indicating a perfect fit and 0 indicating no relationship between the predictors and the target variable. R^2 can help access how well the model explains the variability in the data.

These parameters provide different perspectives on the performance of regression models. MSE, MAE and RMSE focus on the magnitude of errors, while MAPE considers the relative errors. R^2 provides an indication of how well the model fits the data. It's important to consider these parameters collectively to gain a comprehensive understanding of a model's predictive performance.

Let us now illustrate the various accuracy metrics considered for the present study. The performance of the regressor is commendable but not outstanding, as indicated by the various accuracy metrics. The R2 value, which represents the proportion of variability in the target variable that can be accounted for by the regression model, demonstrates that 91% of the observed data can be explained by the model. This outcome is considered favorable.

The Mean Absolute Error (MAE) is determined by taking the average of the absolute errors. In this case, the score indicates that, on average, the model deviates by 1354.63 KWs from the actual record.

The Mean Squared Error (MSE) is computed in a similar manner to the MAE, with the addition of squaring the difference between the actual and predicted values. While the MAE indicates that our model performs well on average, the MSE implies that the model can exhibit significant errors. This is because the squared difference tends to amplify larger mistakes made by the model by incorporating a weighting factor.

Lastly, the Root Mean Squared Error (RMSE) is obtained by taking the square root of the MSE. In this scenario, the square root of 3746515.40 is 1935.59. The RMSE proves to be more informative than the MSE. It's worth noting that 1935 is slightly higher than the MAE of 1354.63, indicating that our model generally predicts well on average. However, it suggests that the model tends to deviate excessively on specific records where the actual and predicted values are considerably distant from each other.

In an ideal situation, we aim to minimize the values of MAE, MSE and RMSE, approaching them as closely as possible to zero. However, it's important to note that we don't want them to be exactly zero, since they could indicate overfitting, which should be avoided. The goal is to strike a balance where the errors are minimized without sacrificing the model's ability to generalize to unseen data. The MAE, MSE, RMSE and R^2 obtained for our study are given below in Figure 24.

Figure 24. XGBoost power consumption results

```
r2:  0.942
MAE:  1052.1826
MSE:  2415661.9171
RMSE:  1554.24
MAPE:  0.9659
```

FUTURE RESEARCH DIRECTIONS

In future projects, there are two primary aspects that can be enhanced: hyper parameter tuning and feature engineering. Additionally, it is often recommended to deploy a Naïve Regressor as a benchmark to compare the accuracy scores. Furthermore, upcoming tasks involve collecting real data on power consumption in Tétouan for recent years and forecasting those records.

CONCLUSION

In conclusion, the analysis of power consumption in Tétouan, Morocco reveals several key findings and considerations. The performance of the regressor model displayed good but not excellent results, as indicated by accuracy metrics such as R^2, MAE, MSE and RMSE. While the model performed well on average, it tended to have significant errors on specific records, indicating room for improvement. Accurate forecasting of energy consumption plays a very vital role in guiding plans, programs, load management, and policies within the institution. A forecasting model that is both accurate and simple can greatly enhance proper planning through informed decisions. In summary, forecasting power consumption using time series data empowers energy companies to optimize resource allocation, reduce costs, balance loads, implement demand response programs, plan infrastructure development, integrate renewable energy, and enhance grid stability. These advantages collectively result in improved operational efficiency, cost savings, and a more sustainable energy system.

REFERENCES

Almeshaiei, E., & Soltan, H. (2011). A methodology for electric power load forecasting. *Alexandria Engineering Journal, 50*(2), 137–144. doi:10.1016/j.aej.2011.01.015

Chujai, P., Kerdprasop, N., & Kerdprasop, K. (2013, March). Time series analysis of household electric consumption with ARIMA and ARMA models. In *Proceedings of the international multiconference of engineers and computer scientists* (Vol. 1, pp. 295-300). Hong Kong: IAENG.

Delima, A. J. P. (2019). Application of Time Series Analysis in Projecting Philippines' Electric Consumption. *International Journal of Machine Learning and Computing, 9*(5), 694–699. doi:10.18178/ijmlc.2019.9.5.860

Dey, S., Chakraborty, A., & Basak, P. (2020). Forecasting electricity consumption in India: An ARDL approach. *Energy Reports, 6*, 300–308.

EI-Latif, A. A. AZobaa, A. FElsonbaty, A. H. (2021). Short-term load forecasting using SARIMA and artificial neural network models. *Electric Power Systems Research, 190*, 106823.

Elsaraiti, M., Ali, G., Musbah, H., Merabet, A., & Little, T. (2021, April). Time series analysis of electricity consumption forecasting using ARIMA model. In 2021 IEEE Green Technologies Conference (GreenTech) (pp. 259-262). IEEE. doi:10.1109/GreenTech48523.2021.00049

Jain, P. K., Quamer, W., & Pamula, R. (2018). Electricity consumption forecasting using time series analysis. In *Advances in Computing and Data Sciences: Second International Conference, ICACDS 2018, Dehradun, India, April 20-21, 2018, Revised Selected Papers, Part II 2* (pp. 327-335). Springer Singapore. 10.1007/978-981-13-1813-9_33

Kalimoldayev, M., Drozdenko, A., Koplyk, I., Marinich, T., Abdildayeva, A., & Zhukabayeva, T. (2020). Analysis of modern approaches for the prediction of electric energy consumption. *Open Engineering, 10*(1), 350–361. doi:10.1515/eng-2020-0028

Lee, Y. W., Tay, K. G., & Choy, Y. Y. (2018). Forecasting electricity consumption using time series model. *International Journal of Engineering & Technology, 7*(4.30), 218.

Li, K., & Zhang, T. (2018). Forecasting electricity consumption using an improved grey prediction model. *Information (Basel), 9*(8), 204. doi:10.3390/info9080204

Liu, X., Huang, L., & Li, Q. (2019). Electricity consumption forecasting in China using seasonal hybrid ARIMA-ETS model. *Energies, 12*(6), 1084.

Morocco-Energy. (2019). Retrieved September 25, 2022, from International Trade Administration | Trade.gov website: https://www.trade.gov/country-commercial-guides/morocco-energy#:~:text=Per%20the%20state%2Downed%20power,%2C%20others%20(2%20percent)

Nichiforov, C., Stamatescu, I., Făgărăşan, I., & Stamatescu, G. (2017, October). Energy consumption forecasting using ARIMA and neural network models. In *2017 5th International Symposium on Electrical and Electronics Engineering (ISEEE)* (pp. 1-4). IEEE. 10.1109/ISEEE.2017.8170657

Nizami, S., Khan, N. A., & Mahmood, K. (2018). ARIMA-based electricity consumption forecasting: A case study of Pakistan. *Journal of Energy and Power Engineering*, *12*(5), 307–313.

Pedregal, D. J., & Trapero, J. R. (2010). Mid-term hourly electricity forecasting based on a multi-rate approach. *Energy Conversion and Management*, *51*(1), 105–111. doi:10.1016/j.enconman.2009.08.028

Popeangă, J., & Lungu, I. (2014). Forecasting Final Energy Consumption using the Centered Moving Average Method and Time Series Analysis. *Database Systems Journal*, *5*(1), 42–50.

Sun, T., Zhang, T., Teng, Y., Chen, Z., & Fang, J. (2019). Monthly electricity consumption forecasting method based on X12 and STL decomposition model in an integrated energy system. *Mathematical Problems in Engineering*, *2019*, 1–16. doi:10.1155/2019/9012543

Tan, C., Tan, Y., & Zhang, Y. (2020). Hybrid ARIMA-LSSVM model for electricity consumption forecasting. *Energy Procedia*, *158*, 4155–4160.

Zhang, Y., Zhang, X., & Wang, Z. (2019). Short-term power load forecasting using ARIMA and LSTM hybrid model. In *Proceedings of the 2019 IEEE International Conference on Smart Grid and Clean Energy Technologies (ICSGCE)* (pp. 122-126). IEEE.

Chapter 8
Modeling of Anti-Islanding Protection for Solar Energy-Based Microgrid

Dharmbir Prasad
https://orcid.org/0000-0002-9010-9717
Asansol Engineering College, India

Rudra Pratap Singh
https://orcid.org/0000-0001-7352-855X
Asansol Engineering College, India

Anis Hakim
Asansol Engineering College, India

Ranadip Roy
https://orcid.org/0000-0003-2111-2581
Sanaka Educational Trust's Group of Institutions, Durgapur, India

Md. Irfan Khan
IAC Electricals Pvt. Ltd., Kolkata, India

ABSTRACT

Effective management of renewable energy resources becomes a key task with an increasing penetration at the distribution level. Renewable plants may be small or medium low voltage networks but consist of several sources and loads having a combined effect. Increasing interest and investment in renewable energy give rise to rapid development of high penetration solar energy. There are multiple ways to interface solar panel arrays with the power grid. A system can switch from a grid tie mode to an island mode of operation in the event of a utility grid fault. The active local passive approach is a point of common coupling taken into consideration for anti-islanding protection in integrated systems. This work analyzes the effect of grid tied PV systems, which can enhance power system reliability. A grid-connected PV system is modeled in PSCAD software to examine its operation and controlling aspects; however, system capacity sizing for the 5.03 kWp microgrid has been performed using PVsyst tools.

DOI: 10.4018/978-1-6684-9130-0.ch008

INTRODUCTION

One of the promising DERs is the solar plant with free and clean source, which needs sunshine to produce electricity. To avoid problems with power outages, the centralized power plant may need to boost the grid's reserve power as a result of the increased use of DER. Additionally, the system may experience problems with anti-islanding protection due to the development of PV penetration. A grid-connected photovoltaic system comprises - PV panels and DC-DC converter which converts DC power from PV panel side voltage to required DC bus voltage. While a DC-AC converter is used as a power interfacing between the DC PV panels and the AC grid. The control strategy of the DC-AC inverter that interfaces the PV array with the utility service needs to achieve the following control objectives in order to assure an efficient energy transfer:

- The control strategy applies to the PV module so as to track the MPP for maximizing the energy capture.
- The efficient conversion of the input DC power into an ac output current which has to be fed into the grid.

The utility must maintain an effective grid-tie DG system to deal with islanding issues (refer to Figure 1). The detection techniques are commonly categorized into two types, active and passive.

Figure 1. Classification of anti-islanding detection methodologies

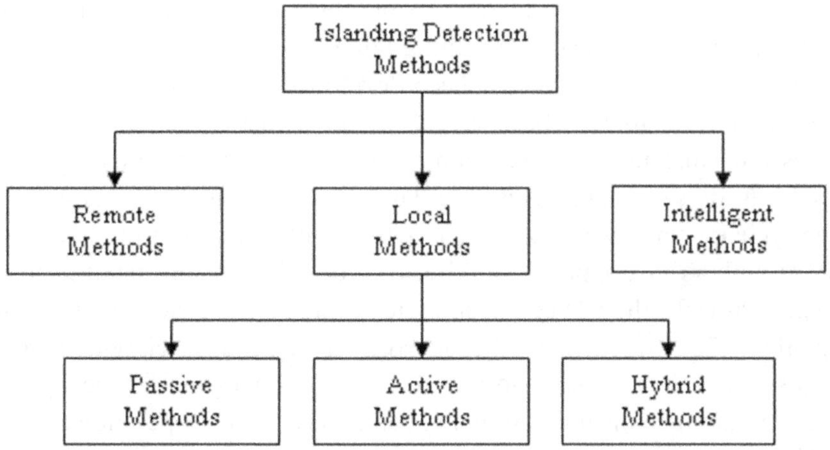

The most basic requirement is expressed in terms of minimum islanding detection time for which quality of the load is an important determinant. The generator must detect the formation of an island and disconnect itself from the grid within the stipulated time period. The following is a list of the most frequent effects of DG connection on utility network security:

- false tripping of feeders
- nuisance tripping of protective devices

- blinding of protection
- fault levels rise or reduction with connection and disconnection of DERs
- unintentional islanding
- automatic reclosing prevention
- re-closing out-of-synchronism devices

Background Review

The distribution system's incorporation of renewable energy sources has significantly increased in recent years. Due to its self-sustaining configuration of distributed energy resources that can also operate in an island mode during grid outages, the microgrid is becoming more and more appealing (Baidya & Nandi, 2022). The occurrence of islanding events is one of the security risks caused by the interconnection of numerous distributed generations. The DG will continue to power the consumers on their own, which is often an unacceptable method of operation. Several techniques have been put forth in the literature to identify islanding quickly and reliably. In order to accurately identify when the mains grid would be disconnected within a reasonable amount of time; a frequency domain technique to analyze the range for the gains needed by an anti-islanding algorithm (John et al., 2003). Major focus in (Xu et al., 2004) study is placed on the slip mode frequency shift technique and the phase jump detection approach, for which it presents a thorough explanation of simulink implementations and their evaluations based on their non-detection zones. Among the various forms of renewable energy generation, small-hydro power generating units make use of the river's naturally flowing water. A controller made for turbines of a different kind that are interfaced with synchronous generators to manage and maintain the power generation when subjected to any variations in load behavior has been reviewed in (Mohamad et al., 2011).

After that, the findings indicate that the resulting voltage sag level may exceed permitted levels of power quality. As a result, a set of indices has been developed to describe how severe the voltage sag is and a strategy based on load flow has been suggested to forecast the problem (Trindade et al., 2013). Abbey et al., have used normal distribution equipment and a recloser controller, a prototype system was built and tested on the utility's distribution test line. The anti-islanding detection time is, according to the results, around one cycle longer than the delay caused by applying the auto-ground (Abbey et al., 2013). In (Datta et al., 2014), authors have applied a multi-criteria decision-making analysis to pinpoint the main difficulties associated with islanding detection and offer a suitable approach for grid-connected solar PV system implementation. Then, researchers introduced an interlocking technique that, by using system impedance estimates, may enhance the performance of rate of change of frequency. For a 9 MW wind power plant, they have examined the relay islanding detection and false operation avoidance capabilities (Jia et al., 2014). The plug-in repeated current control approach for grid-connected inverter systems and the wavelet technique for identifying the condition of the electrical grid are studied in (Pinto & Panda, 2015). In (Saleh et al., 2015), the effectiveness of a passive anti-islanding technique for three-phase cogeneration systems is experimentally examined using the d-q WPT-based anti-islanding approach.

Another research applies a remote islanding detection approach to communicate with circuit breakers to monitor and operate the grid, loads, and PV inverter for a wind-solar hybrid power plant in real time (Bayrak & Kabalci, 2016). Widespread deployment of protection devices is hampered by the grid's sensitivity to severe events, which is a significant problem. The main computational intelligence-based strategies used to solve the issue of islanding in power networks with distributed generators powered by renewable energy are described in (Vyas et al., 2017). Next, a local anti-islanding protection relay is presented as an alternative to the conventional transfer trip in MV feeder applications. It suggests the basic protection for short phase-neutral or phase-phase voltages at PCC of each DG when either of them is crossing zero (Xiao et al., 2017). In (Zhu et al., 2017), a novel active islanding detection method is proposed. It is based on the harmonic impedance obtained from deformed voltage and current waveforms. Analysing the shift in harmonic impedance enables the detection of distributed generator islanding.

In its continuation (Motter & de Melo Vieira, 2018), the pick-up and time-delay settings of passive anti-islanding protection, which are frequently used in distributed generation points of common coupling under frequency/over frequency, rate of change of frequency, vector surge and under voltage/over voltage methodology. Islanding identification is a significant issue that needs more attention for its quicker detection and associated measures among all potential linked issues. A comparison of their technical benefits and drawbacks, as well as practical viability, cost effectiveness, accuracy and simplicity of adoption, has been conducted in (Mishra et al., 2019). In (Ferreira et al., 2019) study, a technique for determining the grid operating conditions viz., utility connected or islanded has been studied. A local scheme that is put near to the protective device is the power electronics based technique. Singh et al., explore several anti-islanding protection strategies before contrasting one of the active strategies with its enhanced variation based on a certain parameter (Singh et al., 2020).

The loading characteristics of the NDZ region, various fault kinds and loads with various power factors have all been taken into consideration with inverter or direct fed islanding detection approaches in microgrids (Khosravi et al., 2021). In (Elshrief et al., 2021), authors suggested using the rate of change of power approach based on the terminal voltage of the PV inverter to detect this anti-islanding event as quickly and precisely as feasible with MATLAB simulink tools. The study of (El Azzaoui, 2021) suggests a novel controller's reference reactive power based on the positive sequence voltage of the point of common coupling and a reactive mismatch coefficient. The criteria also take into account the Discrete Wavelet Transform of the variation in amplitude between nominal and real PCC voltages assessed under various load and disturbance situations. Next, the authors work suggests a quick protection technique based on the Hilbert-Huang transform for the classification and placement of faults and islanding in a distribution system that has been infiltrated by a solar renewable energy source (Shaik et al., 2021).

Furthermore, in order to assess the efficacy of the plan, a standard microgrid power generation model and an inverter control model appropriate for grid-connected and off-grid microgrids are constructed (Wang et al., 2022). The voltage and frequency variations in the two modes are then analyzed. Thereafter, research provides an EMTDC/PSCAD-simulated passive anti-islanding solution based on encrypted communication. This approach includes information on the phase angle difference of superimposed impedance computed at both dispersed generation and the point of common coupling ends (Nale et al., 2022). The effectiveness of a passive anti-islanding protection with low switching losses for three-phase grid-connected solar power systems is presented in (Banu et al., 2023). It was simulated under both grid failures and total disconnections of the PV inverter from the electrical power system.

Paper Layout

Rest of the paper has been organized as follows: Islanding aspects in distribution system, then, solution approach, mathematical formulation, PV grid tie microgrid system integration, system control mechanism, simulation results, finally conclusion has been derived.

ISLANDING IN DISTRIBUTION SYSTEM

The main power utility and local, small-scale distributed energy resources are connected at the distribution voltage level when using renewable distributed generation. As shown in Fig. 2, an island develops when a section of the distribution system separates electrically from the rest of the power system while still being powered by dispersed sources. Islanding may occur accidentally or on purpose. The difference between controlled and uncontrolled island operations must be made. When one or more generators are outfitted with their own control and protection circuitry to ensure safe operation, the controlled island operation is a technique to ensure supply dependability. It is important to prevent the uncontrolled island operation whenever feasible since it is a major issue. Additionally, it's crucial to distinguish between prolonged or long-term island operation, which poses a greater threat to personnel security and produces issues with auto reclosing and may result in severe overvoltage.

Figure 2. Layout of islanding detection system

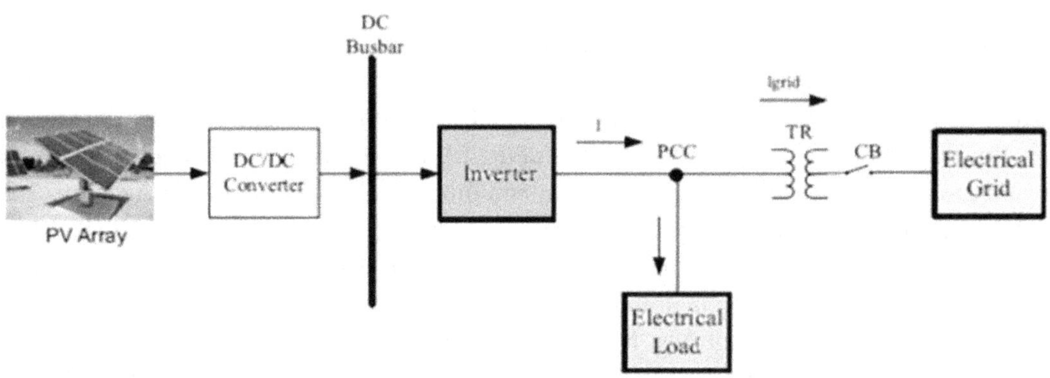

When dispersed power generation systems are present, islanding is desired because local consumers should be served even without power grid connections. People who work on the grid's maintenance are put at risk when islanding happens unintentionally. Protection coordination of the power distribution system is the area that is severely impacted by the penetration of distributed generators. Traditional over current protection is developed for radial distribution systems with fault current flowing in just one direction. However, the single-fed radial networks become complex ones with numerous source locations when PV systems are connected to distribution networks. Due to this, fault currents now travel in both directions instead of just one. For the anti-islanding protection to be managed for healthy load distributions among customers, there are a number of important criteria. The PV array characteristic along with the inverter has been presented in Table 1.

Table 1. Parameters of integral components of solar power based microgrid

Sl.		Components Properties	Specifications
1		*PV module*	
	i	Unit nominal power	335 Wp
	ii	Number of PV modules	15 units
	iii	Nominal power at STC	5.03 KWp
	iv	Modules arrangement	1 string x 15 in series
	v	Modules area	29.1 m^2
	vi	P_{mpp} at 50°C operating condition	4549 Wp
	vii	V_{mpp} at 50°C operating condition	513 V
	viii	I_{mpp} at 50°C operating condition	8.9 A
2		*Inverter*	
	i	Unit nominal power	5.00 kWac
	ii	Number of inverters	1 unit
	iii	Total power	5.0 kWac
	iv	Operating voltage	200-800 V
	v	P_{nom} ratio (DC-AC)	1.01

The proposed system could be achieved under following objectives:

- Grid-tie PV system capacity evaluation using PVsyst tool
- Operation and control modeling of grid-tie PV system islanding using PSCAD tool

SOLUTION APPROACH

Following are the planned execution steps for the undertaken works:

Step-1: *Solar Plant Modeling:* At first, the possibility of installing a 5.03 kWp solar plant as a renewable energy source has been examined using PVsyst tools at Surulia (WB, India) (Latitude: 23.33 °N, Longitude: 86.42 °E) in the Eastern India and also the performance characteristics of the systems has been evaluated (refer to Figure 3-4).

Figure 3. Proposed grid-tie PV system architecture

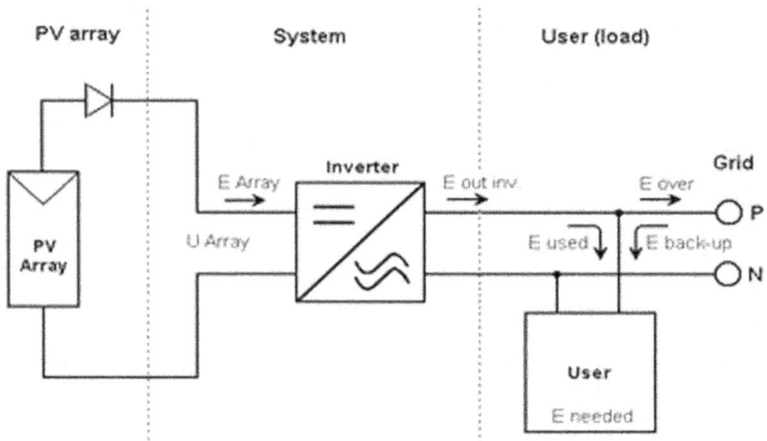

Figure 4. Proposed grid-tie PV system description

System summary

Grid-Connected System		**No 3D scene defined, no shadings**			
PV Field Orientation		**Near Shadings**		**User's needs**	
Fixed plane		No Shadings		Fixed constant load	
Tilt/Azimuth	30 / 0 °			2000 W	
				Global	
				17.52 MWh/Year	
System information					
PV Array			Inverters		
Nb. of modules		15 units	Nb. of units		1 Unit
Pnom total		5.03 kWp	Pnom total		5.00 kWac
			Grid power limit		10.00 kWac
			Grid lim. Pnom ratio		0.503

Step-2: *Modeling of Solar Plant and Distribution Systems:* Then after, modeling of the proposed solar plant containing the PV modules and the inverters has been be conducted using PSCAD tools. The parameters (e.g., *V*, *I*) are adjusted by the DC-DC converter to ensure that the solar plant active power is maximum at a given time and also that the input voltage value of the inverters is getting rated value. The inverters feed the transformers by converting the DC input voltage into three phase AC voltage. By reactive power absorption or generation, the inverters control the voltage level at the common coupling point. The remaining part of the solar power plant design until the common coupling point will be continued including busbar, transformers and cables accessories; and the end grid-tie will be initiated using the PSCAD tools.

Step-3: *Simulation for Islanding Protection for Grid-tie Solar Plant:* In case of islanding scenario, a power network gets separated into controllable/uncontrollable sections containing both sources and loads and operating independently of the main power grid (refer to Figure 5). For grid-mode distributed generation systems, islanding is said to occur when the distributed PV system continues supplying power to the distribution grid even after a portion of the network, that includes the PCC, gets disconnected from the main power system.

Figure 5. Flowchart of passive islanding protection

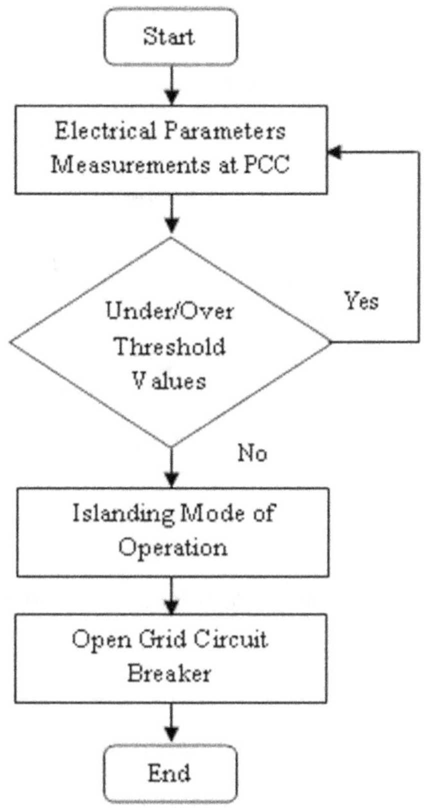

MATHEMATICAL FORMULATION

The proposed system is used to follow certain mathematical formulae, which has been briefed in subsequent sections. The current flowing through the diode (refer to Figure 6) can be derived from an ideal p-n junction diode equation as (1) (Muljadi et al., 2013),

$$I_d = I_{sat}\left[\exp\left(\frac{v_{pv} + i_{pv}R_s}{\eta N V_T}\right) - 1\right] \quad (1)$$

where, η: ideality factor which depends on the fabrication process and semiconductor material of the cell, $Is_{at:}$ reverse saturation current of the on-junction and VT is the thermal voltage of the diode junction which is given by (2) (Shin et al., 2020),

$$V_T = \frac{kT}{q} \tag{2}$$

Here, k: Boltzmann's constant (1.38065×10^{-23} J/K), T: cell's temperature and q: electron's charge (1.6×10^{-19} C).

Figure 6. Representation of PV cell module circuit

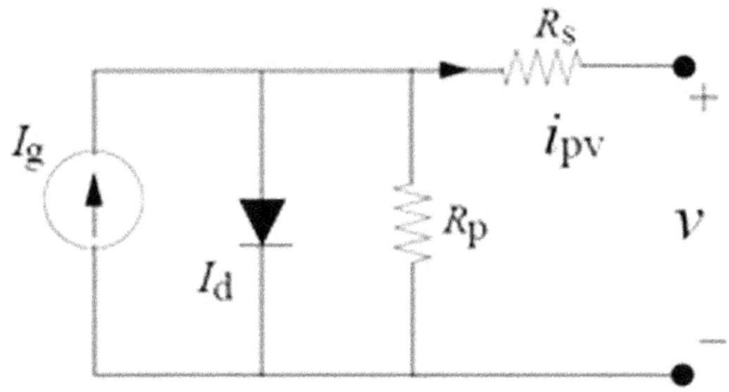

Based on (1) and (2) the mathematical model of solar cell can be described by (3) (Han et al., 2017),

$$i_{pv} = I_g - I_{sat}\left[\exp\left(\frac{u_{pv}}{\eta N V_T}\right) - 1\right] \tag{3}$$

The safe operating voltage and current of the cell can be calculated from the open circuit and short circuit conditions, respectively.

The short-circuit current is the maximum current flowing when the voltage is set to zero for a given temperature and irradiation. This gives $I_{sc}=I_g$. Similarly the open circuit voltage can be calculated for a given temperature and irradiation by setting the cell current is equal to zero (4) (Chowdhury et al., 2009),

$$V_{oc} = \frac{kT}{q}\ln\left(\frac{I_g + I_{sat}}{I_{sat}}\right) \tag{4}$$

The output power of the PV cell is given by multiplying the voltage and current which belongs to the *I-V* curve. For the high efficiency operation, it is desired to extract maximum power from the PV cell. The MPP for a solar cell is given by (5) (Chowdhury et al., 2009),

$$P_{MPP} = V_{MPP} \times I_{MPP} \tag{5}$$

where, V_{MPP} and I_{MPP}: voltage and current corresponding to MPP.

The switching losses in the IGBTs are presented, where the larger energy losses. The losses during switching can be given as in (6) (Bayrak & Kabalci, 2016),

$$P_{switching} = \frac{V_{s(off)} \times f_s}{2} \times \left[t_{on} + t_{off} \right] \times I_{on} \tag{6}$$

where, $P_{switching}$: losses during turn on and turn off (W), $V_{s(off)}$: blocking voltage when switch is off (V), t_{on}: time taken for the switch to turn on (s), t_{off}: time taken for the switch to turn off (s), *f*: switching frequency (Hz) and I_{on}: conduction current (A).

The derivation for sizing the DC link capacitor illustrates the importance of the component due to the difference in AC output power and DC input power. Starting with equations for voltage and current for a single phase of the three phase inverter output (7)-(8) (Bayrak & Kabalci, 2016),

$$V_{LN} = V_{pk}\sin(\omega 0_t) \tag{7}$$

$$I_L = I_{pk}\sin(\omega 0_t + \phi) \tag{8}$$

where, ϕ: difference in phase between the grid voltage and inverter output voltage.

This phase is equal to zero when the current and voltage are in phase. As an approximation, considering a worst case scenario, the three phase instantaneous output power of the inverter can be expressed as (9) (Bayrak & Kabalci, 2016),

$$P_0 = 3V_{pk}I_{pk}\sin(\omega 0_t)\sin(\omega 0t_+\phi) \tag{9}$$

In order to maintain a power balance at the DC bus, a capacitor is used to deliver and store energy (10) (Bayrak & Kabalci, 2016),

$$P0 = PP_V - PP_{Vc}os(2\omega 0t_) \tag{10}$$

The net power into or out of the DC bus could be given by (11) (Bayrak & Kabalci, 2016),

$$P_{Bus} = P_{PV} - P_0 \tag{11}$$

Combining (10) and (11) yields the instantaneous DC bus power that the capacitor will need to sink or source every half cycle (12) (Bayrak & Kabalci, 2016),

$$P_{Bus} = P_{PV}\cos(2\omega 0_t) \tag{12}$$

Integrating (12) for one half cycle and setting it equal to the expression for energy stored in a capacitor, rearranging the result, the equation to size the input capacitor is expressed as (13) (Bayrak & Kabalci, 2016),

$$C = \frac{P_{PV}}{2\pi f_{grid} V_{PV}^2 (\Delta V_{PV-\%})} \tag{13}$$

where, ΔV_{PV}: allowable ripple voltage of V_{PV} expressed as a percentage. For large PV systems the voltage ripple needs to be small so the corresponding power angle.

Then, the instantaneous power injected to the grid is given by (14) (Bayrak & Kabalci, 2016),

$$P_g = 2P\sin^2(\omega g_t) \tag{14}$$

where, P_g is the average power, ωg is the angular grid frequency, and t is the time.

PV GRID TIE MICROGRID SYSTEM INTEGRATION

A system layout for various components developed for the proposed grid-tie PV system as can be seen in Fig. 7. A capacitor is connected in between the maximum power point tracker and photovoltaic array. It is used as an energy buffer for the transient. To fully utilize the photovoltaic energy and connect to the grid, it has to be connected with other electrical devices such as DC/DC boost converter, DC/AC inverter, filter and transformer. The following are major components presented as below:

Figure 7. Modeling layout of grid-tie PV system protection

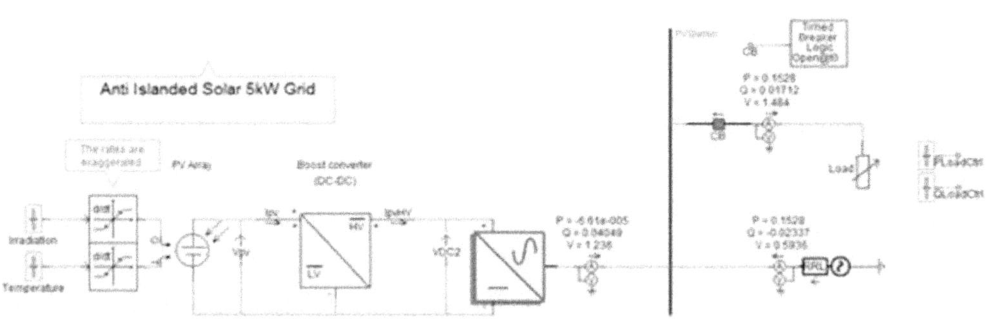

(i). PV System: The PV energy generated from sunlight is captured by PV modules. The power produced by the PV modules is DC electrical power. Integrating the DC power from the PV modules into the existing AC power distribution network can be achieved through grid-tie inverters. The inverters must assure that the PV modules are operating at MPP, which is the operating condition where the most energy is captured. Another function for inverters is to control the current injected into the AC grid at the same phase with the grid voltage at the lowest harmonic distortion levels. Then the electricity emanating from the inverter goes to the load and the excess current goes to the grid. In PSCAD, other than the PV configuration, parameters such as the series resistance, shunt resistance, ideality factor and band gap energy are needed to be set correctly as done in Figure 8-11.

Figure 8. Schematic of PV module block

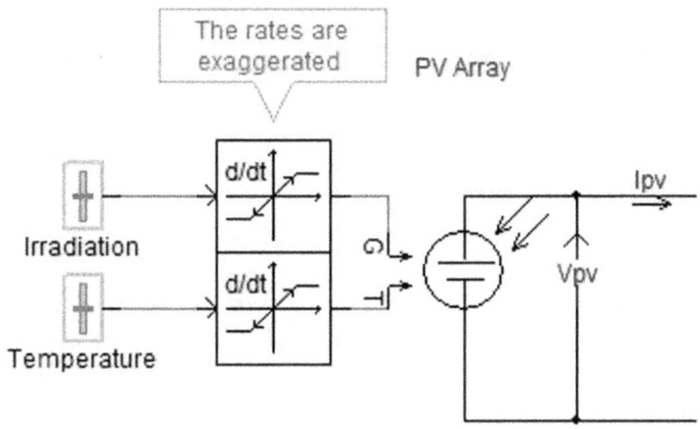

Figure 9. Schematic of PV array parameters

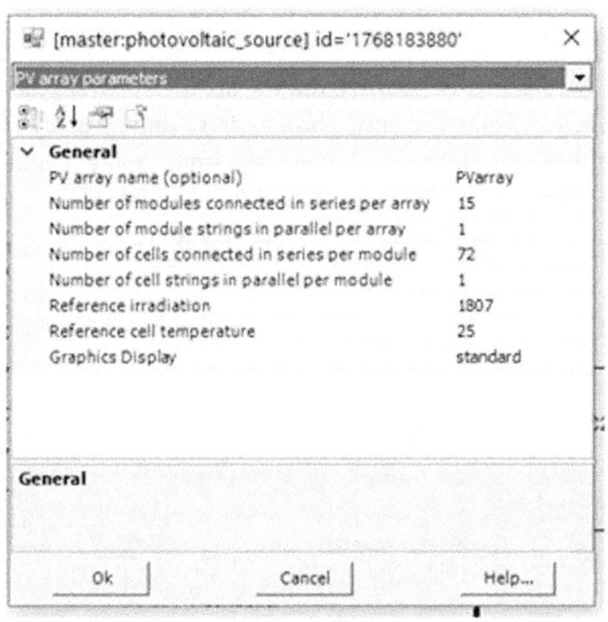

Figure 10. Schematic of PV cell parameters

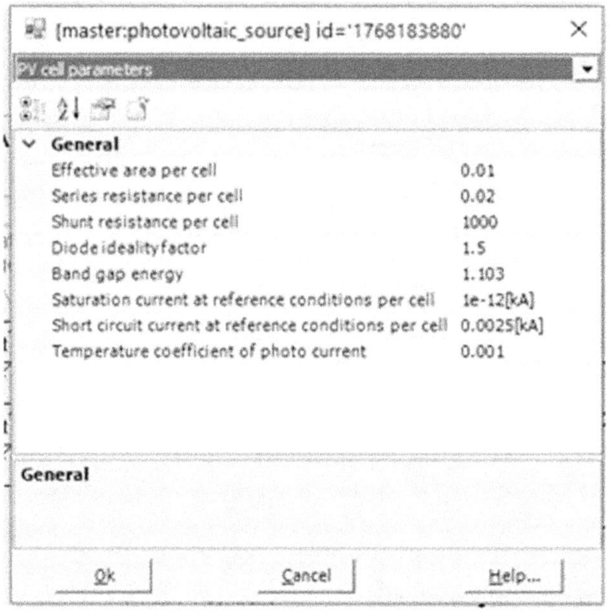

(ii). PV MPPT Controller: MPPT is used to force the PV array to operate the MPP. A DC-DC converter is used to vary the voltage level to the knee point to drive the PV array in order to work at optimum point. The difference between the instantaneous maximum voltage and that of the PV array is fed to the controller. The output of the PI controller will be compared with a high frequency waveform to generate a high duty frequency duty cycle (refer to Figure 12-13).

Figure 11. Summary of PV array losses

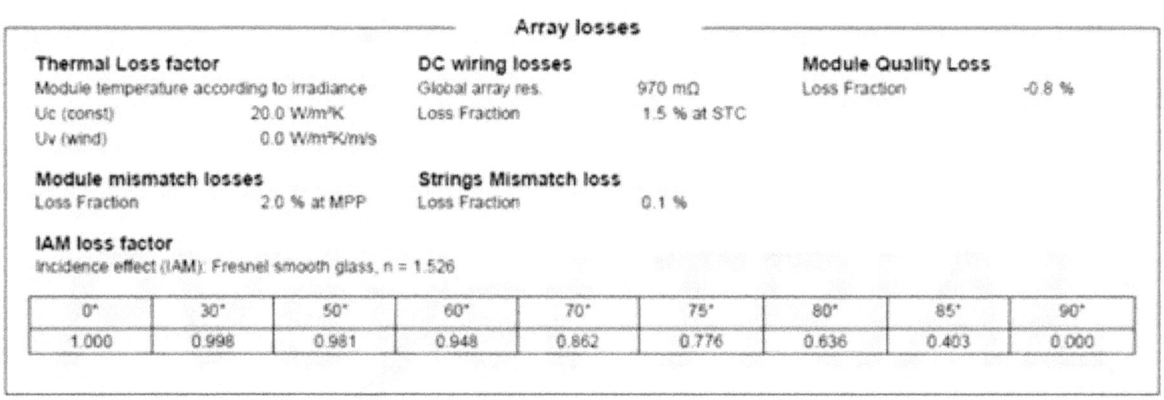

Figure 12. Schematic of PV MPPT controller block

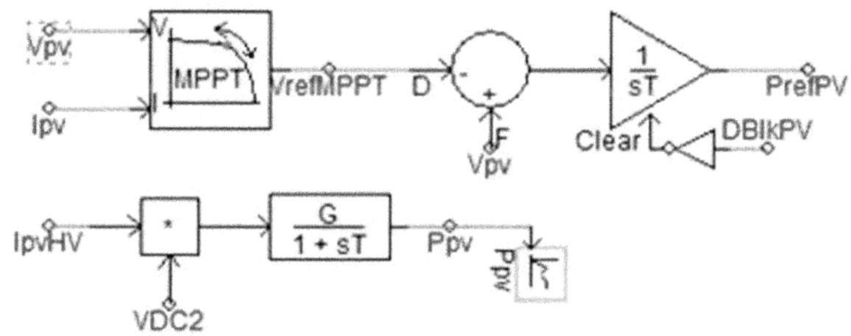

Figure 13. Properties of PV MPPT controller block

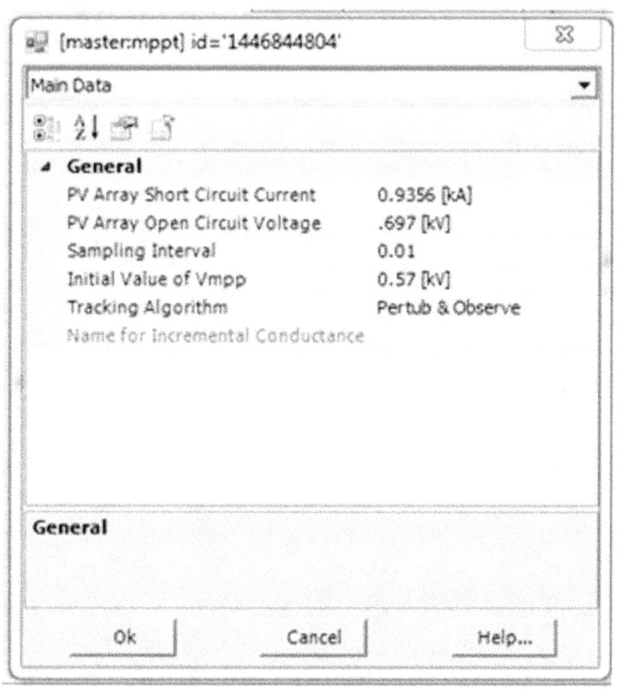

(iii). Phase-Locked Loop: A control system known as a PLL produces an output signal whose phase is correlated to the phase of an input signal. The use of phase-locked loops for synchronization is common. It is usually necessary to know when the loop is out of lock in PLL applications (refer to Figure 14-15). It is utilized in centralized inverters with high power and master-slave architecture that are seen in sizable PV power plants.

Figure 14. Schematic of PLL block for PV modeling

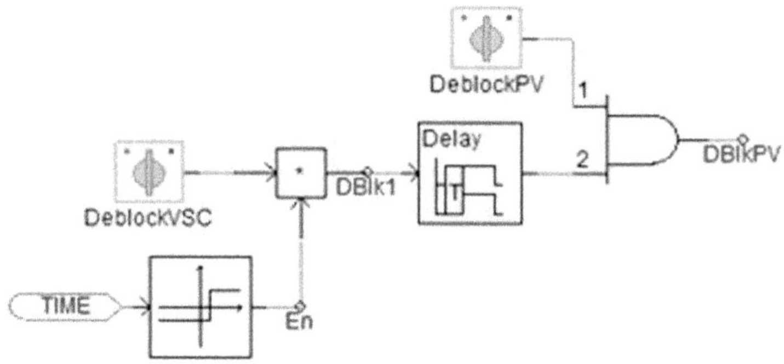

Figure 15. Properties of PLL block

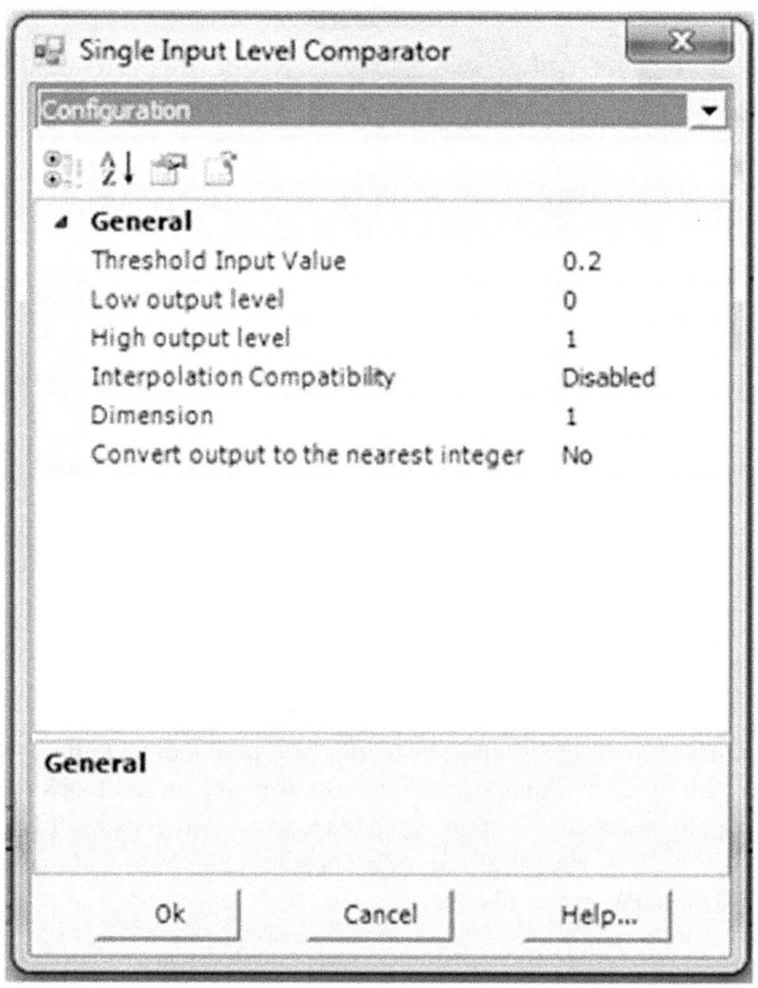

(iv). Boost Converter: The boost converters are commonly used to extract the maximum power from the PV array. These converters produce an output voltage which is higher than the input voltage. When the PV voltage is insufficient, the PV voltage can be boosted with the use of the boost converter. The basic circuit design of a boost converter is shown in Figure 16.

Figure 16. Schematic of boost converter for PV modeling

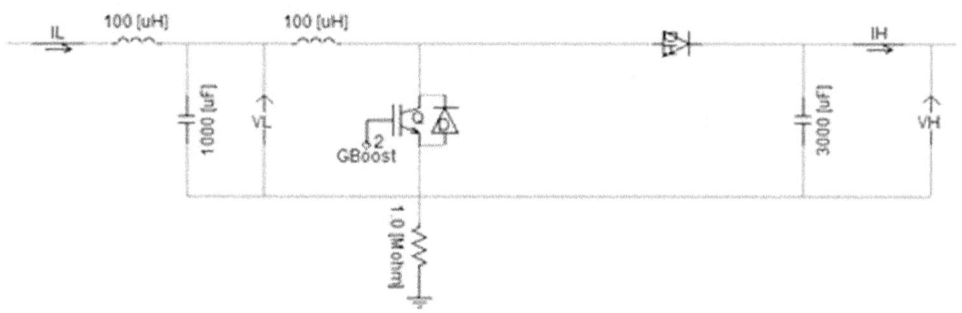

(v). Inverter: Three phase inverter is chosen as the inverter for the PV system. To construct a three phase inverter, three pairs of IGBT based power electronic switches are used as in Figure 17-18, switching losses in the IGBT is significant. A snubber circuit is given to both the diode and switch to protect these devices from damage. Using the Sine wave generator in PSCAD, the amplitude, frequency and phase angle are tuned to meet the user's requirements.

Figure 17. Schematic of IGBT-based inverter for PV modeling

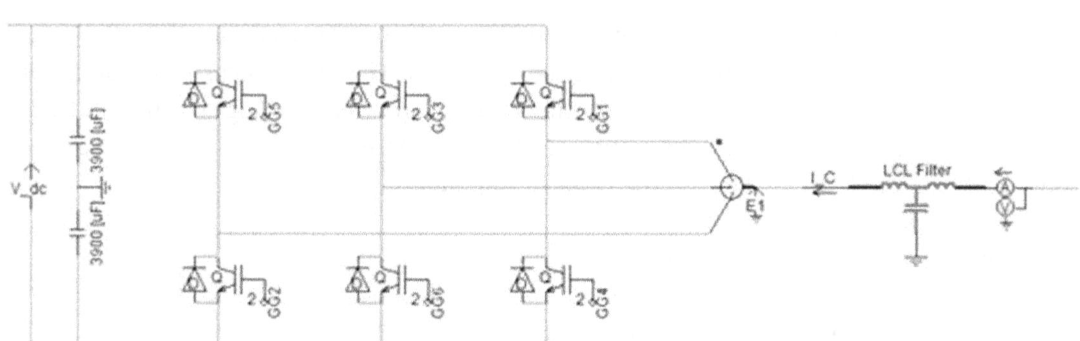

Figure 18. Properties of IGBT-based inverter for PV modeling

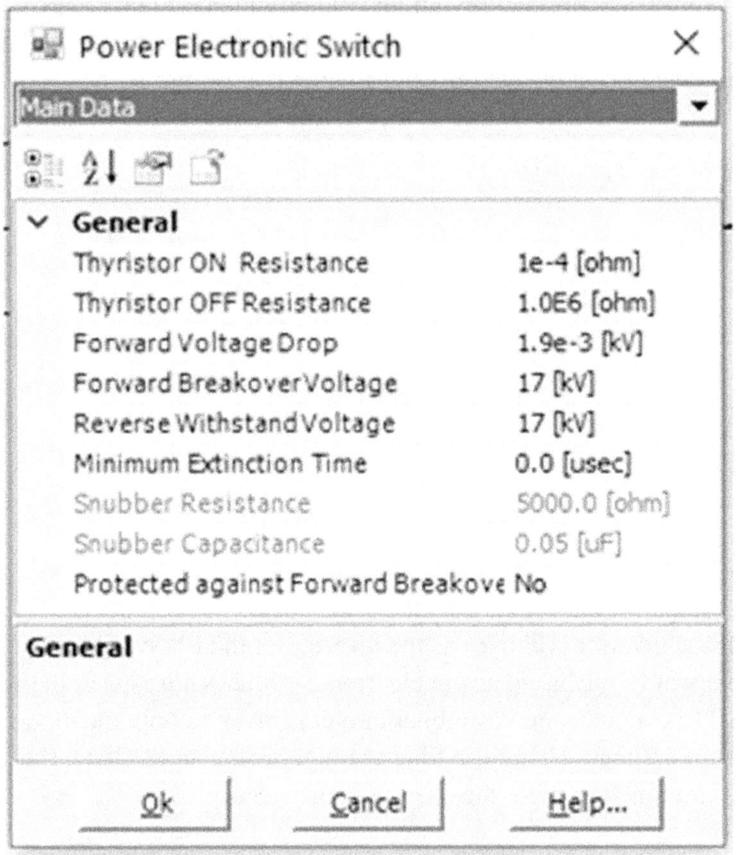

(iv). LCL Filter: Design of these passive filter components requires an accurate estimation of different ripple quantities present in different voltages and currents. The inductance and capacitance are determined by the maximum ripple current at the inductor and ripple voltage on the capacitor, which these components can tolerate (refer to Figure 19). A properly designed filter must also assure the following in addition to attenuating the ripple components of the grid current:

- low electromagnetic interference,
- a high grid power factor,
- a low voltage drop across the filter, and
- a low loss in the damping resistor.

Figure 19. Schematic of LCL filter for PV modeling

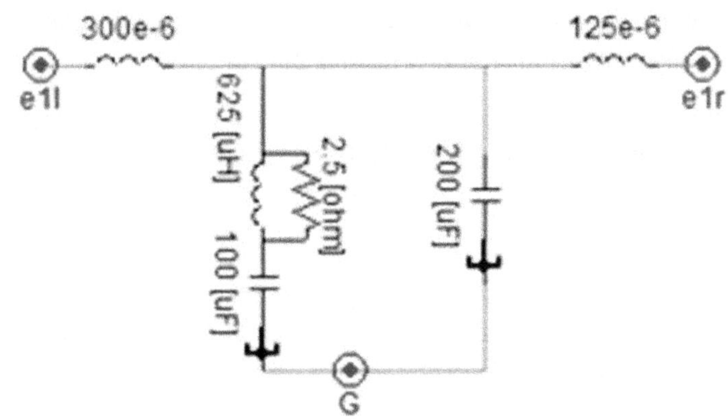

(v). DC Link Capacitor: The bridge between the PV array and the inverter is the DC link capacitor shown in Figure 20. This component is of the utmost importance as it serves three very important functions. The first function is to minimize the voltage ripple across the PV terminals and which results in a ripple of the output power. Secondly, the capacitor acts as a sink and source every half cycle to help create a balance of power on the DC bus.

Figure 20. Properties of DC link capacitor for PV modeling

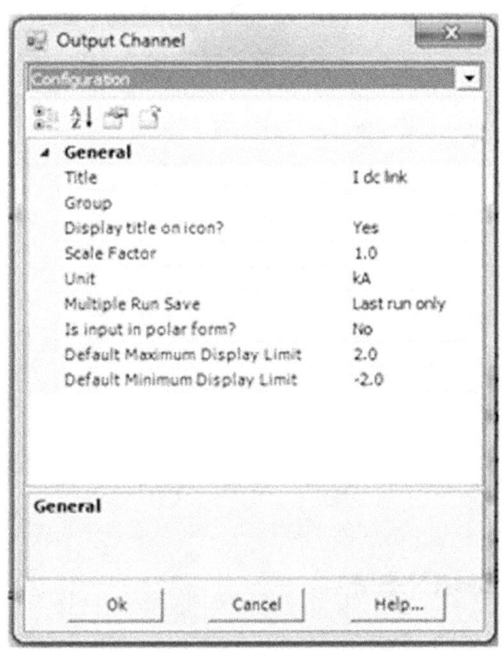

(vi). Transformer: A transformer is used in the PV system to step-up the filtered voltage. The turns ratio, winding type for primary and secondary sides and power rating have to be determined. The power rating of the transformer is designed according to the capacity of generated power from the PV array which has a peak generation (refer to Figure 21-22).

Figure 21. Schematic of various type loads for PV modeling

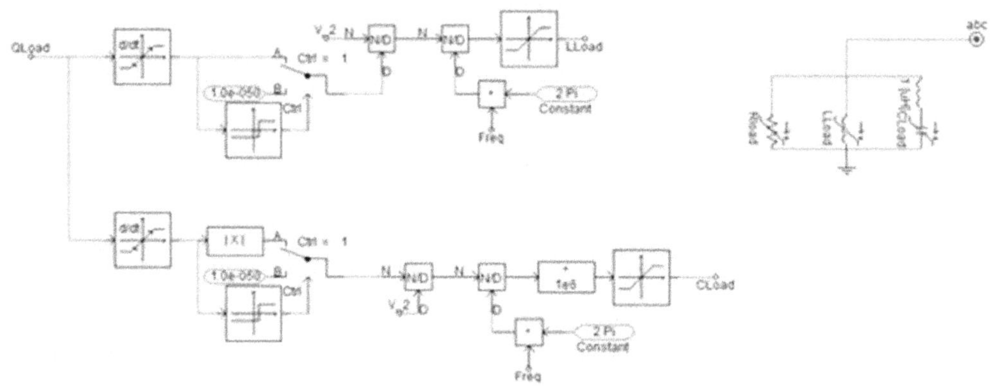

Figure 22. Properties of various types (R,L,C) of loads for PV modeling

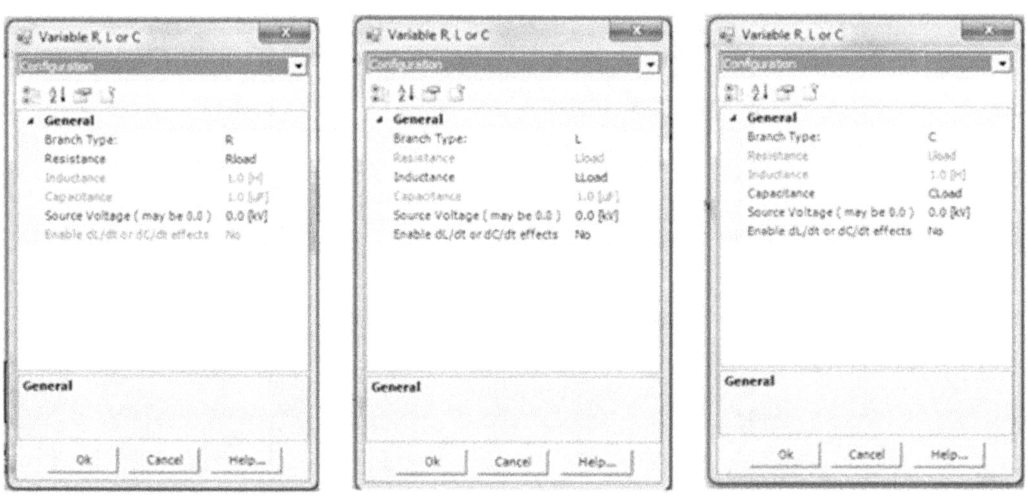

SYSTEM CONTROL MECHANISM

The islanding operation must be stopped when it happens since it might harm both the grid and the DG system. The majority of the time, grid coordination causes changes in the loading of both DGs and storage units. As a result, when a grid failure occurs, the system voltage (V_{dc-ref}: 718V; V_{acL-L}: 415V; $V_{dmax} = V_{qmax}$:

370V; $V_{dmin} = V_{qmin}$: -370V), currents and frequency are also altered. These system characteristics are continually monitored by the relays. The foundation for passive loss of grid protection strategies is this.

Figure 23. Schematic of V_{dc} regulator

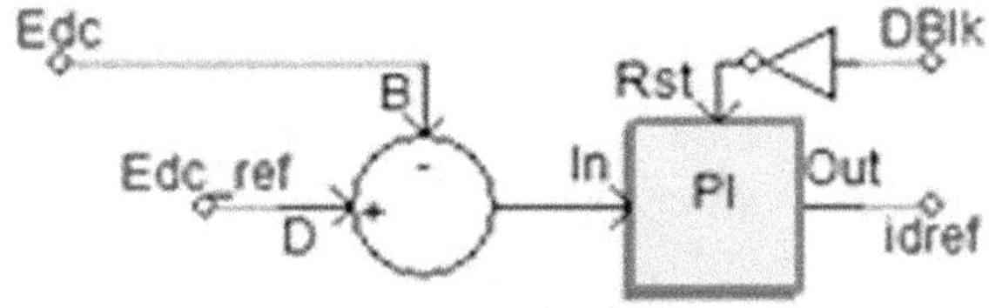

Figure 24. Schematic of I_{dc} regulator

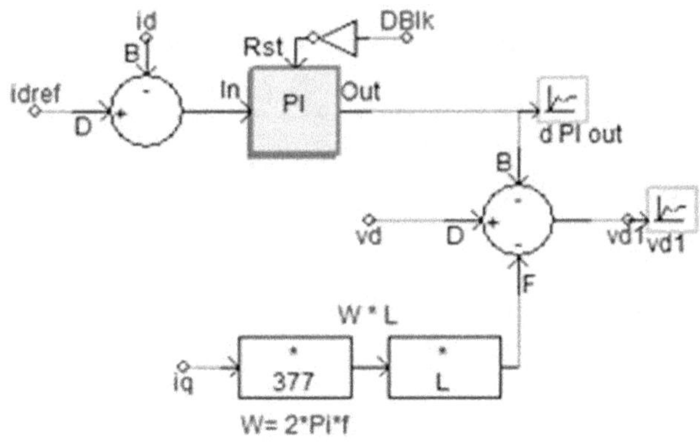

In an islanding condition, the grid voltage and frequency are not steady. In this circumstance, the operation values diverge from the grid reference values. Therefore, during an islanding operation, a CB linked between the grid and the PCC must clear the fault (see Figure 23-27). In the event that CB is unable to be regulated to open the circuit, DG continues to supply power to the nearby load. Voltage unanticipated interruption of the grid is caused by shutdown, equipment failure, and short-circuits situations; these anomalous circumstances lead to islanding operation in a DG system.

Figure 25. Schematic of Q regulator

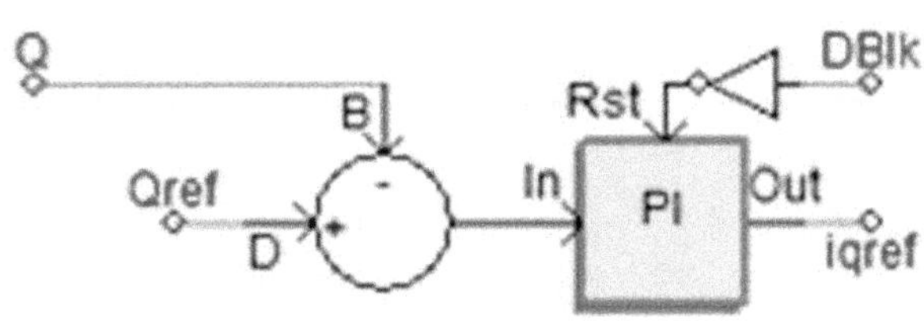

Figure 26. Schematic of I_q regulator

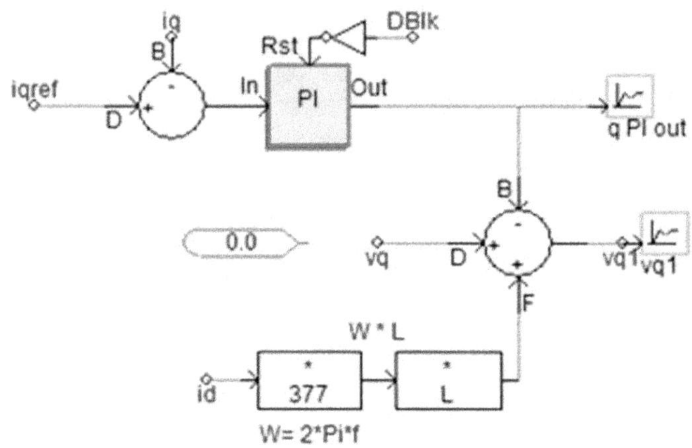

Figure 27. Properties of protective circuit breaker for PV system

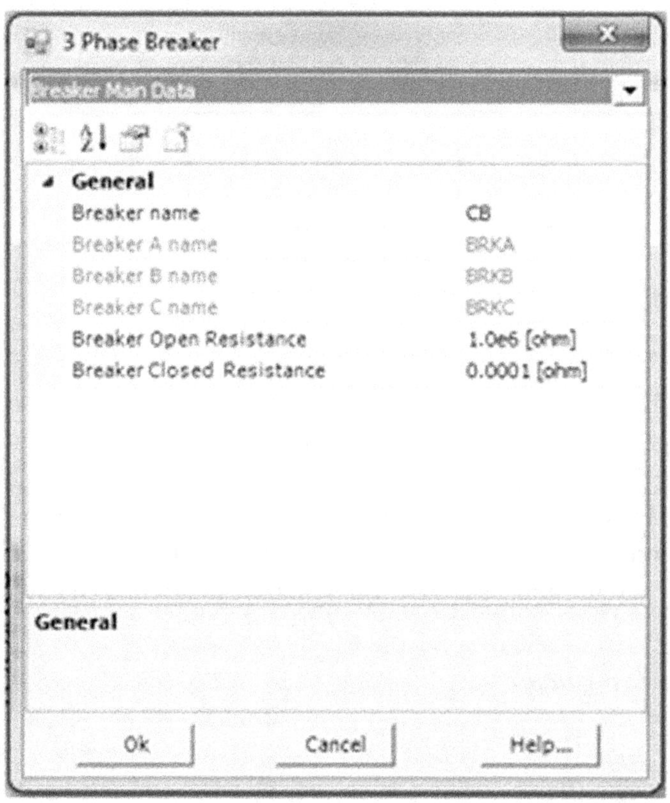

SIMULATION RESULTS

The proposed model has been simulated using PVsyst and PSCAD tools and corresponding results have been as follows (refer to Figure 28-37 and Table 2):

Table 2. Monthly results summary of the proposed Surulia, West Bengal site

Period	$Glob_{Hor}$	$Diff_{Hor}$	T_{amb}	$Glob_{Inc}$	$Glob_{Eff}$	E_{array}	E_{user}	E_{solar}	E_{grid}	EFr_{grid}
January	127.9	49.05	17.93	173.6	170.5	758.2	1488	520.6	217.7	967
February	139.0	54.32	21.89	171.3	168.2	731.9	1344	493.7	219.0	850
March	178.5	73.92	27.04	196.0	192.0	813.7	1488	552.7	238.4	935
April	190.8	79.11	30.73	188.0	183.0	766.3	1440	538.4	206.2	902
May	204.6	92.80	31.83	184.0	178.3	753.2	1488	560.8	170.4	927
June	163.9	97.13	29.71	143.5	138.5	602.2	1440	487.6	95.3	952
July	139.0	85.30	28.53	123.7	119.2	523.8	1488	426.8	77.9	1061
August	140.6	88.55	28.09	131.7	127.5	559.2	1488	450.8	89.6	1037
September	134.6	76.86	27.40	138.2	134.2	585.3	1440	447.6	119.3	992
October	140.6	73.71	26.18	161.5	157.7	686.1	1488	502.5	164.1	985
November	125.6	54.56	21.67	162.0	159.1	697.9	1440	482.0	196.1	958
December	122.3	48.26	17.89	169.3	166.5	742.3	1488	509.4	213.3	979
Year	1807.3	873.55	25.75	1942.7	1894.7	8220.0	17520	5973.0	2007.4	11547

Figure 28. Characteristics of array temperature distribution during running phase

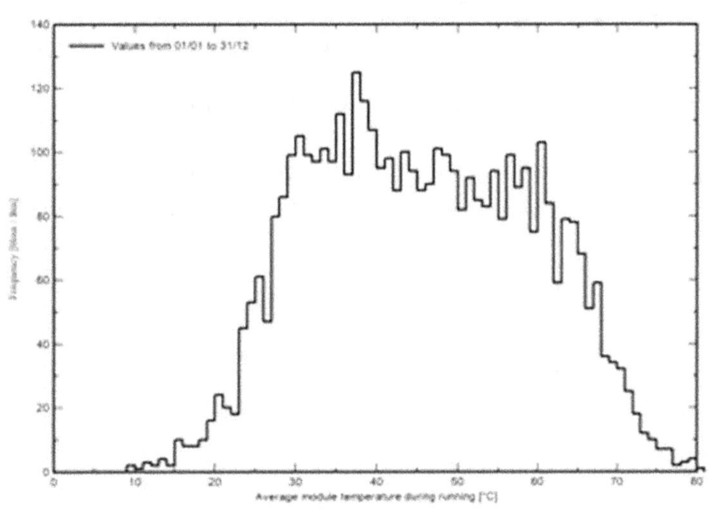

Figure 29. Characteristics of array voltage distribution

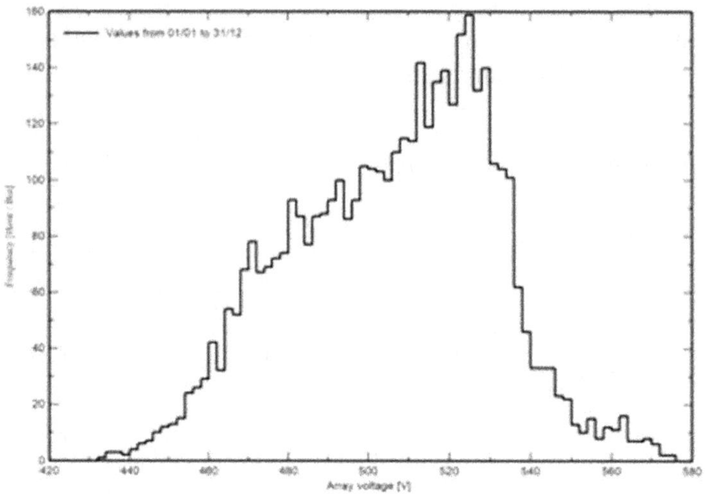

Figure 30. Characteristics of daily system output energy

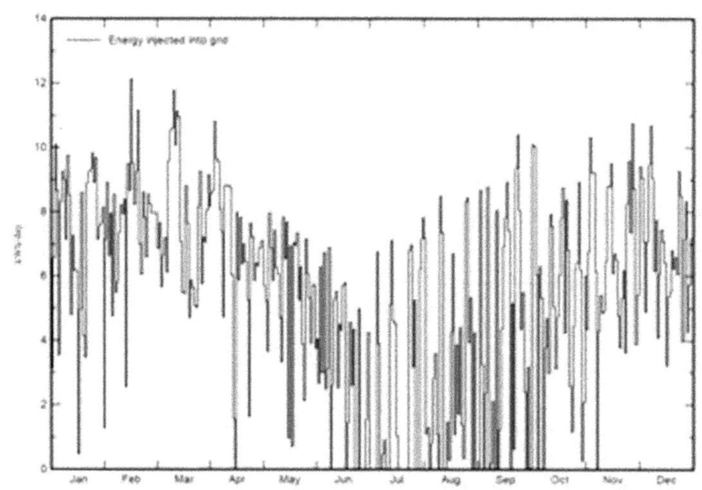

Figure 31. Characteristics of PV performance ratio variations

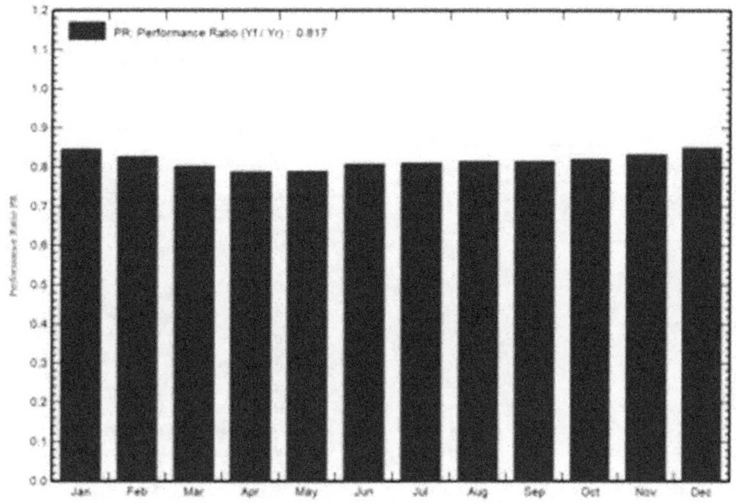

Figure 32. Characteristics of PV installation site

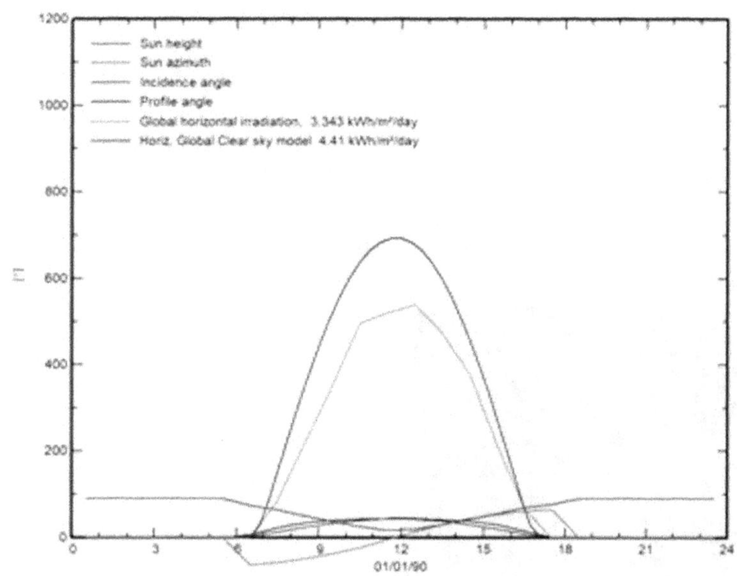

Figure 33. Loss characteristics of undertaken PV system

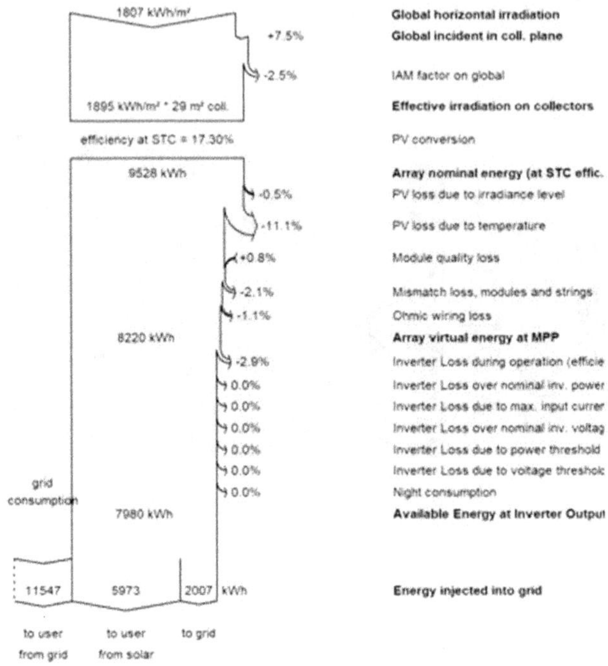

Figure 34. Characteristics of generator voltage and current variations

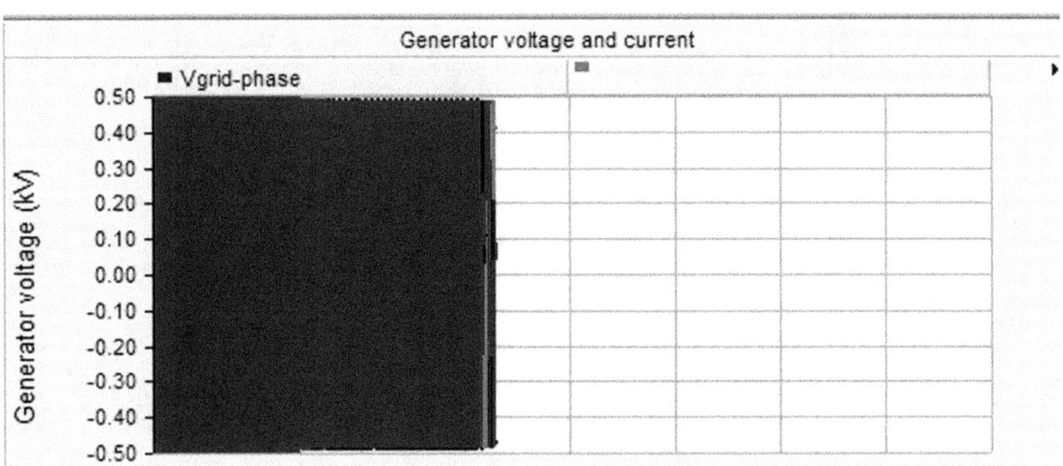

Figure 35. Characteristics of generator reactive power variations

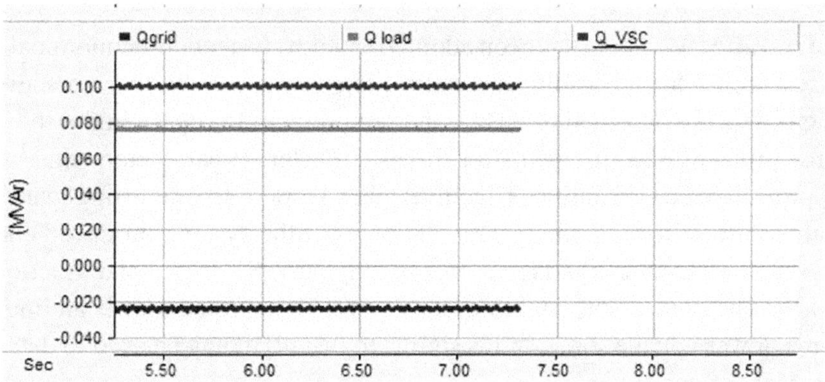

Figure 36. Characteristics of generator real power variations

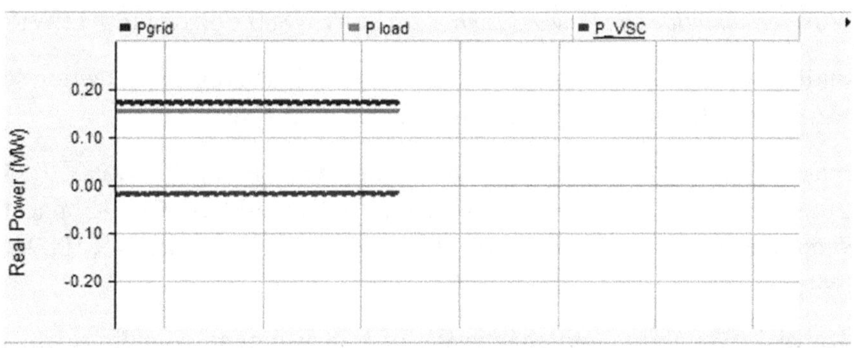

Figure 37. Characteristics of generator current variations feeding to grid

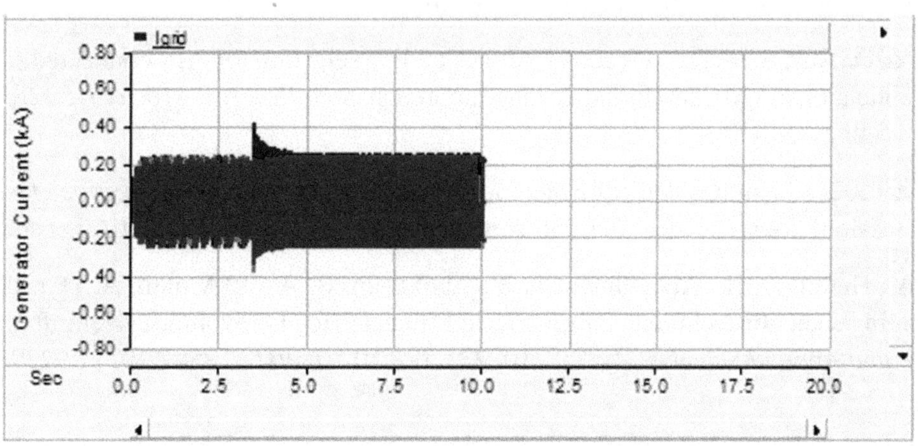

CONCLUSION

In this work, a site specific PV system has been adopted to study. Its techno-economical analysis has been carried using PVsyst tools while operation and controlling mechanism has been achieved using PSCAD tools. Various controllers are designed to ensure the requirement for operating at the MPP imposed by the PV module and protective requirements for the grid. Different parameters viz., - *V*, *I*, *Q*, *P*, *f* have been checked for protective co-ordination. The undertaken work is an interesting learning experience of using new software tools. Following are the key outcomes of the system: produced energy: 7.98 MWh/year; specific production: 1588 kWh/year; performance ratio: 81.75%; solar fraction: 34.09%; fixed constant load: 2 kW. The conclusion could be drawn as the suggested anti-islanding protection has a short detection time, lower voltage stress and is simple to install, it may be used to detect islands in solar systems under a variety of conditions.

REFERENCES

Abbey, C., Brissette, Y., & Venne, P. (2013). An autoground system for anti-islanding protection of distributed generation. *IEEE Transactions on Power Systems*, *29*(2), 873–880. doi:10.1109/TPWRS.2013.2284670

Baidya, S., & Nandi, C. (2022). A comprehensive review on DC Microgrid protection schemes. *Electric Power Systems Research*, *210*, 108051. doi:10.1016/j.epsr.2022.108051

Banu, I. V., Barkat, F., Istrate, M., Guerrero, J. M., Culea, G., Livinti, P., Motas, J. G., Neagu, B., & Andrioaia, D. (2023). Passive anti-Islanding protection for Three-Phase Grid-Connected photovoltaic power systems. *International Journal of Electrical Power & Energy Systems*, *148*, 108946. doi:10.1016/j.ijepes.2023.108946

Bayrak, G., & Kabalci, E. (2016). Implementation of a new remote islanding detection method for wind–solar hybrid power plants. *Renewable & Sustainable Energy Reviews*, *58*, 1–15. doi:10.1016/j.rser.2015.12.227

Chowdhury, S. P., Chowdhury, S., & Crossley, P. A. (2009). Islanding protection of active distribution networks with renewable distributed generators: A comprehensive survey. *Electric Power Systems Research*, *79*(6), 984–992. doi:10.1016/j.epsr.2008.12.012

Datta, A., Saha, D., Ray, A., & Das, P. (2014). Anti-islanding selection for grid-connected solar photovoltaic system applications: A MCDM based distance approach. *Solar Energy*, *110*, 519–532. doi:10.1016/j.solener.2014.09.042

El Azzaoui, M. (2021). Islanding detection method for distributed generation with wavelet based nuisance tripping suppression. *Electric Power Systems Research*, *199*, 107366. doi:10.1016/j.epsr.2021.107366

Elshrief, Y. A., Helmi, D. H., Abd-Elhaleem, S., Abozalam, B. A., & Asham, A. D. (2021). Fast and accurate islanding detection technique for microgrid connected to photovoltaic system. *Journal of Radiation Research and Applied Sciences*, *14*(1), 210–221. doi:10.1080/16878507.2021.1923913

Ferreira, R. R., Colorado, P. J., Grilo, A. P., Teixeira, J. C., & Santos, R. C. (2019). Method for identification of grid operating conditions for adaptive overcurrent protection during intentional islanding operation. *International Journal of Electrical Power & Energy Systems*, *105*, 632–641. doi:10.1016/j.ijepes.2018.09.004

Han, C. H., Ko, B. K., Song, S. Y., & Jang, G. (2017). Application of PV Inverter on distribution system with high penetration of PV. *Journal of International Council on Electrical Engineering*, *7*(1), 263–268. doi:10.1080/22348972.2017.1373726

Jia, K., Bi, T., Liu, B., Thomas, D., & Goodman, A. (2014). Advanced islanding detection utilized in distribution systems with DFIG. *International Journal of Electrical Power & Energy Systems*, *63*, 113–123. doi:10.1016/j.ijepes.2014.05.003

John, V., Ye, Z., & Kolwalkar, A. (2003). *Investigation of antiislanding protection of power converter based distributed generators using frequency domain analysis. In 2003 IEEE Power Engineering Society General Meeting*. IEEE.

Khosravi, H., Samet, H., & Tajdinian, M. (2021). Empirical mode decomposition based algorithm for islanding detection in micro-grids. *Electric Power Systems Research*, *201*, 107542. doi:10.1016/j.epsr.2021.107542

Mishra, M., Chandak, S., & Rout, P. K. (2019). Taxonomy of islanding detection techniques for distributed generation in microgrid. *Renewable Energy Focus*, *31*, 9–30. doi:10.1016/j.ref.2019.09.001

Mohamad, H., Mokhlis, H., & Ping, H. W. (2011). A review on islanding operation and control for distribution network connected with small hydro power plant. *Renewable & Sustainable Energy Reviews*, *15*(8), 3952–3962. doi:10.1016/j.rser.2011.06.010

Motter, D., & de Melo Vieira, J. C. (2018). The setting map methodology for adjusting the DG anti-islanding protection considering multiple events. *IEEE Transactions on Power Delivery*, *33*(6), 2755–2764. doi:10.1109/TPWRD.2018.2816400

Muljadi, E., Singh, M., & Gevorgian, V. (2013). PSCAD modules representing PV generator (No. NREL/TP-5500-58189). National Renewable Energy Lab.

Nale, R., Biswal, M., & Kishor, N. (2022). A passive communication based islanding detection technique for AC microgrid. *International Journal of Electrical Power & Energy Systems*, *137*, 107657. doi:10.1016/j.ijepes.2021.107657

Pinto, S. J., & Panda, G. (2015). Wavelet technique based islanding detection and improved repetitive current control for reliable operation of grid-connected PV systems. *International Journal of Electrical Power & Energy Systems*, *67*, 39–51. doi:10.1016/j.ijepes.2014.11.008

Saleh, S. A., Aljankawey, A. S., Meng, R., Meng, J., Chang, L., & Diduch, C. P. (2015). Apparent power-based anti-islanding protection for distributed cogeneration systems. *IEEE Transactions on Industry Applications*, *52*(1), 83–98. doi:10.1109/TIA.2015.2464307

Shaik, M., Shaik, A. G., & Yadav, S. K. (2022). Hilbert–Huang transform and decision tree based islanding and fault recognition in renewable energy penetrated distribution system. *Sustainable Energy. Grids and Networks*, *30*, 100606.

Shin, H., Chae, S. H., & Kim, E. H. (2020). Design of microgrid protection schemes using PSCAD/EMTDC and ETAP programs. *Energies*, *13*(21), 5784. doi:10.3390/en13215784

Singh, V., Fozdar, M., Singh, A. K., & Singh, S. (2020). Analysis of anti-islanding protection methods integrated in distributed generation. *Intelligent Computing Techniques for Smart Energy Systems Proceedings of ICTSES*, *2018*, 663–671.

Trindade, F. C., do Nascimento, K. V., & Vieira, J. C. (2013). Investigation on voltage sags caused by DG anti-islanding protection. *IEEE Transactions on Power Delivery*, *28*(2), 972–980. doi:10.1109/TPWRD.2012.2237419

Vyas, S., Kumar, R., & Kavasseri, R. (2017). Data analytics and computational methods for anti-islanding of renewable energy based distributed generators in power grids. *Renewable & Sustainable Energy Reviews*, *69*, 493–502. doi:10.1016/j.rser.2016.11.116

Wang, H., Wang, M., Cheng, Q., Lv, S., & Ji, X. (2022). Modeling simulation and inverter control strategy research of microgrid in grid-connected and island mode. *Energy Reports*, *8*, 206–218. doi:10.1016/j.egyr.2022.09.117

Xiao, H. F., Fang, Z., Xu, D., Venkatesh, B., & Singh, B. (2017). Anti-islanding protection relay for medium voltage feeder with multiple distributed generators. *IEEE Transactions on Industrial Electronics*, *64*(10), 7874–7885. doi:10.1109/TIE.2017.2694394

Xu, M., Melnik, R. V., & Borup, U. (2004). Modeling anti-islanding protection devices for photovoltaic systems. *Renewable Energy*, *29*(15), 2195–2216. doi:10.1016/j.renene.2004.04.005

Zhu, K., Wang, Y., Lu, X., & Vieira, J. C. M. (2017). A new DG anti-islanding protection technique based on harmonic impedance measurement. *International Transactions on Electrical Energy Systems*, *27*(2), e2251. doi:10.1002/etep.2251

Chapter 9
Need of 21st Century: A Sustainable Environment Together With a Smart Environment

Shrikant Tiwari
https://orcid.org/0000-0001-6947-2362
Galgotias University, Greater Noida, India

Chidambaram N.
https://orcid.org/0009-0007-1767-385X
CHRIST University, Bangalore, India

Amit Kumar Tyagi
https://orcid.org/0000-0003-2657-8700
National Institute of Fashion Technology, New Delhi, India

ABSTRACT

In this 21st century, the need for a sustainable environment is becoming increasingly essential. This work discusses the importance of adopting sustainable practices and implementing smart technologies to create a useful and eco-friendly living environment. This chapter explains the important need for sustainable practices to mitigate these challenges and ensure a livable future. Smart environment concepts and technologies are then discussed as a means to achieve sustainability. This work discusses the integration of advanced technologies, such as the internet of things (IoT), artificial intelligence (AI), and data analytics, into various aspects of environmental management. Examples include smart energy grids, intelligent waste management systems, and eco-friendly transportation solutions. These technologies enable efficient resource allocation, real-time monitoring, and data-driven decision-making, fostering a sustainable and resilient environment. This work discusses the benefits of adopting a smart environment approach.

DOI: 10.4018/978-1-6684-9130-0.ch009

1. INTRODUCTION

1.1 Background and Importance of Sustainable Environment

A sustainable environment refers to the practice of using resources in a way that meets the needs of the present generation without compromising the ability of future generations to meet their own needs. It involves maintaining the natural balance of ecosystems, conserving resources, and minimizing environmental damage (Rosenzweig et al., 2018). The background and importance of a sustainable environment can be understood from several perspectives:

- Ecological Perspective: The Earth is a complex system where various species and ecosystems are interconnected. A sustainable environment ensures the preservation of biodiversity, protecting habitats, and preventing the extinction of species. It recognizes the intrinsic value of nature and the importance of maintaining a healthy and balanced ecosystem for the well-being of all life forms.
- Human Health and Well-being: A sustainable environment directly impacts human health and well-being. It promotes access to clean air, safe drinking water, and healthy food, which are essential for a good quality of life. By reducing pollution and exposure to harmful substances, a sustainable environment helps prevent diseases and improves overall public health.
- Climate Change Mitigation: Sustainable practices play an essential role in mitigating climate change. By reducing greenhouse gas emissions and promoting renewable energy sources, we can limit global warming and its adverse effects. Sustainable development also involves adapting to the impacts of climate change, such as extreme weather events, sea-level rise, and shifts in ecosystems.
- Resource Conservation: A sustainable environment aims to conserve natural resources such as water, forests, minerals, and energy sources. It encourages responsible consumption and production patterns, reducing waste generation and promoting recycling and reuse. By managing resources efficiently, we can ensure their availability for future generations and reduce our dependence on non-renewable resources.
- Economic Prosperity: Sustainability and economic prosperity are not mutually exclusive. In fact, they are closely intertwined. Sustainable practices foster innovation, create green jobs, and stimulate economic growth in sectors such as renewable energy, clean technology, and eco-tourism. Investments in sustainable infrastructure and practices can lead to long-term economic benefits and resilience.
- Social Equity: A sustainable environment promotes social equity by addressing environmental injustices and ensuring that everyone has equal access to resources and opportunities. It emphasizes the participation and involvement of all consumers/ users, including marginalized communities, in decision-making processes related to environmental policies and projects.
- Global Responsibility: The importance of a sustainable environment goes beyond national boundaries. Issues like deforestation, pollution, and climate change have global implications, requiring international cooperation and collective action. Sustainable development goals, such as those outlined in the United Nations' Agenda 2030, emphasize the need for global partnerships to achieve a sustainable future for all.

In summary, a sustainable environment is important for the well-being of both present and future generations. It encompasses ecological preservation, human health, climate change mitigation, resource conservation, economic prosperity, social equity, and global responsibility. By adopting sustainable practices and policies, we can ensure a healthier planet and a more sustainable future for all.

1.2 Purpose of the Research Towards a Better Sustainable and Smart Environment

The purpose of research towards a better sustainable and smart environment is to generate knowledge, information, and solutions that can contribute to the development and implementation of sustainable practices and technologies. Such research plays an important role in addressing the complex environmental challenges we face and creating a more sustainable and resilient future (Neumann & Emanuel, 2018; Sassen, 2019). Here are some key purposes and objectives of research in this field:

- Understanding Environmental Issues: Research helps in deepening our understanding of various environmental issues, such as climate change, biodiversity loss, pollution, and resource depletion. It allows scientists and experts to analyze the causes, impacts, and interconnections of these challenges, providing a base for effective problem-solving.
- Developing Sustainable Technologies and Innovations: Research enables the development of new technologies, processes, and innovations that support sustainability. It involves exploring alternative energy sources, improving energy efficiency, developing eco-friendly materials, enhancing waste management practices, and promoting circular economy principles. By advancing sustainable technologies, research contributes to reducing environmental impacts and promoting resource conservation.
- Assessing Environmental Impacts: Research helps in assessing and quantifying the environmental impacts of human activities, policies, and technologies. Life cycle assessments, environmental impact assessments, and ecological modeling are some of the research methods used to evaluate the environmental consequences of various actions. This information is essential for making informed decisions and designing strategies that minimize negative impacts.
- Policy Development and Implementation: Research provides evidence and information that inform the development and implementation of environmental policies and regulations. It helps policymakers understand the consequences of different policy options and supports the design of effective strategies for environmental protection and sustainability. Research findings can also help in monitoring and evaluating policy outcomes and adjusting approaches as needed.
- Promoting Behavioral Change: Research plays a role in understanding human behavior and decision-making processes related to environmental issues. It investigates factors that influence individual and collective behaviors and provide ways to promote sustainable practices and behaviors. Research findings can be used to develop educational campaigns, behavioral interventions, and policy incentives that encourage sustainable choices and lifestyles.
- Enhancing Resilience and Adaptation: Research contributes to understanding the impacts of climate change and other environmental stressors on ecosystems and human societies. It helps identify strategies for building resilience and adapting to changing conditions, such as developing climate change adaptation plans, improving disaster preparedness, and enhancing ecosystem-based

approaches to resilience. This research is important for ensuring the long-term sustainability and well-being of communities.

- Stakeholder Engagement and Collaboration: Research facilitates collaboration among stakeholders, including scientists, policymakers, industry leaders, community members, and non-governmental organizations. It provides a platform for dialogue, knowledge sharing, and participatory decision-making processes.

In summary, the purpose of research towards a better sustainable and smart environment is to generate knowledge and solutions that can drive positive change. By advancing our understanding, developing innovative technologies, informing policies, promoting behavior change, and fostering collaboration, research can contribute to building a more sustainable, resilient, and equitable future for our planet.

1.3 Organization of the Work

This chapter is summarized in ten sections.

2. ENVIRONMENTAL CHALLENGES IN THE 21ST CENTURY

2.1 Climate Change and Global Warming

Climate change and global warming are two interconnected phenomena that have become major issues in the 21st century (IPCC, 2018; Sassen, 2019). Here's an overview of their significance and impact:

- Rising Temperatures: Global warming refers to the long-term increase in Earth's average surface temperature, primarily caused by the buildup of greenhouse gases in the atmosphere. Over the past century, the Earth's temperature has risen significantly, with the majority of the warming occurring in the last few decades. This rise in temperatures has far-reaching consequences for ecosystems, weather patterns, and human societies.
- Greenhouse Gas Emissions: The primary driver of global warming is the emission of greenhouse gases (GHGs) into the atmosphere, mainly carbon dioxide (CO2) from burning fossil fuels and deforestation. Other GHGs, such as methane (CH4) and nitrous oxide (N2O), also contribute to warming. These gases trap heat in the atmosphere, leading to the greenhouse effect and a subsequent increase in global temperatures.
- Extreme Weather Events: Climate change intensifies the frequency and severity of extreme weather events, including heatwaves, droughts, hurricanes, floods, and wildfires. Warmer temperatures provide more energy to fuel these events, leading to greater destruction and risks to human lives, infrastructure, and ecosystems. Changes in precipitation patterns can also result in water scarcity or excessive rainfall, further impacting agriculture, water resources, and vulnerable communities.
- Sea-Level Rise: As global temperatures rise, ice sheets and glaciers melt, contributing to sea-level rise. This threatens coastal areas and low-lying islands, increasing the risk of flooding, coastal erosion, and salinization of freshwater sources. It also endangers ecosystems like coral reefs and coastal wetlands, which provide important habitats and act as natural buffers against storms.

- Biodiversity Loss: Climate change poses a significant threat to global biodiversity. Rising temperatures and altered precipitation patterns disrupt ecosystems, affecting the distribution and behavior of species. Many species are facing habitat loss, reduced reproductive success, and increased vulnerability to disease and invasive species. These impacts have consequences for ecological balance, ecosystem services, and the survival of several plant and animal species.
- Human Health Impacts: Climate change has direct and indirect impacts on human health. Heatwaves and increased temperatures can lead to heat-related illnesses and deaths. Changes in disease vectors and the spread of infectious diseases are influenced by climate conditions. Additionally, extreme weather events, displacement, and food and water insecurity resulting from climate change can cause physical and mental health issues, particularly among vulnerable populations.
- Socioeconomic Challenges: Climate change has wide-ranging socioeconomic implications. Disruptions in agriculture and food production, water scarcity, and natural disasters can lead to economic losses, food insecurity, and increased migration. Industries reliant on natural resources, such as tourism and fisheries, are vulnerable to climate impacts. Addressing climate change also requires significant investments in renewable energy, infrastructure, and sustainable practices, creating both economic challenges and opportunities.
- Global Cooperation and Mitigation Efforts: Addressing climate change requires global cooperation and mitigation efforts. The Paris Agreement, an international treaty signed in 2015, aims to limit global warming well below 2 degrees Celsius above pre-industrial levels and pursue efforts to limit the temperature increase to 1.5 degrees Celsius. This agreement emphasizes the reduction of greenhouse gas emissions, adaptation measures, and financial support for developing nations.

In summary, climate change and global warming present urgent and complex challenges that require immediate action. Mitigating the impacts of climate change, transitioning to a low-carbon economy, promoting sustainable practices, and investing in adaptation measures are essential for safeguarding the planet's ecosystems, human well-being, and future generations

2.2 Resource Depletion and Environmental Degradation

Resource depletion and environmental degradation have been significant issues in the 21st century, driven by population growth, unsustainable consumption patterns, and industrial activities (Rees & Wackernagel, 2013; Song & Zegeye, 2017). Here are some key aspects of these challenges:

- Deforestation: Large-scale deforestation continues to be a major cause of environmental degradation. Forests play an important role in carbon sequestration, biodiversity conservation, and regulating the water cycle. Deforestation for agriculture, logging, and urban expansion leads to habitat loss, soil erosion, increased greenhouse gas emissions, and the loss of important ecosystem services.
- Water Scarcity: Increasing demands for water, coupled with inefficient use and pollution, contribute to water scarcity in many regions. Over-extraction of groundwater for irrigation and industrial purposes depletes aquifers, leading to the sinking of land (subsidence) and the loss of freshwater resources. Pollution from industrial activities, agriculture, and inadequate sanitation also compromises water quality, posing risks to human health and ecosystems.

- Loss of Biodiversity: Human activities, such as habitat destruction, pollution, overexploitation, and climate change, have caused a rapid decline in biodiversity. Species extinction rates are now several orders of magnitude higher than natural background rates. The loss of biodiversity disrupts ecosystem functioning, reduces resilience, and diminishes the provision of important ecosystem services, such as pollination, nutrient cycling, and natural pest control.
- Soil Degradation: Unsustainable agricultural practices, deforestation, and urbanization contribute to soil degradation. Soil erosion, nutrient depletion, and contamination reduce agricultural productivity and can lead to desertification in vulnerable regions. Degraded soils are less capable of supporting plant growth, exacerbating food insecurity and threatening livelihoods.
- Air Pollution: Industrial activities, transportation, and energy generation contribute to air pollution, posing risks to human health and ecosystems. Emissions of pollutants such as particulate matter, nitrogen oxides, and sulfur dioxide contribute to respiratory illnesses, cardiovascular problems, and premature deaths. Additionally, air pollution impacts the quality of soil, water, and vegetation, affecting agricultural productivity and ecosystem health.
- Overfishing and Marine Degradation: Overfishing, destructive fishing practices, and pollution pose significant threats to marine ecosystems. Unsustainable fishing practices lead to the depletion of fish stocks, disrupt food webs, and impact the livelihoods of coastal communities. Pollution from land-based activities, such as agricultural runoff and industrial waste, contributes to marine pollution, including harmful algal blooms, dead zones, and the accumulation of plastic waste.
- Mineral Resource Depletion: The increasing demand for minerals, metals, and rare earth elements for industrial processes, electronics, and infrastructure has led to resource depletion. Mining activities have significant environmental impacts, including habitat destruction, soil erosion, water pollution, and the release of greenhouse gases. Recycling and responsible mining practices are essential to reduce the environmental and social impacts associated with mineral extraction.

Hence, addressing resource depletion and environmental degradation requires a shift towards sustainable practices and policies. This includes promoting sustainable land management, reforestation efforts, sustainable agriculture, efficient water use, waste reduction and recycling, renewable energy adoption, and the protection of biodiversity-rich areas. Transitioning to a circular economy, where resources are conserved, reused, and recycled, is also important for reducing the strain on natural resources and minimizing environmental impacts.

2.3 Pollution and Its Impact on Ecosystems and Human Health

Pollution is a widespread environmental issue in the 21st century, resulting from various human activities (Rocha, 2018). It poses significant threats to ecosystems and human health. Here are some key aspects of pollution and its impacts:

- Air Pollution: The release of pollutants into the air from industrial activities, transportation, and energy generation leads to air pollution. Particulate matter, nitrogen oxides, sulfur dioxide, and volatile organic compounds are among the common air pollutants. Prolonged exposure to air pollution can lead to respiratory problems, cardiovascular diseases, allergies, and lung cancer. It also damages vegetation, contributes to climate change, and affects the quality of ecosystems.

- Water Pollution: Pollution of water bodies, including rivers, lakes, and oceans, occurs due to the discharge of industrial waste, agricultural runoff, sewage, and improper waste disposal. Water pollution has detrimental effects on aquatic life, leading to the decline of fish populations, disruption of ecosystems, and harmful algal blooms. Contaminated water sources pose risks to human health, causing waterborne diseases and long-term exposure to toxic substances.
- Soil Pollution: Soil pollution occurs when chemicals, heavy metals, or hazardous substances are deposited into the soil through industrial activities, improper waste disposal, and agricultural practices. Contaminated soil affects the quality of agricultural land, reduces crop productivity, and can lead to the accumulation of pollutants in the food chain. Soil pollution also impacts soil biodiversity and microorganisms, affecting soil health and ecosystem functioning.
- Noise Pollution: Noise pollution, primarily generated by transportation, industrial activities, and urbanization, can have adverse effects on human health and ecosystems. Prolonged exposure to high noise levels can lead to stress, hearing loss, sleep disturbances, and impaired cognitive function. Noise pollution can also disrupt animal communication, migration patterns, and breeding behaviors, impacting wildlife populations and ecological balance.
- Light Pollution: The excessive and misdirected artificial lighting in urban areas contributes to light pollution. It affects nocturnal wildlife, disrupting their behavior, migration, and reproductive patterns. Light pollution also interferes with ecosystems' natural cycles, such as disrupting the breeding patterns of certain species and altering the natural darkness necessary for proper functioning. Moreover, it can have implications for human health, including disruptions to sleep patterns and the disturbance of circadian rhythms.
- Chemical Pollution: Chemical pollution encompasses the release of harmful substances, such as pesticides, heavy metals, pharmaceuticals, and industrial chemicals, into the environment. These pollutants can accumulate in ecosystems, affecting biodiversity and ecosystem functions. They can also enter the food chain and pose risks to human health, causing developmental issues, hormonal imbalances, and chronic diseases.
- Climate Change: Pollution, particularly the emission of greenhouse gases, contributes to climate change. The resulting global warming affects ecosystems and human health in several ways. It leads to rising sea levels, increased frequency and intensity of extreme weather events, altered precipitation patterns, and shifts in ecosystems. These changes pose risks to biodiversity, agriculture, water resources, and human livelihoods.

Hence, pollution requires collective efforts, including strict regulations, technological advancements, and behavioral changes. Implementing pollution control measures, transitioning to cleaner and sustainable energy sources, promoting recycling and waste management, and adopting environmentally friendly practices are important for mitigating the impacts of pollution on ecosystems and human health in the 21st century.

3. SUSTAINABLE PRACTICES FOR A BETTER FUTURE

3.1 Concept of Sustainability: Now and Before

The concept of sustainability has evolved over time, with a growing recognition of the need to balance economic development, social well-being, and environmental stewardship (Batty, 2018). Here's a comparison of the concept of sustainability in the past and its current understanding:

Before:

- Limited Environmental Focus: In the past, the emphasis was primarily on economic growth and development, with little consideration for the environmental consequences. The exploitation of natural resources and pollution were often seen as necessary byproducts of industrialization and progress.
- Short-Term Perspective: The focus was often on immediate gains and benefits without considering the long-term impacts. Natural resources were considered limitless, and little thought was given to the finite nature of resources or the potential for irreversible damage to ecosystems.
- Sectoral Approach: Sustainability was often seen as an issue for environmentalists or specific sectors such as conservation or renewable energy. It was not widely integrated into broader decision-making processes or across different sectors of society.

Now:

- Holistic Approach: The understanding of sustainability now recognizes the interconnections between social, economic, and environmental factors. It emphasizes the need for a holistic approach that considers the long-term well-being of people and the planet.
- Environmental Stewardship: There is a greater understanding of the importance of protecting and preserving the natural environment. Sustainability now includes the promotion of environmental stewardship, conservation of biodiversity, and mitigation of pollution and climate change.
- Long-Term Perspective: The concept of sustainability now recognizes the need to consider the long-term impacts of actions and decisions. This includes assessing the potential environmental, social, and economic consequences of activities and striving for solutions that meet the needs of the present without compromising the ability of future generations to meet their own needs.
- Integration and Collaboration: Sustainability is now seen as a cross-cutting issue that requires collaboration and integration across different sectors, including government, business, academia, and civil society. It involves incorporating sustainability principles and practices into various aspects of decision-making, policy development, and planning processes.
- Social Equity and Well-being: The current understanding of sustainability emphasizes the importance of social equity, inclusivity, and ensuring a high quality of life for all people. It recognizes that sustainable development should address social issues such as poverty, inequality, and access to basic needs like education, healthcare, and clean water.
- Systems Thinking: The concept of sustainability now embraces systems thinking, recognizing that social, economic, and environmental systems are interconnected and interdependent. This approach involves considering the broader impacts and feedback loops of actions, as well as identifying and addressing root causes rather than just symptoms.

In summary, the concept of sustainability has evolved from a narrow focus on economic growth to a clearer approach that integrates environmental, social, and economic dimensions. It emphasizes the need for long-term thinking, collaboration, and holistic decision-making to ensure a sustainable and resilient future for generations to come.

3.2 Sustainable Development Goals (SDGs) for Today's Generation

The Sustainable Development Goals (SDGs) are a set of 17 interconnected goals (refer figure 1) adopted by the United Nations in 2015 (Angel et al., 2011; Batty, 2018; IPCC, 2018; Neumann & Emanuel, 2018; Pickett et al., 2018; Rees & Wackernagel, 2013; Rocha, 2018; Rosenzweig et al., 2018; Sassen, 2019; Song & Zegeye, 2017). They provide a framework for global development efforts and aim to address key social, economic, and environmental challenges. Here are the SDGs that are particularly relevant for today's generation:

- No Poverty (SDG 1): End poverty in all its forms and dimensions, including extreme poverty, by implementing social protection systems and promoting sustainable economic growth.
- Zero Hunger (SDG 2): Achieve food security, improve nutrition, and promote sustainable agriculture to ensure everyone has access to safe, nutritious, and sufficient food.
- Good Health and Well-being (SDG 3): Ensure healthy lives and promote well-being for all at all ages, focusing on areas such as reducing child and maternal mortality, combating communicable diseases, and strengthening healthcare systems.
- Quality Education (SDG 4): Ensure inclusive and equitable quality education and promote lifelong learning opportunities for all, aiming to provide access to affordable and quality education from early childhood to tertiary level.
- Gender Equality (SDG 5): Achieve gender equality and empower all women and girls by eliminating discrimination, violence, and harmful practices and ensuring equal opportunities in all areas of life.
- Clean Water and Sanitation (SDG 6): Ensure availability and sustainable management of water and sanitation for all, aiming to provide access to safe drinking water, sanitation facilities, and proper hygiene practices.
- Affordable and Clean Energy (SDG 7): Ensure access to affordable, reliable, sustainable, and modern energy for all, while promoting renewable energy sources and increasing energy efficiency.
- Decent Work and Economic Growth (SDG 8): Promote sustained, inclusive, and sustainable economic growth, full and productive employment, and decent work for all, aiming to eradicate forced labor and child labor.
- Industry, Innovation, and Infrastructure (SDG 9): Build resilient infrastructure, promote inclusive and sustainable industrialization, and foster innovation to support sustainable development and enhance connectivity.
- Reduced Inequalities (SDG 10): Reduce inequalities within and among countries, addressing income inequality, social inclusion, and equal opportunities for all, irrespective of age, gender, disability, race, ethnicity, or economic status.
- Sustainable Cities and Communities (SDG 11): Make cities and human settlements inclusive, safe, resilient, and sustainable, focusing on issues such as affordable housing, sustainable transport, and access to green spaces.

- Responsible Consumption and Production (SDG 12): Ensure sustainable consumption and production patterns, promoting resource efficiency, reducing waste, and minimizing the environmental impact of consumption and production processes.
- Climate Action (SDG 13): Take urgent action to combat climate change and its impacts, implementing policies and practices to reduce greenhouse gas emissions, enhance climate resilience, and promote sustainable development.
- Life Below Water (SDG 14): Conserve and sustainably use the oceans, seas, and marine resources for sustainable development, addressing issues such as marine pollution, overfishing, and ocean acidification.
- Life on Land (SDG 15): Protect, restore, and sustainably manage terrestrial ecosystems, promoting the conservation of biodiversity, combating desertification, and halting deforestation.
- Peace, Justice, and Strong Institutions (SDG 16): Promote peaceful and inclusive societies for sustainable development, provide access to justice, and build effective, accountable, and inclusive institutions at all levels.

Figure 1. SDG goals

- Partnerships for the Goals (SDG 17): Strengthen the means of implementation and revitalize the global partnership for sustainable development, fostering collaboration among governments, civil society, and the private sector to achieve the SDGs.

4. SMART ENVIRONMENT: INTEGRATING TECHNOLOGY FOR SUSTAINABILITY

4.1 Role of Internet of Things (IoT) in Smart Environment

In this smart era, Internet of Things (IoT) plays an essential role in enabling and enhancing the functionality of smart environments. IoT refers to the network of interconnected physical devices embedded with sensors, software, and connectivity that enables them to collect and exchange data. Here's how IoT contributes to the development and operation of smart environments:

- Connectivity and Data Collection: IoT enables devices and sensors within a smart environment to connect and communicate with each other, creating a network of data collection points. These devices can collect real-time data on various parameters such as temperature, humidity, occupancy, energy consumption, air quality, and more. The data collected provides important information for optimizing processes, resource management, and decision-making.
- Automation and Control: IoT facilitates automation and control in smart environments. Devices and systems can be interconnected, allowing for automated responses and actions based on the collected data. For example, in a smart home, IoT devices can automatically adjust lighting and temperature settings based on occupancy or user preferences. In a smart building, IoT systems can optimize energy usage by controlling lighting, HVAC systems, and other equipment based on occupancy patterns and real-time data.
- Monitoring and Maintenance: IoT enables continuous monitoring of various aspects within a smart environment. Sensors and devices can track performance, detect anomalies, and identify maintenance needs in real-time. For example, in a smart grid system, IoT sensors can monitor power distribution, detect faults, and trigger maintenance actions to ensure efficient and reliable operation. Predictive maintenance can be implemented using IoT data analytics to identify equipment failures before they occur, reducing downtime and costs.
- Energy Efficiency and Sustainability: IoT plays a significant role in improving energy efficiency and promoting sustainability in smart environments. Connected devices and sensors can monitor energy usage, identify inefficiencies, and optimize energy consumption based on real-time data. IoT enables smart energy management systems to control and adjust lighting, heating, cooling, and other energy-consuming devices, leading to reduced energy waste and lower environmental impact.
- Enhanced User Experience: IoT enhances the user experience within smart environments by providing personalized and adaptive services. Smart devices and applications can learn user preferences, adapt settings accordingly, and provide customized experiences. For example, in a smart home, IoT-enabled voice assistants can control various devices, such as lighting, entertainment systems, and security, providing convenience and seamless integration for users.
- Safety and Security: IoT strengthens safety and security measures within smart environments. Connected sensors and cameras can monitor and detect potential hazards, such as fire, gas leaks, or unauthorized access. IoT-enabled security systems can send alerts, trigger alarms, and notify authorities or occupants in case of emergencies. Real-time data analytics and predictive algorithms can help identify and prevent security breaches or anomalies, enhancing overall safety.

- Decision-Making and Optimization: The large amount of data collected through IoT devices and sensors in smart environments provides important information for decision-making and optimization. Data analytics, combined with AI and machine learning techniques, can analyze patterns, identify trends, and optimize processes. This allows for data-driven decision-making, resource allocation, and continuous improvement of operations within the smart environment.

Hence, IoT acts as the backbone of connectivity, data collection, and automation in smart environments. It enables the seamless integration of various devices, systems, and applications, transforming physical spaces into intelligent and responsive environments that enhance efficiency, sustainability, safety, and user experience.

4.3 Artificial Intelligence (AI) and Machine Learning (ML) for Environmental Management

Artificial Intelligence (AI) and Machine Learning (ML) have significant potential for transforming environmental management practices. They can help analyze large amount of data, identify patterns and trends, make predictions, and support decision-making processes. Here's how AI and ML can be applied in environmental management:

- Environmental Monitoring and Data Analysis: AI and ML algorithms can process large volumes of environmental data collected from sensors, satellites, and other sources. They can identify patterns and anomalies in the data to monitor environmental conditions, such as air quality, water quality, biodiversity, and deforestation rates. This data analysis can provide important information for understanding environmental trends and informing decision-making processes.
- Predictive Modeling and Risk Assessment: AI and ML techniques can be used to develop predictive models for environmental changes and associated risks. For example, they can predict the impact of climate change on ecosystems, assess the likelihood of natural disasters, or forecast air pollution levels. These predictive models help in proactive planning, risk assessment, and designing appropriate mitigation strategies.
- Environmental Impact Assessment: AI and ML can support the assessment of environmental impacts of various projects and activities. They can analyze project data, historical records, and environmental parameters to evaluate potential impacts on ecosystems, habitats, and natural resources. This enables better decision-making and helps in identifying measures to mitigate or minimize adverse effects.
- Ecosystem Management and Conservation: AI and ML techniques can aid in ecosystem management and conservation efforts. They can analyze complex ecological data, identify species distribution patterns, detect habitat degradation, and assess biodiversity levels. This information can guide conservation strategies, including habitat restoration, species protection, and identifying areas of high conservation value.
- Natural Resource Management: AI and ML can optimize the management of natural resources, such as water and energy. They can analyze usage patterns, identify inefficiencies, and develop models for resource allocation and optimization. For example, ML algorithms can predict water demand, optimize irrigation systems, and suggest energy-saving measures in buildings.

- Wildlife Protection and Anti-Poaching Efforts: AI and ML can contribute to wildlife protection and anti-poaching efforts. They can analyze data from camera traps, drones, and other sources to detect and track wildlife, identify illegal activities, and detect poaching threats. This can assist in real-time monitoring, early warning systems, and effective enforcement strategies.
- Sustainable Agriculture and Precision Farming: AI and ML techniques can support sustainable agriculture practices by optimizing crop management, irrigation scheduling, and pest control. They can analyze soil data, weather patterns, and crop health indicators to provide personalized recommendations for farmers. This helps reduce resource waste, increase yields, and minimize environmental impact.
- Environmental Compliance and Regulation: AI and ML can assist in monitoring compliance with environmental regulations and detecting violations. They can analyze large amounts of data, such as industry emissions, pollution levels, and regulatory records, to identify non-compliance or potential environmental risks. This enables more effective enforcement and regulatory decision-making.
- Decision Support Systems: AI and ML can be integrated into decision support systems for environmental management. They can analyze multiple variables, scenarios, and trade-offs to provide information and recommendations for complex decision-making processes. This enhances the effectiveness and efficiency of decision-making, particularly in areas where multiple factors and uncertainties need to be considered.

Note that AI and ML technologies hold great promise for addressing environmental challenges and enabling more effective and sustainable environmental management practices. However, it is important to ensure responsible and ethical use of these technologies, considering privacy, bias, and the need for human oversight in decision-making processes.

4.2 Data Analytics and Big Data in Sustainable Decision-Making

Data analytics and big data play an essential role in sustainable decision-making by providing information and actionable information for addressing environmental, social, and economic challenges. Here's how data analytics and big data contribute to sustainable decision-making:

- Data-driven Information: Data analytics techniques enable the extraction of important information from large and complex datasets. By analyzing environmental, social, and economic data, organizations can gain a deeper understanding of the underlying trends, patterns, and relationships. This information help inform decision-making processes and enable evidence-based strategies for sustainable development.
- Identifying Key Performance Indicators (KPIs): Big data analytics allows organizations to identify and measure key performance indicators related to sustainability. By collecting and analyzing relevant data, organizations can track their progress towards sustainability goals and objectives. This enables better monitoring, evaluation, and reporting of sustainability performance.
- Predictive Analytics: Predictive analytics uses historical data and statistical models to make predictions about future events and trends (Agarwal & Tripathi, 2022; Alfen & Ten Brink, 2018; Bai & Imura, 2016; Coaffee et al., 2018; Giffinger et al., 2007; Glaeser, 2011; Dutta et al., 2023; Midha et al., 2017; Midha & Tripathi, 2019; Hura et al., 2021; Midha et al., 2021; Ranchhodbhai

& Tripathi, 2019; Sai, Tripathi, & Tyagi, 2023; Arunachalam et al., 2022; Somisetti et al., 2020; Subasree & Sakthivel, 2022; Tulumello & Hernández-Molina, 2018; Tyagi, Kukreja, Nair et al, 2022; Varsha et al., 2021; Vishnuram et al., 2022). In the context of sustainability, predictive analytics can forecast environmental impacts, identify emerging risks, and project the outcomes of different interventions. This helps decision-makers anticipate challenges and make proactive choices to achieve long-term sustainability.

- Resource Optimization: Big data analytics helps optimize the use of resources by identifying inefficiencies and opportunities for improvement. By analyzing data on resource consumption, such as energy usage, water usage, or waste generation, organizations can identify areas where conservation measures can be implemented, leading to reduced costs and environmental impact.
- Risk Assessment and Management: Big data analytics enables organizations to assess and manage risks associated with sustainability. By integrating various datasets, including environmental data, social indicators, and economic factors, organizations can identify potential risks, vulnerabilities, and impacts. This helps in developing risk management strategies and implementing measures to mitigate adverse effects.
- Real-time Monitoring and Feedback: Big data analytics enables real-time monitoring of sustainability-related metrics. By collecting and analyzing data in real-time or near real-time, organizations can track performance, identify deviations, and take immediate corrective actions. This allows for more agile and responsive decision-making, especially in dynamic and rapidly changing environments.
- Lifecycle Analysis and Circular Economy: Data analytics facilitates lifecycle analysis, which assesses the environmental impact of products and processes throughout their lifecycle. By analyzing data on raw material sourcing, production, use, and disposal, organizations can identify opportunities for resource efficiency, waste reduction, and the adoption of circular economy practices.
- Decision Support Systems: Data analytics and big data can be integrated into decision support systems to provide decision-makers with relevant information and recommendations. These systems use algorithms and models to process data, analyze scenarios, and generate information for decision-making. By adding these systems, organizations can make more informed and sustainable decisions.

In summary, data analytics and big data play an essential role in sustainable decision-making by providing information, predictive capabilities, and evidence-based strategies. By adding the power of data, organizations can optimize resource use, assess risks, engage consumers, and monitor progress towards sustainability goals.

5. BENEFITS OF A SMART ENVIRONMENT

A smart environment provides a wide range of benefits across various domains, including urban planning, energy management, healthcare, transportation, and more. These benefits are driven by the integration of advanced technologies, data analytics, and connectivity to create more efficient, sustainable, and responsive ecosystems. Here are some key benefits of a smart environment:

A. Sustainability: In this regard, we will have the following benefits;

- Resource Efficiency: Smart environments optimize the use of resources, such as energy, water, and materials, reducing waste and environmental impact.
- Renewable Energy Integration: Smart grids and energy management systems enable the efficient integration of renewable energy sources, reducing reliance on fossil fuels and lowering greenhouse gas emissions.

B. Efficient Urban Planning: In this regard, we will have the following benefits;

- Traffic Management: Smart transportation systems improve traffic flow, reduce congestion, and minimize commuting times, leading to fuel savings and reduced air pollution.
- Infrastructure Optimization: Smart cities use data to optimize the allocation of resources for infrastructure maintenance and development, reducing costs and improving service delivery.

C. Enhanced Public Services: In this regard, we will have the following benefits;

- Healthcare: Remote monitoring and telehealth services in smart healthcare environments provide timely care, reduce hospital admissions, and improve patient outcomes.
- Education: Smart classrooms and digital learning tools enhance the quality of education by personalizing learning experiences and promoting engagement.

D. Improved Safety and Security: In this regard, we will have the following benefits;

- Public Safety: Smart surveillance, emergency response systems, and disaster management tools enhance public safety and reduce response times during emergencies.
- Cybersecurity: Advanced cybersecurity measures protect important infrastructure and sensitive data from cyber threats.

E. Quality of Life: In this regard, we will have the following benefits;

- Health and Wellness: Smart homes support health and wellness by monitoring health parameters, providing medication reminders, and assisting with daily tasks for the elderly and individuals with disabilities.
- Convenience: Smart devices and automation simplify daily routines and enhance convenience, from voice-activated assistants to home automation systems.

F. Economic Benefits: In this regard, we will have the following benefits;

- Job Creation: The development and implementation of smart technologies create job opportunities in various sectors, including technology, data analytics, and urban planning.
- Economic Growth: Efficiency gains and cost savings in smart environments can stimulate economic growth and innovation.

G. Environmental Conservation: In this regard, we will have the following benefits;

- Biodiversity Preservation: Smart environmental monitoring helps protect and conserve ecosystems and wildlife habitats.
- Waste Reduction: Smart waste management systems optimize collection routes, reduce litter, and promote recycling.

H. Data-Driven Decision-Making: In this regard, we will have the following benefits;

- Informed Policies: Data analytics in smart environments provide insights for policymakers to make informed decisions and allocate resources effectively.
- Continuous Improvement: Data-driven feedback loops enable continuous improvement in various systems, from transportation to healthcare.

I. Resilience and Disaster Preparedness: In this regard, we will have the following benefits;

- Resilience: Smart technologies enhance a community's ability to withstand and recover from natural disasters and emergencies.
- Early Warning Systems: Advanced sensors and monitoring systems provide early warnings for events like floods, earthquakes, and wildfires.

Hence, a smart environment provides the potential to improve quality of life, reduce environmental impact, enhance public safety, and drive economic growth. These benefits make smart environments an essential strategy for addressing the complex challenges of the 21st century.

6. IMPORTANT CHALLENGES AND OPEN ISSUES IN IMPLEMENTING SMART ENVIRONMENT

Implementing a smart environment, where technology is harnessed to create sustainable, efficient, and responsive ecosystems, presents a range of important challenges and open issues. These challenges cut across various domains, from urban planning and energy management to data privacy and cybersecurity. Here are some of the key challenges and open issues in implementing a smart environment:

A. Data Privacy and Security: We will face following challenges and issues;

- Challenge: Collecting and managing large amount of data in smart environments raises significant issues about data privacy and security (Madhav & Tyagi, 2023; Nair & Tyagi, 2021; Sheth, 2022; Tyagi et al., 2021; Tyagi & Sreenath, 2021; Tyagi, Chandrasekaran, & Sreenath, 2022; Tyagi et al., 2020; Tyagi & Sreenath, 2023a; Tyagi & Sreenath, 2023b; Tyagi & Sreenath, 2023c). Unauthorized access or data breaches can have severe consequences.
- Open Issue: Developing robust data encryption, access control, and authentication mechanisms while complying with evolving privacy regulations.

B. Interoperability and Standards: We will face following challenges and issues;

- Challenge: The integration of diverse smart devices and systems from different manufacturers often results in interoperability issues. Lack of common standards can hinder seamless data exchange and cooperation between components.
- Open Issue: Establishing and adhering to open standards for smart environment components to ensure compatibility and facilitate scalability.

C. Sustainability and Resource Management: We will face following challenges and issues;

- Challenge: Smart environments should prioritize sustainability, but their energy and resource consumption can be significant, especially data centers and energy-intensive devices.
- Open Issue: Finding ways to optimize resource use and energy efficiency while maintaining the functionality and performance of smart systems.

D. Digital Divide and Inclusion: We will face following challenges and issues;

- Challenge: The benefits of smart environments should be accessible to all, but the digital divide can exacerbate disparities in access and participation.
- Open Issue: Ensuring equitable access to smart technologies and services, particularly in underserved or marginalized communities.

E. Scalability and Infrastructure: We will face following challenges and issues;

- Challenge: As smart environments grow, the demand on infrastructure and network capacity increases, potentially leading to bottlenecks and service interruptions.
- Open Issue: Designing scalable architectures that can accommodate the growth of smart technologies without sacrificing performance.

F. Environmental Impact: We will face following challenges and issues;

- Challenge: The production and disposal of smart devices and infrastructure can have environmental consequences, including e-waste and resource depletion.
- Open Issue: Developing sustainable materials, recycling programs, and circular economy practices for smart technology.

G. Public Acceptance and Trust: We will face following challenges and issues;

- Challenge: Public trust is important for the success of smart environments. Issues about surveillance, data misuse, or loss of privacy can hinder adoption.
- Open Issue: Fostering transparency, educating the public, and implementing robust ethical guidelines to build and maintain trust.

H. Regulatory and Legal Frameworks: We will face following challenges and issues;

- Challenge: The rapid development of smart technologies often outpaces regulatory and legal frameworks, creating legal uncertainties and challenges.
- Open Issue: Updating and adapting regulations to address the unique challenges posed by smart environments, including data governance and liability.

I. Community Engagement: We will face following challenges and issues;

- Challenge: Effective implementation of smart environments requires active engagement with local communities and consumers.
- Open Issue: Developing mechanisms for involving residents, businesses, and organizations in the planning, implementation, and evaluation of smart environment initiatives.

Hence, these important challenges and open issues require collaboration among governments, industries, academia, and civil society. So, a multidisciplinary and inclusive approach is essential to ensure that smart environments benefit society while addressing potential pitfalls and ethical issues.

7. ECONOMIC OPPORTUNITIES AND JOB CREATION

7.1 Green Economy and Sustainable Business Practices

The concept of a green economy involves transitioning economic systems and business practices towards sustainability, with a focus on minimizing environmental impact, promoting resource efficiency, and fostering social inclusiveness. Sustainable business practices are integral to achieving a green economy. Here are some key aspects of green economy and sustainable business practices:

- Environmental Stewardship: Sustainable businesses prioritize environmental stewardship by minimizing their ecological footprint. They adopt practices that reduce energy consumption, conserve water, minimize waste generation, and promote the use of renewable resources. This includes implementing energy-efficient technologies, adopting clean production processes, and investing in renewable energy sources.
- Circular Economy: Sustainable businesses embrace the principles of a circular economy, aiming to eliminate waste and maximize resource efficiency. They design products and services with a focus on reducing, reusing, and recycling materials throughout their life cycles. This includes implementing take-back programs, adopting eco-design principles, and promoting the use of recycled or biodegradable materials.
- Responsible Supply Chain Management: Sustainable businesses prioritize responsible supply chain management by considering the environmental and social impacts of their entire value chain. They engage suppliers that adhere to sustainable practices, promote fair trade, and ensure the responsible sourcing of raw materials. Supply chain transparency and traceability are important to identify and address potential environmental and social risks.
- Social Responsibility: Sustainable businesses recognize their role in promoting social well-being and inclusiveness. They prioritize fair labor practices, promote diversity and inclusion, and support local communities. This includes providing safe and healthy working conditions, fair wages,

and opportunities for skill development. Sustainable businesses also contribute to community development initiatives and engage in philanthropic activities.

- Green Innovation and Research: Sustainable businesses prioritize innovation and research to develop new technologies, products, and services that have a positive environmental and social impact. They invest in research and development efforts to improve resource efficiency, develop sustainable alternatives, and drive technological advancements that support the transition to a green economy.
- Measurement and Reporting: Sustainable businesses establish systems for measuring, monitoring, and reporting their environmental and social performance. They set goals and targets aligned with sustainability objectives, track progress, and disclose relevant information to consumers. Transparent reporting enables accountability, facilitates benchmarking, and encourages continuous improvement.
- Collaboration and Advocacy: Sustainable businesses collaborate with industry peers, government agencies, and non-governmental organizations to drive sustainability initiatives at a broader scale. They participate in industry associations, sustainability networks, and multi-user platforms to share best practices, advocate for supportive policies, and collectively address sustainability challenges.
- Financial Viability: Sustainable businesses understand that long-term financial viability is essential for driving sustainable practices. They integrate sustainability considerations into their business strategies, recognizing that sustainable practices can lead to cost savings, enhanced reputation, improved customer loyalty, and access to new markets. They incorporate sustainability into their financial decision-making processes and consider the economic benefits of adopting green practices.

Hence, by adopting green economy principles and implementing sustainable business practices, organizations can contribute to environmental protection, social well-being, and long-term economic prosperity. The integration of sustainability into business models is essential for creating a more resilient and equitable future.

7.2 Innovation and Entrepreneurship in Smart Environmental Solutions

Innovation and entrepreneurship play an important role in developing and implementing smart environmental solutions. They drive the creation of new technologies, business models, and approaches that address environmental challenges and promote sustainability. Here are some key aspects of innovation and entrepreneurship in the context of smart environmental solutions:

- Technological Advancements: Innovation and entrepreneurship drive technological advancements that enable smart environmental solutions. This includes the development of Internet of Things (IoT) devices, sensors, data analytics tools, artificial intelligence (AI), and machine learning algorithms that help monitor, analyze, and optimize environmental processes (Sengodan et al., 2022; Tyagi et al., 2023; Gomathi et al., 2023; Tyagi, Abraham, Kaklauskas et al, 2022; Malik et al., 2022; Mapanga et al., 2017; Midha & Tripathi, 2020; Pal et al., 2022; Pramod, 2022; Sai, Tyagi, & Sreenath, 2023; Tyagi, 2022a; Tyagi, 2022b; Tyagi & Aswathy, 2021; Tyagi & Chahal, 2020; Tyagi, 2021; Tyagi, Agarwal, & Sreenath, 2022; Tyagi & Bansal, 2023; Tyagi & Sreenath,

2023d). Technological innovations provide the base for creating smart systems and infrastructure that improve resource efficiency, reduce environmental impacts, and enhance sustainability.

- Startups and Sustainable Ventures: Entrepreneurs and startups are at the forefront of developing innovative solutions for environmental challenges. They identify market gaps and opportunities, and add their creativity and risk-taking abilities to develop and commercialize smart environmental technologies and services. These ventures often bring fresh perspectives, agility, and disruptive ideas that can drive transformative change in the environmental sector.
- Collaborative Platforms and Incubators: Collaborative platforms and incubators provide a supportive environment for innovation and entrepreneurship in the smart environmental space. These platforms bring together entrepreneurs, technologists, researchers, and investors, fostering collaboration, knowledge exchange, and access to resources. They provide mentorship, funding opportunities, and networking platforms that accelerate the development and scaling of smart environmental solutions.
- Cross-Sector Partnerships: Innovation and entrepreneurship in smart environmental solutions often require collaboration across sectors. Partnerships between technology companies, government entities, academia, and non-profit organizations can add diverse expertise and resources to address complex environmental challenges. These collaborations facilitate the exchange of knowledge, access to funding, and the co-creation of innovative solutions that have a broader impact.
- Business Model Innovation: In addition to technological advancements, innovation in business models is important for the successful implementation of smart environmental solutions. Entrepreneurs and innovators provide new approaches to financing, pricing, and delivering smart environmental technologies and services. This includes subscription models, pay-per-use schemes, and performance-based contracts that incentivize resource efficiency and promote sustainable consumption and production.
- User-Centric Design: Innovation and entrepreneurship in smart environmental solutions prioritize user-centric design principles. By understanding the needs, preferences, and behaviors of end-users, entrepreneurs can develop solutions that are intuitive, user-friendly, and capable of driving behavior change. User-centric design enhances the adoption and acceptance of smart environmental technologies and encourages sustainable practices among individuals and organizations.
- Scalability and Replicability: Successful smart environmental solutions are scalable and replicable, meaning they can be implemented in different contexts and scaled up to have a broader impact. Entrepreneurs and innovators focus on designing solutions that can be easily adapted to different locations, sectors, and user groups. This allows for widespread adoption and accelerates the transition to a sustainable and smart environment.

Hence, innovation and entrepreneurship are essential drivers of change in the environmental sector. They enable the development and implementation of smart environmental solutions that address pressing challenges such as climate change, resource depletion, and pollution. By fostering a culture of innovation and supporting entrepreneurial initiatives, societies can unlock the potential for transformative change and create a sustainable and smart environment for future generations.

7.3 Job Creation in Green Industries

The transition to a green economy and the adoption of sustainable practices creates several/ multiple job opportunities across various sectors. Green industries, also known as the clean or renewable energy sector, focus on environmentally friendly practices and technologies that reduce carbon emissions and promote sustainability. Here are some key areas that contribute to job creation in green industries:

- Renewable Energy: The shift towards renewable energy sources such as solar, wind, hydro, and geothermal power has created a significant number of jobs. These include roles in the installation, operation, and maintenance of renewable energy systems, as well as in manufacturing components such as solar panels and wind turbines.
- Energy Efficiency: Energy efficiency initiatives, including building retrofits, insulation, and efficient lighting systems, require skilled workers to carry out installation and maintenance activities. Energy auditors and efficiency consultants are also in demand to identify energy-saving opportunities in buildings and industrial processes.
- Green Construction: Green building practices, which focus on energy efficiency, sustainable materials, and environmental impact reduction, create employment opportunities for architects, engineers, construction workers, and technicians specialized in green construction techniques.
- Sustainable Transportation: The transition to sustainable transportation systems, including electric vehicles (EVs), charging infrastructure, and public transportation, creates jobs in manufacturing, installation, maintenance, and operation of EVs, charging stations, and related infrastructure.
- Waste Management and Recycling: The growing emphasis on waste reduction, recycling, and waste-to-energy technologies creates jobs in waste collection, sorting, recycling facilities, and the development and operation of innovative waste management systems.
- Water and Wastewater Management: The demand for clean water and efficient wastewater treatment drives job opportunities in water conservation, water treatment technologies, and wastewater management. Roles in water monitoring, purification, and distribution systems contribute to job creation in this sector.
- Sustainable Agriculture and Food Systems: Sustainable agriculture practices, organic farming, and the promotion of local and seasonal food production generate employment in farming, food processing, distribution, and marketing of sustainably produced food.
- Environmental Consulting and Services: The need for environmental consulting services, environmental impact assessments, sustainability reporting, and compliance with environmental regulations create jobs for environmental scientists, engineers, and consultants.
- Research and Development: Green industries require research and development (R&D) efforts to drive innovation, improve technologies, and develop new sustainable solutions. R&D roles in renewable energy, energy storage, clean technologies, and sustainable materials contribute to job creation and drive the advancement of green industries.
- Green Business and Entrepreneurship: The growth of green industries fosters opportunities for green businesses and startups. Entrepreneurs and innovators can establish sustainable ventures that provide environmentally friendly products, services, and solutions, creating employment opportunities in various sectors.

Note that the transition to a green economy also involves job transformation and reskilling. Some existing jobs may require upskilling or transitioning to align with the changing needs of the green industries. Governments, educational institutions, and training providers play an important role in provideing training and skill development programs to equip individuals with the necessary skills and knowledge for green jobs. In summary, the shift towards green industries and sustainable practices provides significant job creation potential, fostering economic growth while addressing environmental challenges. It presents opportunities for both traditional and emerging sectors, contributing to a more sustainable and inclusive economy.

10. CONCLUSION

The need for a sustainable environment together with the pursuit of a smart environment defines an important imperative for the 21st century. As our world faces unprecedented environmental challenges, the integration of sustainability and smart technologies becomes not just a goal but a necessity. In this 21st century, the dual pursuit of sustainability and smartness is not only possible but also essential. It's a call to action that transcends borders and ideologies, uniting us in the shared goal of protecting our planet while embracing the opportunities of a technologically advanced future. Through innovation, cooperation, and collective responsibility, we can build a sustainable, smart environment that benefits both current and future generations.

REFERENCES

Abhishek, B., & Tyagi, A. K. (2022). An Useful Survey on Supervised Machine Learning Algorithms: Comparisons and Classifications. In T. Sengodan, M. Murugappan, & S. Misra (Eds.), *Advances in Electrical and Computer Technologies. Lecture Notes in Electrical Engineering* (Vol. 881). Springer. doi:10.1007/978-981-19-1111-8_24

Agarwal, D., & Tripathi, K. (2022). A Framework for Structural Damage detection system in automobiles for flexible Insurance claim using IOT and Machine Learning. *2022 International Mobile and Embedded Technology Conference (MECON),* 5-8. 10.1109/MECON53876.2022.9751889

Alfen, H. W., & Ten Brink, P. (Eds.). (2018). *Nature-based solutions for urban resilience*. Springer.

Angel, S., Sheppard, S., & Civco, D. L. (Eds.). (2011). *The dynamics of global urban expansion*. World Bank Publications.

Bai, X., & Imura, H. (Eds.). (2016). *Urbanization and sustainability: Linking urban ecology, environmental justice and global environmental change*. Springer.

Batty, M. (2018). *The new science of cities*. MIT Press.

Coaffee, J., Healey, P., & O'Brien, P. (2018). *Resilient cities: Rethinking urban protection*. Springer.

Deshmukh, A., Patil, D. S., Soni, G., & Tyagi, A. K. (2023). Cyber Security: New Realities for Industry 4.0 and Society 5.0. In A. Tyagi (Ed.), *Handbook of Research on Quantum Computing for Smart Environments* (pp. 299–325). IGI Global. doi:10.4018/978-1-6684-6697-1.ch017

Giffinger, R., Fertner, C., Kramar, H., Kalasek, R., Pichler-Milanović, N., & Meijers, E. (2007). Smart cities: Ranking of European medium-sized cities. Centre of Regional Science (SRF*).* Vienna University of Technology. 10.1145/1947940.1947974

Glaeser, E. L. (2011). *Triumph of the city: How our greatest invention makes us richer, smarter, greener, healthier, and happier*. Penguin.

Gomathi, L., Mishra, A. K., & Tyagi, A. K. (2023). Industry 5.0 for Healthcare 5.0: Opportunities, Challenges and Future Research Possibilities. *2023 7th International Conference on Trends in Electronics and Informatics (ICOEI),* 204-213. 10.1109/ICOEI56765.2023.10125660

IPCC. (2018). *Global warming of 1.5°C: An IPCC special report on the impacts of global warming of 1.5°C above pre-industrial levels and related global greenhouse gas emission pathways, in the context of strengthening the global response to the threat of climate change*. Cambridge University Press.

Jajula, S. K., Tripathi, K., & Bajaj, S. B. (2023). Review of Detection of Packets Inspection and Attacks in Network Security. In P. Dutta, S. Chakrabarti, A. Bhattacharya, S. Dutta, & V. Piuri (Eds.), *Emerging Technologies in Data Mining and Information Security. Lecture Notes in Networks and Systems* (Vol. 491). Springer. doi:10.1007/978-981-19-4193-1_58

Madhav, A. V. S., & Tyagi, A. K. (2022). The World with Future Technologies (Post-COVID-19): Open Issues, Challenges, and the Road Ahead. In A. K. Tyagi, A. Abraham, & A. Kaklauskas (Eds.), *Intelligent Interactive Multimedia Systems for e-Healthcare Applications*. Springer. doi:10.1007/978-981-16-6542-4_22

Madhav, A. V. S., & Tyagi, A. K. (2023). Explainable Artificial Intelligence (XAI): Connecting Artificial Decision-Making and Human Trust in Autonomous Vehicles. In *Proceedings of Third International Conference on Computing, Communications, and Cyber-Security. Lecture Notes in Networks and Systems* (vol. 421). Springer. 10.1007/978-981-19-1142-2_10

Malik, S., Bansal, R., & Tyagi, A. K. (Eds.). (2022). *Impact and Role of Digital Technologies in Adolescent Lives*. IGI Global. doi:10.4018/978-1-7998-8318-0

Mapanga, Kumar, Makondo, Kushboo, Kadebu, & Chanda. (2017). Design and implementation of an intrusion detection system using MLP-NN for MANET. *2017 IST-Africa Week Conference (IST-Africa),* 1-12. . doi:10.23919/ISTAFRICA.2017.8102374

Midha, S., Kaur, G., & Tripathi, K. (2017). Cloud deep down — SWOT analysis. *2017 2nd International Conference on Telecommunication and Networks (TEL-NET),* 1-5. 10.1109/TEL-NET.2017.8343560

Midha, S., & Tripathi, K. (2019). Extended TLS security and Defensive Algorithm in OpenFlow SDN. *2019 9th International Conference on Cloud Computing, Data Science & Engineering (Confluence),* 141-146. 10.1109/CONFLUENCE.2019.8776607

Midha, S., & Tripathi, K. (2020). Remotely Triggered Blackhole Routing in SDN for Handling DoS. In *Proceedings of International Conference on IoT Inclusive Life (ICIIL 2019), NITTTR Chandigarh, India. Lecture Notes in Networks and Systems* (vol. 116). Springer. 10.1007/978-981-15-3020-3_1

Midha, S., & Tripathi, K. (2021). Extended Security in Heterogeneous Distributed SDN Architecture. In G. Hura, A. Singh, & L. Siong Hoe (Eds.), *Advances in Communication and Computational Technology. Lecture Notes in Electrical Engineering* (Vol. 668). Springer. doi:10.1007/978-981-15-5341-7_75

Midha, S., Tripathi, K., & Sharma, M. K. (2021, April). Practical Implications of Using Dockers on Virtualized SDN. *Webology., 18*(1), 312–330. doi:10.14704/WEB/V18SI01/WEB18062

Mishra, S., & Tyagi, A. K. (2022). The Role of Machine Learning Techniques in Internet of Things-Based Cloud Applications. In S. Pal, D. De, & R. Buyya (Eds.), *Artificial Intelligence-based Internet of Things Systems. Internet of Things (Technology, Communications and Computing).* Springer. doi:10.1007/978-3-030-87059-1_4

Nair, M. M., & Tyagi, A. K. (2021). Privacy: History, Statistics, Policy, Laws, Preservation and Threat Analysis. Journal of Information Assurance & Security, 16(1), 24-34.

Neumann, M. B., & Emanuel, K. (2018). The nature of future risks: Changing threats in the 21st century. *Global Environmental Change, 53*, 1–9.

Pickett, S. T., Cadenasso, M. L., & Grove, J. M. (Eds.). (2018). *Resilient cities: Meaning, models, and metaphor for integrating the ecological, socio-economic, and planning realms.* Springer.

Pramod, A. (2022). *Emerging Innovations in the Near Future Using Deep Learning Techniques, Book: Advanced Analytics and Deep Learning Models.* Wiley Scrivener. doi:10.1002/9781119792437.ch10

Ranchhodbhai, P.N., & Tripathi, K. (2019). Identifying and Improving the Malicious Behavior of Rushing and Blackhole Attacks using Proposed IDSAODV Protocol. *International Journal of Recent Technology and Engineering, 8*(3), 6554-6562.

Rees, W., & Wackernagel, M. (2013). The shoe fits, but the footprint is larger than Earth: Urban ecological footprints show that cities can be unsustainable. *Environment and Urbanization, 25*(2), 481–499.

Rocha, M. F. (2018). The future of cities in the 21st century: Implications of urban energy use and greenhouse gas emissions. *Energy Procedia, 147*, 72–77.

Rosenzweig, C., Solecki, W., Romero-Lankao, P., Mehrotra, S., & Dhakal, S. (Eds.). (2018). *Climate change and cities: Second assessment report of the Urban Climate Change Research Network.* Cambridge University Press. doi:10.1017/9781316563878

Sai, G. H., Tripathi, K., & Tyagi, A. K. (2023). Internet of Things-Based e-Health Care: Key Challenges and Recommended Solutions for Future. In *Proceedings of Third International Conference on Computing, Communications, and Cyber-Security. Lecture Notes in Networks and Systems* (vol. 421). Springer. 10.1007/978-981-19-1142-2_37

Sai, G. H., Tyagi, A. K., & Sreenath, N. (2023). Biometric Security in Internet of Things Based System against Identity Theft Attacks. *2023 International Conference on Computer Communication and Informatics (ICCCI),* 1-7. 10.1109/ICCCI56745.2023.10128186

Sassen, S. (2019). *Cities in a world of cities: The globalization of urban governance*. Yale University Press.

Sheth, H. S. K. (2022). Deep Learning, Blockchain based Multi-layered Authentication and Security Architectures. *2022 International Conference on Applied Artificial Intelligence and Computing (ICAAIC)*, 476-485. 10.1109/ICAAIC53929.2022.9793179

Shreyas Madhav, A. V., Ilavarasi, A. K., & Tyagi, A. K. (2022). The Heroes and Villains of the Mix Zone: The Preservation and Leaking of USer's Privacy in Future Vehicles. In V. Arunachalam & K. Sivasankaran (Eds.), *Microelectronic Devices, Circuits and Systems. ICMDCS 2022. Communications in Computer and Information Science* (Vol. 1743). Springer. doi:10.1007/978-3-031-23973-1_12

Somisetti, K., Tripathi, K., & Verma, J. K. (2020). Design, Implementation, and Controlling of a Humanoid Robot. *2020 International Conference on Computational Performance Evaluation (ComPE)*, 831-836. 10.1109/ComPE49325.2020.9200020

Song, Y., & Zegeye, A. (2017). Urbanization and the environment: An overview. In *The Urbanization and the Environment* (pp. 1–16). Springer.

Subasree, S., & Sakthivel, N. K. (2022). Combining the advantages of radiomic features based feature extraction and hyper parameters tuned RERNN using LOA for breast cancer classification. *Biomedical Signal Processing and Control, 72*(A). doi:10.1016/j.bspc.2021.103354

Tulumello, S., & Hernández-Molina, G. (2018). Smart cities as techno-social assemblages: A framework for understanding the rise of smart urbanism. *Frontiers in Sociology*, *3*, 30.

Tyagi. (2022a). *Handbook of Research on Technical, Privacy, and Security Challenges in a Modern World*. IGI Global. doi:10.4018/978-1-6684-5250-9

Tyagi. (2022b). *Using Multimedia Systems, Tools, and Technologies for Smart Healthcare Services*. IGI Global. doi:10.4018/978-1-6684-5741-2

Tyagi & Aswathy. (2021). Autonomous Intelligent Vehicles (AIV): Research statements, open issues, challenges and road for future. *International Journal of Intelligent Networks, 2*, 83-102. doi:10.1016/j.ijin.2021.07.002

Tyagi, Aswathy, Aghila, & Sreenath. (2021, October). AARIN: Affordable, Accurate, Reliable and INnovative Mechanism to Protect a Medical Cyber-Physical System using Blockchain Technology. *IJIN*, *2*, 175–183.

Tyagi, & Chahal. (2020). *Artificial Intelligence and Machine Learning Algorithms. In Challenges and Applications for Implementing Machine Learning in Computer Vision*. IGI Global. doi:10.4018/978-1-7998-0182-5.ch008

Tyagi & Sreenath. (2021). Cyber Physical Systems: Analyses, challenges and possible solutions. *Internet of Things and Cyber-Physical Systems, 1*, 22-33. doi:10.1016/j.iotcps.2021.12.002

Tyagi, A., Kukreja, S., Nair, M. M., & Tyagi, A. K. (2022). Machine Learning: Past, Present and Future. *NeuroQuantology : An Interdisciplinary Journal of Neuroscience and Quantum Physics*, *20*(8). Advance online publication. doi:10.14704/nq.2022.20.8.NQ44468

Tyagi, A. K. (Ed.). (2021). *Multimedia and Sensory Input for Augmented, Mixed, and Virtual Reality*. IGI Global. doi:10.4018/978-1-7998-4703-8

Tyagi, A. K., Agarwal, D., & Sreenath, N. (2022). SecVT: Securing the Vehicles of Tomorrow using Blockchain Technology. *2022 International Conference on Computer Communication and Informatics (ICCCI)*, 1-6. 10.1109/ICCCI54379.2022.9740965

Tyagi, A. K., & Bansal, R. (2023). A Step-To-Step Guide to Write a Quality Research Article. In Intelligent Systems Design and Applications. ISDA 2022. Lecture Notes in Networks and Systems (vol. 717). Springer. doi:10.1007/978-3-031-35510-3_36

Tyagi, A. K., Chandrasekaran, S., & Sreenath, N. (2022). Blockchain Technology:– A New Technology for Creating Distributed and Trusted Computing Environment. *2022 International Conference on Applied Artificial Intelligence and Computing (ICAAIC)*, 1348-1354. 10.1109/ICAAIC53929.2022.9792702

Tyagi, A. K., Nair, M. M., Niladhuri, S., & Abraham, A. (2020). Security, Privacy Research issues in Various Computing Platforms: A Survey and the Road Ahead. Journal of Information Assurance & Security, 15(1), 1-16.

Tyagi, A. K., & Sreenath, N. (2023a). Security, Privacy, and Trust Issues in Intelligent Transportation System. In *Intelligent Transportation Systems: Theory and Practice. Disruptive Technologies and Digital Transformations for Society 5.0*. Springer. doi:10.1007/978-981-19-7622-3_8

Tyagi, A. K., & Sreenath, N. (2023b). Artificial Intelligence—Internet of Things-Based Intelligent Transportation System. In *Intelligent Transportation Systems: Theory and Practice. Disruptive Technologies and Digital Transformations for Society 5.0*. Springer. doi:10.1007/978-981-19-7622-3_10

Tyagi, A. K., & Sreenath, N. (2023c). Future Intelligent Vehicles: Open Issues, Critical Challenges, and Research Opportunities. In *Intelligent Transportation Systems: Theory and Practice. Disruptive Technologies and Digital Transformations for Society 5.0*. Springer. doi:10.1007/978-981-19-7622-3_15

Tyagi, A. K., & Sreenath, N. (2023d). Intelligent Transportation System: Past, Present, and Future. In *Intelligent Transportation Systems: Theory and Practice. Disruptive Technologies and Digital Transformations for Society 5.0*. Springer. doi:10.1007/978-981-19-7622-3_2

Varsha, R., Nair, S. M., Tyagi, A. K., & Aswathy, S. U. (2021). The Future with Advanced Analytics: A Sequential Analysis of the Disruptive Technology's Scope. In Hybrid Intelligent Systems. HIS 2020. Advances in Intelligent Systems and Computing (vol. 1375). Springer. doi:10.1007/978-3-030-73050-5_56

Vishnuram, G., Tripathi, K., & Kumar Tyagi, A. (2022). Ethical Hacking: Importance, Controversies and Scope in the Future. *2022 International Conference on Computer Communication and Informatics (ICCCI)*, 1-6. 10.1109/ICCCI54379.2022.9740860

Chapter 10
Revolution Towards DC Microgrids: A History

Rinki Maurya
https://orcid.org/0000-0002-0652-9613
Indian Institute of Information Technology, Allahabad, India

Priya Sharma
Indian Institute of Information Technology, Allahabad, India

Ashutosh Kumar Singh
Indian Institute of Information Technology, Allahabad, India

Surya Prakash
Indian Institute of Information Technology, Allahabad, India

ABSTRACT

A DC microgrid is a decentralized electricity system that uses direct current (DC) power for distribution and consumption. It is an alternative to the traditional alternating current (AC) microgrid, which is the dominant electrical distribution system used in most parts of the world. The adoption of DC microgrids has been gaining momentum in recent years due to their numerous advantages over AC microgrids, including improved energy efficiency, reduced power loss during transmission, and the ability to integrate renewable energy sources more easily. In this chapter, the authors review the technical and economic benefits of all types of possible grids and microgrids and discuss the challenges and opportunities associated with their deployment. Overall, the analysis suggests that the transition to DC microgrids is a promising approach for addressing the energy challenges of the 21st century and could play a significant role in the ongoing energy transition.

DOI: 10.4018/978-1-6684-9130-0.ch010

1. INTRODUCTION

Electricity demand is increasing day by day due to the world's population increment and lifestyle improvements. Now, the conventional AC grid is insufficient to fill the gap between electricity generation and consumption because these grids are mainly run by the combustion of limited fossil fuels, which intensify the concentration of pollutants in the green environment (like PM2.5, SO_2, NO_x, CO_2, Mercury (Hg), PM_{10}, (CO), and Unburned hydrocarbons, soot, and sulfur oxides (Munawer, 2018) and escalate global warming, which results in severe health risks like heart & Respiratory, Lung Cancer, Coughing, Asthma, skin, & stomach diseases etc. So, now many nations are switching to environmentally friendly alternatives to fulfill the rising electricity demand. According to the NASA 2021 report (Lin et al., 2023), the global CO_2 level in the atmosphere was nearly 417.41 ppm by December 2021, reaching the greatest trend despite the COVID-19 pandemic. In 2021, the electricity generated by coal was approximately 27% worldwide (Buonocore et al., 2021).

Firstly, the concept of a small AC grid was introduced for remote areas where electricity was generated by diesel, petrol, etc. After the improvements in solar photovoltaics, an AC-DC hybrid microgrid (Saeed et al., 2021)came into existence, which is used to generate electricity for remote areas. PV panels generate DC electricity which requires an inverter for prevailing appliances to convert it from DC to AC which has frequency and voltage synchronization issue with the AC grid (Malik et al., 2017). In this type of system, hybrid AC/DC microgrids are used, yet two other issues, high power losses and synchronization comes. As of now, most of the appliances are operated on DC, which tends the conventional grid toward DC microgrids. Many researchers are working for a zero-carbon emission environment with fewer electricity bills using this, in which there is enormous scope for improvement (Ahmad et al., 2020).Various challenges for designing DC microgrids and their lasting solutions are discussed in this manuscript.

In this paper, section 2 explains the electricity generation by the AC grid and the reason for finding other sources to generate electricity. Section 3 describes the classification of microgrids based on several factors like type, mode of operation, the polarity of the voltage, structure, type of energy resources, and their scenario, etc. Section 4 describes the conclusion followed by the future scope.

2. STANDARD AC GRID

An energy source is referred to as a conventional source of energy if it cannot be utilized again after being used once. Traditional energy sources are unrenewable and environmentally damaging. Conventional AC grids as shown in Fig.1, are mostly run by the use of various types of coal like bituminous coal, sub-bituminous coal, or lignite. For electricity generation, the coals are burned and used to heat the water at a very high temperature approx. 1600 degrees centigrade, causing water vapors to be created. These water vapors are forced towards turbines, which generate electricity, So in TPPs, thermal energy is converted into electrical energy, but with many proposed ideas the annual CO_2 emission will decline to zero by the year 2050, according to the NZE (Net Zero Emission) scenario (Lin et al., 2023). Global warming also increases with these TPPs (Finkelman et al., 2021).

Figure 1. Conventional AC grid block diagram

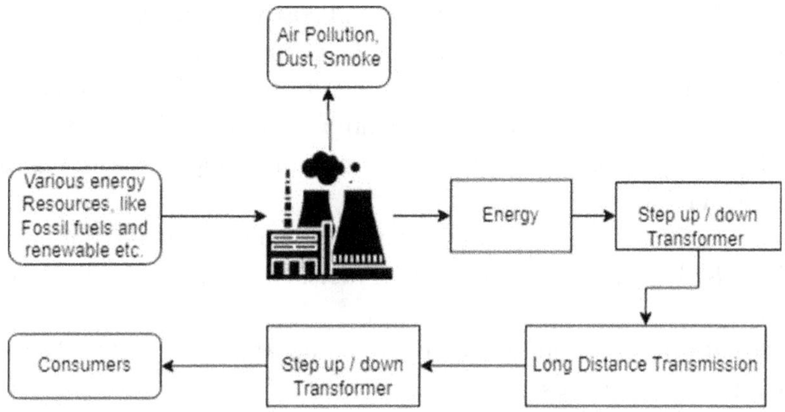

Merits of the Conventional AC Grids

There are several merits of the conventional AC grid (Modu et al., 2023; Planas et al., 2015; Sabzehgar, 2017; *World Energy Outlook 2022*, n.d.), as identified in the literature:

1. **Widely accepted and established**: AC power has been the dominant electrical distribution system for over a century and is widely accepted and used around the world.
2. **Versatility**: AC power can be easily transformed to different voltages and frequencies, which makes it flexible and adaptable to different types of loads and transmission distances.
3. **High power transmission capacity:** It can transmit large amounts of power over long distances with minimal loss.
4. **Efficiency:** While AC transmission lines do experience some power loss, they are still generally considered to be efficiently compared to DC transmission lines.
5. **Cost-effective:** AC transmission lines and related infrastructure have been in use for a long time and are well-established, which makes them cost-effective to install and maintain on a large scale.
6. **Safety:** AC power is generally considered to be safe, as it is less likely to cause electrical shocks compared to DC power due to Zero crossing of current.
7. **Less number of Plants needed:** A smaller number of power plant installations is needed to cover the entire electricity demand.
8. **Easy transportation of Raw material:** Raw Materials like, Coal, oil, and natural gasses can be efficiently transported by trains or ships.

Demerits of the Conventional AC Grid

There are several issues with the conventional AC grid grid (Modu et al., 2023; Planas et al., 2015; Sabzehgar, 2017; *World Energy Outlook 2022*, n.d.),that have been identified in the literature:

1. **Inefficiency:** AC transmission lines experience power loss due to resistance, which results in reduced efficiency.

2. **Difficulty integrating renewable energy sources:** sources, such as photovoltaic panels and wind turbines, produce DC power. It can be difficult and costly to convert this power to AC for use on the grid. This can limit the potential for renewable energy to contribute to the grid.
3. **Voltage fluctuation:** AC voltage can fluctuate due to various factors, such as load demand and changes in the power grid. These voltage fluctuations can cause damage to electrical equipmen.
4. **Complexity:** The AC grid can be complex, with multiple components and layers of infrastructure that need to be maintained.
5. **Security:** The AC grid is vulnerable to cyber-attacks and physical disruptions.
6. **Reduce fossil fuels:** It totally depends on fossil fuels, which are limited in nature and day by day decreases also.
7. **Produce air pollution:** Air pollution due to the combustion of fossil fuels causes an increase in global warming and health issues.
8. **Synchronization issue:** Synchronization issues like voltage and frequency, happen when combined with other AC grids.
9. **Load shedding issues:** Unable to fulfill the current electricity demand due to continuous increase in the population as well as their need. Power blackouts can happen at any time.

A Possible Solution to the Conventional Grid

1. Smart grids (Fang et al., 2012)
2. Distributed energy resources (Winardi et al., 2021)
3. Energy storage
4. Advanced transmission and distribution systems
5. Enhanced cybersecurity measures: etc.

3. MICROGRID

“A microgrid is a small-scale decentralized power grid that can operate independently or collaboratively with other small power grids” (Alam et al., 2023). It can be classified into some sections depending on various factors like the type of grid, mode of operation, the polarity of voltage, the structure of the grid, type of energy resource, and scenario of the grid (Maurya et al., 2022; Subburaj et al., 2022).

Classification of the Microgrid

Microgrids can be classified in various ways as shown in Figure 2, depending on the following criteria used.

Figure 2. Classification of microgrids

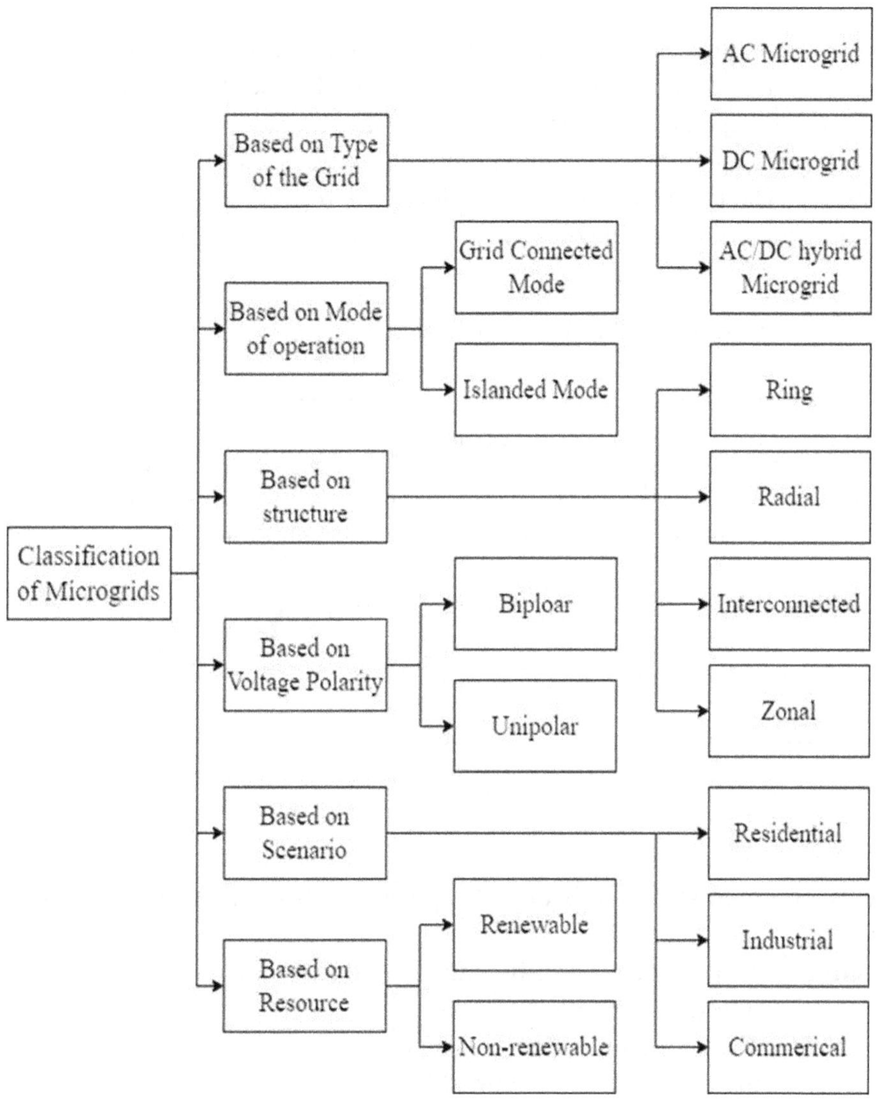

3.1 Microgrid Based on the "Type of the Power System Grid"

The "Type" of the grid is categorized based on the main bus voltage, if the main bus has an AC type voltage, is called an AC microgrid, has a DC type of voltage, is called DC microgrid, and are two bus voltages AC and DC than it is called as hybrid AC-DC (Abdolmaleki & Bergna-Diaz, 2022; Altaf et al., 2022; Vegunta et al., 2021) voltage. Microgrids can be classified based on the type of power system they use, including AC, DC, and hybrid AC-DC systems (Sabzehgar, 2017; Sahoo et al., 2018).

A. AC Microgrids

AC microgrids use alternating current (AC) power for distribution and consumption. AC microgrids can be connected to the main power grid or operate independently, and they can range in size from small, village-scale systems to larger, regional-scale systems. The generation of electricity in an AC microgrid can be done by any of the methods like photovoltaic cells, wind turbines, renewable energy sources, fuel cells, tidal plants, etc. The main complications with AC microgrids shown in Fig.3, (Altaf et al., 2022) are the synchronization of line parameters with the conventional AC grid.

Figure 3. General block diagram of an AC microgrid

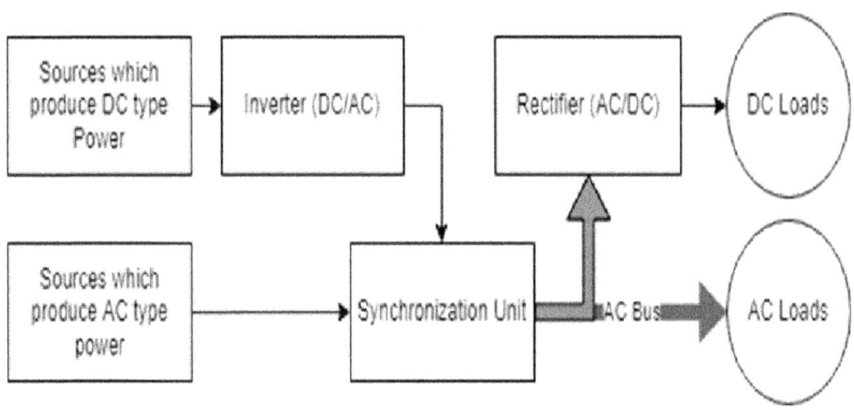

- **Merits of the AC microgrid**
    1. **Widely accepted and established:** AC power has been the dominant electrical distribution system is widely accepted.
    2. **Versatility:** AC power can be easily transformed to different voltages and frequencies, which makes it flexible and adaptable.
    3. **High power transmission capacity:** AC transmission lines can transmit large amounts of power over long distances with minimal loss.
    4. **Cost-effective:** Well established infrastructure of AC transmission lines system makes them cost-effective to install and maintain.
    5. **Safety:** AC power is generally considered to be safe, as it is less likely to cause electrical shocks compared to DC power.
- **Demerits of the AC microgrid** (Abdolmaleki & Bergna-Diaz, 2022; Altaf et al., 2022; Vegunta et al., 2021).
    1. **Inefficiency:** AC transmission lines experience power loss due to resistance, which can reduce the efficiency of the system.
    2. Difficulty integrating renewable energy sources: Because of DC output power, it can be difficult and costly to convert this power to AC.
    3. **Voltage fluctuation:** It can damage to electrical equipment and affect the stability of the microgrid.

4. **Complexity:** AC microgrids can be complex, with multiple components and layers of infrastructure that need to be maintained.
5. **Security:** AC microgrids are vulnerable to cyber-attacks and physical disruptions.

Overall, AC microgrids face many of the same challenges as the conventional AC grid (Abdolmaleki & Bergna-Diaz, 2022).

B. DC Microgrids

A small version of the grid, where the main line has only a DC voltage parameter is known as a DC microgrid, to power these devices from an AC source, an inverter is required to change the AC supply into a DC.. They have several advantages over AC microgrids (Al-Ismail, 2021). The general block diagram of DC microgrids is shown in Fig.4.

Figure 4. General block diagram of DC microgrid

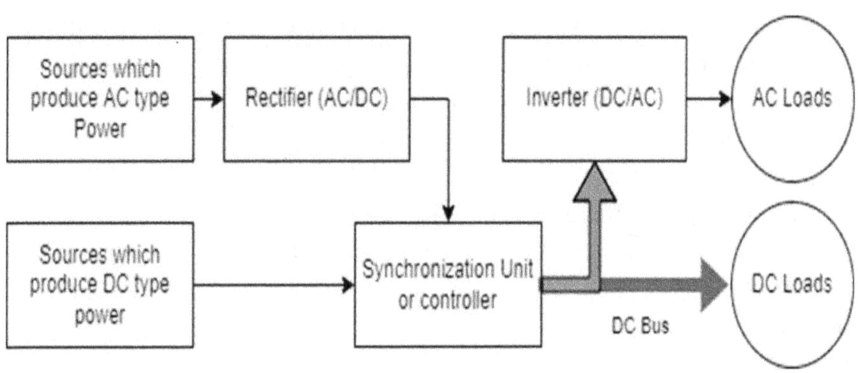

Nevertheless, if we use a DC supply for these DC loads, this conversion is not required. Therefore, the island's DC microgrid is the subject of intense research.(Alam et al., 2023).

- **Merits of the DC microgrid**: There are several merits and demerits bof DC microgrid systems (Abhishek et al., 2020; Al-Ismail, 2021; Unamuno & Barrena, 2015a, 2015b) as identified in the literature:
  1. **Energy efficiency:** DC microgrids can be more energy efficient than AC, as they experience less power loss during transmission.
  2. **Renewable energy integration:** DC microgrids can integrate DC type sources more easily, without the need for expensive conversion equipment. This can make it easier to incorporate renewable energy into the microgrid.
  3. **Improved reliability:** It is more reliable as they are less affected by voltage fluctuations and less susceptible to disruptions caused by power quality issues.
  4. **Easy to control:** DC microgrids can be easily controlled and managed, as they do not require complex control systems or infrastructure.

5. **Safety:** DC power is generally considered to be safer than AC power, as it is less likely to cause electrical shocks..
6. Air pollution can be minimized.
7. Reduction of fossil fuels can be saved.

- **Demerits of the DC microgrid**:
  1. **Limited use of legacy equipment:** Most electrical equipment is designed to work with AC power, and it may not be compatible with DC power. This can make it difficult to incorporate legacy equipment into a DC microgrid (Malik et al., 2017)
  2. **The complexity of voltage conversion:** DC microgrids may require the use of voltage conversion equipment to power AC loads or to interface with the main power grid. This can add complexity and cost to the system.
  3. **Limited transmission distance:** DC transmission lines experience more power loss than AC transmission lines over long distances, which limits the transmission distance of DC power. This can make it more challenging to transmit electricity from remote renewable energy sources to consumers.
  4. **Limited voltage range:** DC voltage is more limited in range than AC voltage, which can make it more difficult to transmit power over long distances or to power certain types of loads.
  5. **Security:** They are vulnerable to physical disruptions, such as electrical faults. This can have serious consequences, including power outages and economic losses.
  6. Voltage/Current control
  7. Constant and pulse power load
  8. Maximum power point tracking
  9. Synchronization with electrical or conventional grid
  10. Power quality
  11. Anti-isolation protection
  12. Energy storage
  13. Monitoring of PV modules

Overall, DC microgrids face some unique challenges that need to be addressed in order to realize their full potential.

C. Hybrid AC-DC Microgrids

It (Xiong Liu et al., 2010) combine elements of both AC and DC power systems and can offer the benefits of both. They can use AC power for distribution and consumption, while also integrating DC sources, such as renewable energy sources. As shown in Fig.5, can be particularly useful in situations where both AC and DC loads need to be supplied.

Figure 5. General block diagram of AC/DC hybrid microgrid

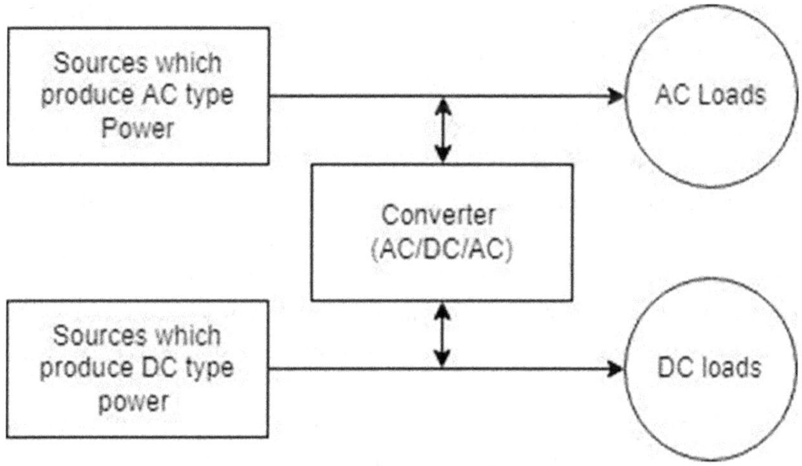

There are two different kinds of sources: first produces AC-type electricity, and second produces DC. Currently, certain appliances function for AC, while others for DC in the end. Therefore, in an AC/DC hybrid microgrid, resources that provide AC power are utilized to support AC-type loads, whilst resources that produce DC power are used to support DC-type loads (Gao et al., 2020).

- **Merits of the hybrid AC-DC microgrid**: There are several merits and demerits (Gao et al., 2020; Xia et al., 2018; Xiong Liu et al., 2010) of AC-DC hybrid microgrid systems, as identified in the literature:
    1. **Energy efficiency:** It can be more energy efficient than AC microgrids, as they can integrate DC renewable energy sources more easily, reducing power loss during transmission.
    2. **Renewable energy integration:** It can integrate both AC and DC renewable energy sources, allowing them to take advantage of the benefits of both types of power systems. This can increase the potential for renewable energy to contribute to the microgrid.
    3. **Improved reliability:** It can be more reliable than AC microgrids, This can improve the stability and resilience of the microgrid.
    4. **Easy to control:** It can be easily controlled and managed, as they do not require complex control systems or infrastructure.
    5. **Safety:** It can offer the safety benefits of both AC and DC power systems, depending on the specific configuration and applications of the system.
    6. No need for an ADC or DAC.
    7. Air pollution is reduced but less than in the comparison of DC microgrids alone.
    8. Higher efficiency is achieved when compared with individual AC or DC microgrids.
    9. This type of system can be operated in grid-connected or islanded mode.
- **Demerits of the hybrid AC-DC microgrid:** Hybrid AC-DC microgrids have several potential drawbacks that may limit their widespread adoption. Some of the potential demerits of hybrid AC-DC microgrids include:-

1. **Complexity:** They are more complex than traditional AC microgrids, because of addition of DC system, and the management of power flows in multiple domains. This can increase the design and operational complexity of the microgrid.
2. **Compatibility:** It may not be compatible with certain types of loads or generation sources that are designed to operate in one domain only. This can limit the types of resources that can be integrated into the microgrid.
3. **Losses:** There are additional losses associated with the conversion of power between AC/DC, which can reduce the overall efficiency of the microgrid. These losses can be minimized with high-efficiency converters, but they may still be significant in some cases.
4. **Interoperability:**It may not be fully interoperable with traditional AC microgrids, which could limit their integration into larger power systems or the ability to share resources with other microgrids.
5. **Regulation:** It may be subject to different regulatory frameworks or standards than traditional AC microgrids, which could create additional challenges for their implementation and operation.
6. The reduction of fossil fuels is not controlled permanently due to the use of fossil fuels in the AC grid.
7. Management between rated AC frequencies and rated DC voltages of different AC and DC subgrids is complicated.

3.2 Microgrid Based on the "Mode of Operation or Connectivity to the Main Grid"

The microgrid is separated into two forms as mentioned below, depending on how it connects to the traditional grid (Augustine et al., 2018; D'silva et al., 2020; Talapur et al., 2018).

A. Grid-connected microgrids
B. Islanded/standalone microgrids

A. Grid-connected

A grid-connected microgrid is a type of microgrid that is connected to a larger electrical grid and can operate in both grid-connected and island modes. The key benefit of this is, it can supply power when microgrids cannot.

- **Merits of the Grid-connected microgrid**: Some of the potential merits or advantages of a grid-connected microgrid include:
  1. **Reliability:** It can provide backup power to its local loads in the event of a grid outage, improvs the reliability.
  2. **Flexibility:** They can take advantage of the grid as a source of power when local generation is insufficient, and can also sell excess power back to the grid when it is not needed by the local loads. This can provide the microgrid with greater flexibility in terms of energy management.
  3. **Efficiency:** It can take advantage of the economies of scale and efficiency of a larger electrical grid, and can also reduce the need for expensive on-site energy storage systems.

4. **Sustainability:** It can help reduce greenhouse gas emissions and support the integration of renewable energy sources by allowing the local generation of clean energy and reducing the need for fossil fuel-based power from the grid.

- **Demerits of the Grid-connected microgrid**: Some of the potential demerits or drawbacks of a grid-connected microgrid include:
  1. **Dependency:** It is dependent on the larger electrical grid for power, and may be vulnerable to outages or disruptions in the grid.
  2. **Complexity:** It may be more complex to design and operate than a standalone microgrid, due to the need to manage the interaction with the grid and coordinate power flows in both directions.
  3. **Regulation:** It may be subject to different regulatory frameworks or standards than standalone microgrids, which could create additional challenges for its implementation and operation.
  4. **Cost:** This is costly to implement due to additional equipment and infrastructure required.

B. Islanded Mode

When operating in an island mode (Augustine et al., 2018; D'silva et al., 2020; Talapur et al., 2018) only microgrids can contribute all of the energy needed because there is no way to connect to a traditional AC grid. An islanded microgrid is a type of microgrid that operates independently of the larger electrical grid and is not connected to it. An islanded microgrid is designed to provide electricity to a local area or group of loads and can operate in either grid-connected or island mode. Fig. 6 shows the grid-connected and islanded type microgrid.

Figure 6. Grid-connected and Islanded microgrid

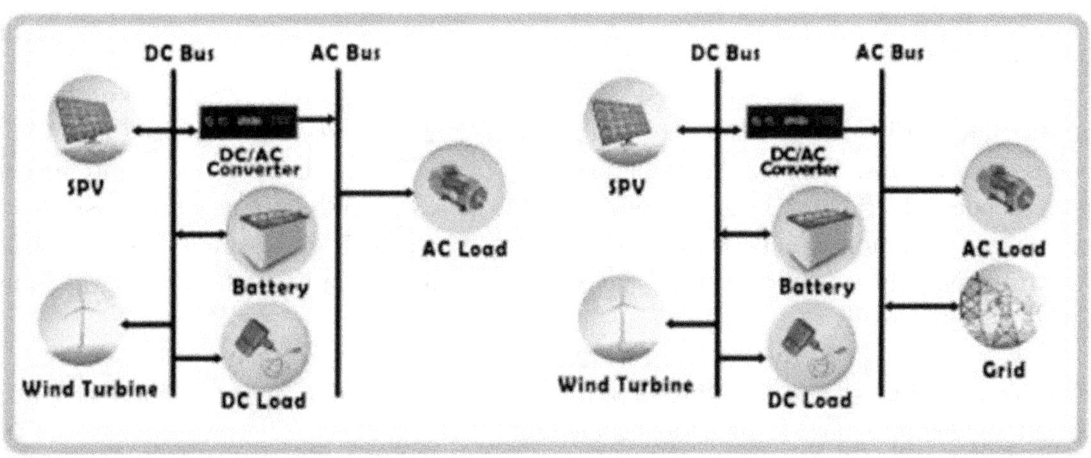

- **Merits of an Islanded microgrid**: Some of the potential benefits of an islanded microgrid include:
  1. **Reliability:** It can provide a reliable source of electricity to its local loads, even in the event of a grid outage or disruption.

2. **Resiliency**: It can be designed to be resilient to natural disasters or other disruptions, as it is not dependent on the larger electrical grid.
3. **Sustainability**: It can support the integration of renewable energy sources, and can reduce greenhouse gas emission.
4. **Independence:** It can provide electricity to remote areas that are not served by the larger grid, enabling greater energy independence and self-sufficiency.

- **Demerits of an Islanded microgrid**: However, there are also some potential drawbacks to an islanded microgrid, including:
  1. **Cost:** It is expensive to implement and operate than a grid-connected, due to the need for on-site generation and storage systems and the lack of access to the economies of scale and efficiency of a larger grid.
  2. **Complexity:** It is more complex to design and operate than a grid-connected microgrid, due to the need to manage the local generation and storage resources.
  3. **Limited capacity:** It have limited capacity and may not be able to meet the electricity needs of larger loads.

3.3 Microgrid Based on the "Structure of the Grid"

According to the internal connection between source and loads, microgrids (Priyadarshini Padhi & Deepa, 2021; L. Xu et al., 2022) are divided into four types explained below.

A. Ring microgrid
B. Radial microgrid
C. Interconnected microgrid
D. Zonal microgrid

A. Ring Structure

A ring structure (Orji et al., 2017; Yadav & Tummuru, 2023)microgrid is organized around a ring-shaped network of transmission and distribution lines. In this, the loads are connected to the network through secondary distribution lines, and power is distributed from the primary transmission lines to the secondary distribution lines through transformers. It (L. Xu et al., 2022) allows for the distribution of power from multiple generation sources and enables the microgrid to operate in both grid-connected and island modes.

- **Merits of a Ring Microgrid:** Ring structure microgrids can offer several potential benefits compared to other microgrid architectures, including:
  1. **Reliability:** It allows for the distribution of power from multiple sources, which can improve the reliability of the power supply.
  2. **Flexibility:** It enables the microgrid to operate in both grid-connected and island modes, providing greater flexibility in terms of energy management.
  3. **Efficiency:** It allows for the efficient distribution of power from the generation sources to the loads, reducing transmission losses and improving the overall efficiency of the microgrid.

- **Demerits of a Ring Microgrid:** However, there are also some potential drawbacks or limitations to ring structure microgrids, including:
    1. **Complexity:** It may be more complex to design and operate than other microgrid architectures, due to the need to manage the transmission and distribution network and coordinate power flows in multiple domains.
    2. **Cost:** It may be more expensive to implement and operate than other microgrid architectures, due to the need for a transmission and distribution network and the use of transformers.
    3. **Regulation:** It may be subject to different regulatory frameworks or standards than other microgrid architectures, which could create additional challenges for their implementation and operation.
    4. The main demerit of this type of structure is when any one of the generation or consumption units is not in working condition then the overall system gets affected because the closed loop is not maintained in that situation.

B. Radial Structure

Radial microgrid (Kant & Gupta, 2022; Kishor et al., 2022; L. Xu et al., 2022) is organized around a network of distribution lines that radiate out from a central generation source. In this, loads are connected to the distribution lines through secondary lines, and power is distributed from the central generation source to the loads through transformers.

It allows for the distribution of power from a single-generation source and typically enables the microgrid to operate in grid-connected mode only. In grid-connected mode, the microgrid can take advantage of the larger electrical grid as a source of power and can sell excess power back to the grid when it is not needed by the local loads.

- **Merits of Radial Microgrid**: It can offer several potential benefits compared to other microgrid architectures, including.
    1. **Simplicity:** It can be simpler to design and operate than other microgrid architectures, as they do not require a complex transmission and distribution network and can rely on a single central generation source.
    2. **Efficiency:** It can be more efficient than other microgrid architectures, as they eliminate the need for multiple generation sources and reduce transmission losses.
    3. **Cost:** They may be less expensive to implement and operate than other microgrid architectures, due to the simplicity of their design and the lack of transmission losses.
- **Demerits of Radial Microgrids**: However, there are also some potential drawbacks or limitations to radial structure microgrids, including
    1. **Dependency:** They are dependent on a single central generation source and may be vulnerable to outages or disruptions in the power supply.
    2. **Limited capacity:** It may have limited capacity, as they are typically designed to serve smaller loads or communities.
    3. **Regulation:** It may be subject to different regulatory frameworks or standards than other microgrid architectures, which could create additional challenges for their implementation and operation.

C. Interconnected Mesh Structure

An interconnected mesh microgrid (Nawaz et al., 2022; Zolfaghari et al., 2022) is a type of microgrid that is organized around a network of transmission and distribution lines that are connected in a mesh configuration. In this, loads are connected to the network through secondary distribution lines, and power is distributed from the primary transmission lines to the secondary distribution lines through transformers.

- **Merits of an Interconnected Mesh Microgrid**: It can offer several potential benefits compared to other microgrid architectures, including:
  1. **Reliability:** It allows for the distribution of power from multiple sources, which can improve the reliability of the power supply.
  2. **Flexibility:** It enables the microgrid to operate in both grid-connected and island modes, providing greater flexibility in terms of energy management.
  3. **Resiliency:** It can be more resilient to outages or disruptions, as it allows for the distribution of power through multiple paths and can enable the microgrid to continue operating even if part of the network is damaged.
- **Demerits of an Interconnected Mesh Microgrid**: However, there are also some potential drawbacks or limitations to interconnected mesh microgrids, including:
  1. **Complexity:** It may be more complex to design and operate than other microgrid architectures, due to the need to manage the transmission and distribution network and coordinate power flows in multiple domains.
  2. **Cost:** It may be more expensive to implement and operate than other microgrid architectures, due to the need for a transmission and distribution network and the use of transformers.
  3. **Regulation:** It may be subject to different regulatory frameworks or standards than other microgrid architectures, which could create additional challenges for their implementation and operation.

D. Zonal Microgrid

A zonal microgrid (Nawaz et al., 2022; Zolfaghari et al., 2022) is a small-scale electrical grid that operates within a defined geographic zone. It is typically connected to the larger, central grid, but can operate independently if necessary. The primary function of a zonal microgrid is to provide a reliable and secure source of power to a specific area, such as a neighborhood, campus, or industrial park.

It can be powered by a variety of energy sources, including renewable sources such as solar panels, wind turbines, and hydroelectric generators, as well as more traditional sources such as fossil fuels.

- **Merits of a Zonal Microgrid:** There are several potential benefits and drawbacks to zonal microgrids:
  1. **Improved reliability:** By being able to operate independently, it can provide a reliable source of power even if the central grid experiences an outage or disruption.
  2. **Energy security:** They can help to ensure that a specific area has a secure source of energy, which can be important in regions that are vulnerable to disruptions in the central grid.
  3. **Cost savings:** By generating and using energy locally, It can reduce the need for long-distance transmission of electricity, which can be expensive and inefficient.

4. **Improved energy efficiency:** They can help to optimize energy use by matching energy supply with demand, which can help to reduce waste and improve overall energy efficiency.
5. **Reduced carbon emissions:** By promoting the use of renewable energy sources, It can help to reduce carbon emissions and contribute to a cleaner, more sustainable energy system.

- **Demerits of a Zonal Microgrid**: There are several potential benefits and drawbacks to zonal microgrids:
  1. **Initial costs:** Setting up and maintaining a It can be expensive, particularly if it involves installing new infrastructure or upgrading existing systems.
  2. **Complexity:** They can be complex systems that require specialized expertise to operate and maintain.
  3. **Interoperability:** It may not always be compatible with the central grid, which can make it difficult to seamlessly integrate them into the overall energy system.
  4. **Regulation:** It may be subject to various regulations and policies, which can vary by location and can impact their development and operation.

3.4 Microgrids Based on "Voltage Polarity"

Based on the different voltage polarities (Ferrera et al., 2015; Guo et al., 2022; Mohammadi et al., 2019; Wang et al., 2017) as shown in Fig.7, the different microgrids, can be classified into two groups.

A. Unipolar microgrid
B. Bipolar microgrid

A. Unipolar Microgrid

When microgrids have only one level of voltage in the whole grid, known as unipolar microgrids, like VDC. In this system, all the components i.e., loads and generation sources are connected via two-wire lines i.e., positive, and negative, and power is supplied directly to the loads using DC-to-DC converters. It is simple, but it cannot provide redundancy and is weak to defend against even a single fault.

- **Merits of a Unipolar microgrid**: They can offer several potential benefits compared to traditional alternating current (AC) microgrids, which use multiple voltage levels and AC-to-DC converters to supply power to the loads. Some of the potential advantages of unipolar microgrids include:
  1. **Simplicity:** It can be simpler to design and operate than traditional AC microgrids, as they do not require multiple voltage levels or the use of AC-to-DC converters.
  2. **Efficiency:** It can be more efficient than AC microgrids, as they eliminate the losses associated with AC-to-DC conversion and the need for transformers to step down the voltage.
  3. **Compatibility:** It can be compatible with a wider range of loads and generation sources, as they do not require the use of AC-to-DC converters to interface with the microgrid.
  4. **Cost:** They may be less expensive to implement and operate than AC microgrids, due to the simplicity of their design and the lack of AC-to-DC conversion losses.
  5. **No need for a voltage balancer:** Due to single voltage availability, no need for a voltage balancer circuitry.

- **Demerits of a Unipolar Microgrid**: However, They may also have some potential drawbacks or limitations, including
  1. **Limited capacity:** They may have limited capacity, as they are typically designed to serve smaller loads or communities.
  2. **Interoperability:** They may not be fully interoperable with traditional AC microgrids, which could limit their integration into larger power systems or the ability to share resources with other microgrids.
  3. **Regulation:** It may be subject to different regulatory frameworks or standards than traditional AC microgrids, which could create additional challenges for their implementation and operation.
  4. **Less flexibility:** No flexibility in voltage level is allowed.
  5. **Highly Sensitive:** Highly sensitive to a single fault.

Figure 7. Unipolar and bipolar microgrid

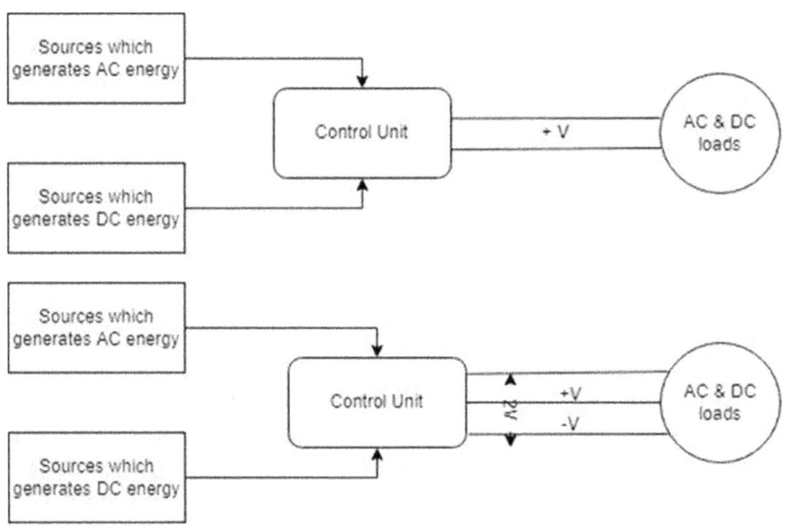

B. Bipolar Microgrid

A bipolar microgrid (Guo et al., 2022; Mohammadi et al., 2019; Wang et al., 2017) is a type of microgrid that uses two or more direct currents (DC) voltage levels for the distribution of electricity within the microgrid like $+V_{DC}$, $-V_{DC}$, or $2V_{DC}$, etc. In this, the loads and generation sources are connected to one of two DC buses, with a DC-to-DC converter used to interface between the two voltage levels.

- **Merits of a Bipolar Microgrid**: Some of the potential advantages of bipolar microgrids include:
  1. **Compatibility:** It can be compatible with a wider range of loads and generation sources because three voltage levels are available in this configuration, as they can use different voltage levels to match the voltage requirements of different resources.

2. **Efficiency:** It can be more efficient than traditional alternating current (AC) microgrids, as they eliminate the losses associated with AC-to-DC conversion and the need for transformers to step down the voltage.
3. **Cost:** It may be less expensive to implement and operate than AC microgrids, due to the simplicity of their design.

- **Demerits of a Bipolar Microgrid**: However, They may also have some potential drawbacks or limitations, including
    1. **Complexity:** It may be more complex to design and operate than traditional AC microgrids, as they involve the management of power flows in multiple voltage domains.
    2. **Interoperability:** It may not be fully interoperable with traditional AC microgrids, which could limit their integration into larger power systems or the ability to share resources with other microgrids.
    3. **Regulation:** It may be subject to different regulatory frameworks or standards than traditional AC microgrids, which could create additional challenges for their implementation and operation.

3.5 Microgrids Based on the "Scenario or End Use Applications"

When microgrids are designed for some specific area or purposes, are comes under the category of 'According to scenario' and are generally divided into three groups (Ashok Kumar & Amutha Prabha, 2022).

A. Residential Microgrids

A residential microgrid (Ali et al., 2021; Golmohamadi, 2022; Haider et al., 2016) is a small-scale electrical grid that serves a single residential community, such as a neighborhood or apartment complex. It is typically connected to the larger, central grid, but can operate independently if necessary. The primary function of a residential microgrid(*Artificial Intelligence Framework for Smart City Microgrids: State of the Art, Challenges, and Opportunities*, n.d.; Rodriguez-Diaz et al., 2015) is to provide a reliable and secure source of power to the homes and businesses within its service area.

- **Merits of Residential microgrid**:
    1. **Improved reliability:** By being able to operate independently, It can provide a reliable source of power even if the central grid experiences an outage or disruption. This can be particularly important in emergency situations.
    2. **Energy security:** It can help to ensure that a specific community has a secure source of energy, which can be important in regions that are vulnerable to disruptions in the central grid.
    3. **Cost savings:** By generating and using energy locally, It can reduce the need for long-distance transmission of electricity, which can be expensive and inefficient.
    4. **Improved energy efficiency:** It can help to optimize energy use by matching energy supply with demand, which can help to reduce waste and improve overall energy efficiency.
    5. **Reduced carbon emissions:** By promoting the use of renewable energy sources, It can help to reduce carbon emissions and contribute to a cleaner, more sustainable energy system.
- **Demerits of Residential microgrid**:

1. **Initial costs:** Setting up and maintaining a It can be expensive, particularly if it involves installing new infrastructure or upgrading existing systems.
2. **Complexity:** It can be complex systems that require specialized expertise to operate and maintain.
3. **Interoperability:** It may not always be compatible with the central grid, which can make it difficult to seamlessly integrate them into the overall energy system.
4. **Regulation:** It may be subject to various regulations and policies, which can vary by location and can impact their development and operation.

B. Industrial Microgrid

An industrial microgrid (Hatziargyriou et al., 2007; Pan et al., 2014; Yamashita et al., 2020)is a small-scale electrical grid that serves a specific industrial area or facility. It is typically connected to the larger, central grid, but can operate independently if necessary. The primary function of an industrial microgrid (Z. Xu et al., 2018) is to provide a reliable and secure source of power to the industrial facilities within its service area.

- **Merits of Industrial microgrids**: There are several potential benefits and drawbacks to industrial microgrids:
    1. **Improved reliability:** By being able to operate independently, It can provide a reliable source of power even if the central grid experiences an outage or disruption. This can be particularly important for critical industrial facilities, such as factories or hospitals, which require a stable and secure power supply.
    2. **Energy security:** It can help to ensure that a specific industrial area has a secure source of energy, which can be important in regions that are vulnerable to disruptions in the central grid.
    3. **Cost savings:** By generating and using energy locally, It can reduce the need for long-distance transmission of electricity, which can be expensive and inefficient.
    4. **Improved energy efficiency:** It can help to optimize energy use by matching energy supply with demand, which can help to reduce waste and improve overall energy efficiency.
    5. **Reduced carbon emissions:** By promoting the use of renewable energy sources It can help to reduce carbon emissions and contribute to a cleaner, more sustainable energy system.
- **Demerits of Industrial microgrid**:
    1. **Initial costs:** Setting up and maintaining It can be expensive, particularly if it involves installing new infrastructure or upgrading existing systems.
    2. **Complexity:** It can be complex systems that require specialized expertise to operate and maintain.
    3. **Interoperability** It may not always be compatible with the central grid, which can make it difficult to seamlessly integrate them into the overall energy system.
    4. **Regulation** It may be subject to various regulations and policies, which can vary by location and can impact their development and operation.

C. The Commercial Microgrid

A commercial microgrid (Jain & Saxena, 2023; Meegahapola et al., 2014; Tao et al., 2011; Tavakoli et al., 2018; Zachar & Daoutidis, 2018; Zhang et al., 2015)is a small-scale electrical grid that serves a specific commercial area or facility, such as a business park or shopping center. It is typically connected to the larger, central grid, but can operate independently if necessary. The primary function is to provide a reliable and secure source of power to businesses and other commercial facilities within its service area.

- **Merits of the commercial microgrid**:
    1. **Improved reliability**: By being able to operate independently It can provide a reliable source of power even if the central grid experiences an outage or disruption. This can be particularly important for businesses that rely on a stable and secure power supply to maintain their operations.
    2. **Energy security:** It can help to ensure that a specific commercial area has a secure source of energy, which can be important in regions that are vulnerable to disruptions in the central grid.
    3. **Cost savings:** By generating and using energy locally, It can reduce the need for long-distance transmission of electricity, which can be expensive and inefficient. They can also help to reduce energy costs by allowing commercial customers to choose their own energy sources and control their energy use.
    4. **Improved energy efficiency:** It can help to optimize energy use by matching energy supply with demand, which can help to reduce waste and improve overall energy efficiency.
    5. **Reduced carbon emissions:** By promoting the use of renewable energy sources, It can help to reduce carbon emissions and contribute to a cleaner, more sustainable energy system.
- **Demerits of the commercial microgrid**:
    1. **Initial costs**: Setting up and maintaining a It can be expensive, particularly if it involves installing new infrastructure or upgrading existing systems.
    2. **Complexity**: It can be complex systems that require specialized expertise to operate and maintain.
    3. **Interoperability** It may not always be compatible with the central grid, which can make it difficult to seamlessly integrate them into the overall energy system.

3.6 Microgrids Based on the "Generation Source Type"

According to the material or sources used to generate energy, microgrids can be classified into two groups (Jirdehi et al., 2020; Rezkallah et al., 2023):

A. Renewable energy resources microgrid
B. Non-renewable energy resources microgrid

A. Renewable Energy Sources Type Microgrid

Renewable energy sources (Fazal et al., 2023; Hai et al., 2023; Rezkallah et al., 2023) are sources of energy that are naturally replenished and can be used indefinitely without running out. Some examples of renewable energy sources include solar, wind, hydroelectric, geothermal, and bioenergy. These energy sources are typically used to generate electricity and can be an important component of microgrids, particularly in areas where it is difficult or impossible to connect to the central grid.

- **Merits of Renewable energy resources microgrid**:
  1. **Sustainable:** It are naturally replenished and can be used indefinitely, which makes them a more sustainable source of energy than non-renewable sources.
  2. **Clean:** It tends to have low or zero emissions, which makes them a cleaner source of energy than non-renewable sources, which often produce air pollution and greenhouse gases.
  3. **Cost-effective:** In the long term It can be more cost-effective than non-renewable sources, as they do not require the extraction and processing of finite resources, which can be expensive and environmentally damaging.
  4. **Localized:** It can often be generated and used locally, which can help to reduce the need for long-distance transmission of electricity and improve energy security.
  5. **Community ownership:** It can be owned and managed by the community or region in which they are located, which can help to foster a sense of ownership and control over the energy system.
- **Demerits of Renewable energy resources microgrid:**
  1. **Initial costs:** Installing this systems can be expensive, particularly if it involves building new infrastructure or upgrading existing systems.
  2. **Intermittency:** Some renewable energy sources, such as solar and wind, are intermittent, which means that their generation can vary depending on weather conditions. This can make it more difficult to match energy supply with demand and may require the use of energy storage systems.
  3. **Land use:** Some renewable energy sources, such as large-scale solar and wind projects, may require significant amounts of land, which can be a concern in areas with limited land resources or environmental sensitivity.
  4. **Regulation:** Renewable energy projects may be subject to various regulations and policies, which can vary by location and can impact their development and operation (Adefarati & Bansal, 2019).

B. Non-Renewable Energy Resources Microgrids

Non-renewable energy sources are sources of energy that are finite and cannot be replenished at a rate that is equal to or greater than their rate of consumption. Some examples of non-renewable energy sources include (Abu-Elzait & Parkin, 2019; Husein & Chung, 2018) fossil fuels such as coal, oil, and natural gas, as well as nuclear energy. These energy sources are typically used to generate electricity, and are often the primary source of energy for traditional electrical grids.

- **Merits of Non-renewable energy microgrid**:
  1. **Widely available:** Non-renewable energy sources, such as fossil fuels, are widely available and have been extensively developed and tested, which makes them relatively easy to access and use.

2. **High energy density:** They tend to have a high energy density, which means that they can provide a lot of energy per unit of weight or volume. This can be particularly useful in small, portable systems such as microgrids.
3. **Established infrastructure:** They are often supported by established infrastructure, including pipelines, transmission lines, and storage facilities, which can make them easier to transport and use.

- **Demerits of Non-renewable energy microgrid**:
    1. **Limited supply:** They are finite and will eventually run out, which means that they are not a sustainable source of energy in the long term.
    2. **Environmental impacts:** The extraction and use it, can have negative environmental impacts, such as air pollution and greenhouse gas emissions, which can contribute to climate change.
    3. **Increasing cost:** The cost of extracting and processing non-renewable energy sources tends to increase over time, which can make them more expensive than renewable energy sources in the long term.
    4. **Dependence on external sources:** They are often imported from other countries, which can make a microgrid dependent on external sources and vulnerable to disruptions in supply.
    5. **Political instability:** In some cases, the extraction and use of non-renewable energy sources can be associated with political instability or conflict, which can create risks for microgrids that rely on these sources.

To produce energy, two different sorts of resources are utilized. Given that these resources are abundantly available in nature, the cost of raw materials is typically zero when electricity is produced using renewable energy sources. The resources on Earth that are utilized to create energy are frequently divided into two groups. The first one is a renewable resource, but the others are not. In Table.1, various resource kinds are distinguished.

Table 1. Type of resources for generation of electricity

S.N.	Non-Renewable Energy Resources	Renewable Energy Resources
1	Coal	Solar energy
2	Petroleum products	Wind energy
3	Crude oil	Hydropower
4	Non-renewable waste	Geothermal Power
5	Derived gasses	Biomass energy (ethanol, wood, sevage)
6	Natural gasses	Tidal energy
7	Bitumens	Wave energy
8	Groundwater	Biodiesel
9	Nuclear energy	

4. CONCLUSION AND FUTURE SCOPE

In the study that is being presented, a thorough history of the power generation transition towards DC-type microgrids is explained. The issues with the traditional AC grid and the difficulty of providing electricity to all small units or remote areas are also covered. Researchers have recommended microgrids in response to the issue. Whereas DC microgrids are the best option in the current situation. However, in this manuscript, the drawbacks of every generation system are briefly explained, and the benefits of every generation method are also covered. Presently most of the appliances are working in the DC nature like lighting systems, DC vehicles, etc.

This document can serve as a platform to identify the precise research point to benefit the country by obtaining information on the benefits and drawbacks of all types of generation systems. The easiest way to deal with the current issue is to employ a DC microgrid with DC loads that operate at voltages lower than 50 V. In addition to safety for human beings, improved efficiency, good power quality, a decrease in air pollution, and global warming are also benefits of choosing below 50V.

REFERENCES

Abdolmaleki, B., & Bergna-Diaz, G. (2022). Distributed Control and Optimization of DC Microgrids: A Port-Hamiltonian Approach. *IEEE Access : Practical Innovations, Open Solutions*, *10*, 64222–64233. doi:10.1109/ACCESS.2022.3183209

Abhishek, A., Ranjan, A., Devassy, S., Kumar Verma, B., Ram, S. K., & Dhakar, A. K. (2020). Review of hierarchical control strategies for DC microgrid. *IET Renewable Power Generation*, *14*(10), 1631–1640. doi:10.1049/iet-rpg.2019.1136

Abu-Elzait, S., & Parkin, R. (2019). Economic and Environmental Advantages of Renewable-based Microgrids over Conventional Microgrids. *2019 IEEE Green Technologies Conference(GreenTech)*, 1–4. 10.1109/GreenTech.2019.8767146

Adefarati, T., & Bansal, R. C. (2019). Reliability, economic and environmental analysis of a microgrid system in the presence of renewable energy resources. *Applied Energy*, *236*, 1089–1114. doi:10.1016/j.apenergy.2018.12.050

Ahmad, F., Alam, M. S., & Alsaidan, I. S. (2020). Campus Microgrid: A Case Study. *2020 IEEE/PES Transmission and Distribution Conference and Exposition (T&D)*, 1–6. 10.1109/TD39804.2020.9299998

Al-Ismail, F. S. (2021). DC Microgrid Planning, Operation, and Control: A Comprehensive Review. *IEEE Access : Practical Innovations, Open Solutions*, *9*, 36154–36172. doi:10.1109/ACCESS.2021.3062840

Alam, M., Al-Ismail, F. S., Al-Sulaiman, F. A., & Abido, M. A. (2023). Energy management in DC microgrid with an efficient voltage compensation mechanism. *Electric Power Systems Research*, *214*, 108842. doi:10.1016/j.epsr.2022.108842

Ali, S., Zheng, Z., Aillerie, M., Sawicki, J.-P., Péra, M.-C., & Hissel, D. (2021). A Review of DC Microgrid Energy Management Systems Dedicated to Residential Applications. *Energies*, *14*(14), 4308. doi:10.3390/en14144308

Altaf, M. W., Arif, M. T., Islam, S. N., & Haque, M. (2022). Microgrid Protection Challenges and Mitigation Approaches–A Comprehensive Review. *IEEE Access : Practical Innovations, Open Solutions, 10*, 38895–38922. doi:10.1109/ACCESS.2022.3165011

Ashok Kumar, A., & Amutha Prabha, N. (2022). A comprehensive review of DC microgrid in market segments and control technique. *Heliyon, 8*(11), e11694. doi:10.1016/j.heliyon.2022.e11694 PMID:36425418

Augustine, S., Quiroz, J. E., Reno, M. J., & Brahma, S. (2018). *DC Microgrid Protection: Review and Challenges.* doi:10.2172/1465634

Buonocore, J. J., Salimifard, P., Michanowicz, D. R., & Allen, J. G. (2021). A decade of the U.S. energy mix transitioning away from coal: Historical reconstruction of the reductions in the public health burden of energy. *Environmental Research Letters, 16*(5), 054030. doi:10.1088/1748-9326/abe74c

D'silva, S., Shadmand, M., Bayhan, S., & Abu-Rub, H. (2020). Towards Grid of Microgrids: Seamless Transition between Grid-Connected and Islanded Modes of Operation. *IEEE Open Journal of the Industrial Electronics Society, 1*, 66–81. doi:10.1109/OJIES.2020.2988618

Fang, X., Misra, S., Xue, G., & Yang, D. (2012). Smart Grid — The New and Improved Power Grid: A Survey. *IEEE Communications Surveys and Tutorials, 14*(4), 944–980. doi:10.1109/SURV.2011.101911.00087

Fazal, S., Enamul Haque, M., Taufiqul Arif, M., Gargoom, A., & Oo, A. M. T. (2023). Grid integration impacts and control strategies for renewable based microgrid. *Sustainable Energy Technologies and Assessments, 56*, 103069. doi:10.1016/j.seta.2023.103069

Ferrera, M. B., Litran, S. P., Duran, E., & Andujar, J. M. (2015). A SEPIC-Cuk converter combination for bipolar DC microgrid applications. *2015 IEEE International Conference on Industrial Technology (ICIT)*, 884–889. 10.1109/ICIT.2015.7125209

Finkelman, R. B., Wolfe, A., & Hendryx, M. S. (2021). The future environmental and health impacts of coal. *Energy Geoscience, 2*(2), 99–112. doi:10.1016/j.engeos.2020.11.001

Gao, F., Wang, X., Yang, P., Kou, S., & Sun, M. (2020). Research and Simulation of Hybrid AC/DC Microgrid. *2020 4th International Conference on HVDC (HVDC)*, 1276–1280. 10.1109/HVDC50696.2020.9292793

Golmohamadi, H. (2022). Demand-Side Flexibility in Power Systems: A Survey of Residential, Industrial, Commercial, and Agricultural Sectors. *Sustainability (Basel), 14*(13), 7916. doi:10.3390/su14137916

Guo, C., Liao, J., & Zhang, Y. (2022). Adaptive droop control of unbalanced voltage in the multi-node bipolar DC microgrid based on fuzzy control. *International Journal of Electrical Power & Energy Systems, 142*, 108300. doi:10.1016/j.ijepes.2022.108300

Hai, T., Zhou, J., & Muranaka, K. (2023). Energy management and operational planning of renewable energy resources-based microgrid with energy saving. *Electric Power Systems Research, 214*, 108792. doi:10.1016/j.epsr.2022.108792

Haider, H. T., See, O. H., & Elmenreich, W. (2016). A review of residential demand response of smart grid. *Renewable & Sustainable Energy Reviews, 59*, 166–178. doi:10.1016/j.rser.2016.01.016

Hatziargyriou, N., Asano, H., Iravani, R., & Marnay, C. (2007). Microgrids. *IEEE Power & Energy Magazine*, *5*(4), 78–94. doi:10.1109/MPAE.2007.376583

Husein, M., & Chung, I.-Y. (2018). Optimal design and financial feasibility of a university campus microgrid considering renewable energy incentives. *Applied Energy*, *225*, 273–289. doi:10.1016/j.apenergy.2018.05.036

Jain, D., & Saxena, D. (2023). Comprehensive review on control schemes and stability investigation of hybrid AC-DC microgrid. *Electric Power Systems Research*, *218*, 109182. doi:10.1016/j.epsr.2023.109182

Jirdehi, M. A., Tabar, V. S., Ghassemzadeh, S., & Tohidi, S. (2020). Different aspects of microgrid management: A comprehensive review. *Journal of Energy Storage*, *30*, 101457. doi:10.1016/j.est.2020.101457

Kant, K., & Gupta, O. H. (2022). DC Microgrid: A Comprehensive Review on Protection Challenges and Schemes. *IETE Technical Review*, 1–17. doi:10.1080/02564602.2022.2136271

Kishor, Y., Kamesh Rao, C. H., & Patel, R. N. (2022). Comprehensive Analysis on DC-Microgrid Application for Remote Electrification. In *Renewable Energy Technologies* (pp. 435–475). Wiley. doi:10.1002/9781119827634.ch13

Lin, Z., Wang, P., Ren, S., & Zhao, D. (2023). Comprehensive Impact Assessment of Carbon Neutral Pathways and Air pollution Control Policies in Shaanxi Province of China. *Resources, Conservation & Recycling Advances*, *200143*. Advance online publication. doi:10.1016/j.rcradv.2023.200143

Liu, X., Wang, P., & Loh, P. C. (2010). A hybrid AC/DC micro-grid. *2010 Conference Proceedings IPEC*, 746–751. 10.1109/IPECON.2010.5697024

Malik, S. M., Ai, X., Sun, Y., Zhengqi, C., & Shupeng, Z. (2017). Voltage and frequency control strategies of hybrid AC/DC microgrid: A review. *IET Generation, Transmission & Distribution*, *11*(2), 303–313. doi:10.1049/iet-gtd.2016.0791

Maurya, R., Prakash, S., & Singh, A. K. (2022). Challenges, Configuration, Control, and Scope of DC Microgrid Systems: A Review. *Journal of Electrical Engineering & Technology*. Advance online publication. doi:10.1007/s42835-022-01304-y

Meegahapola, L. G., Robinson, D., Agalgaonkar, A. P., Perera, S., & Ciufo, P. (2014). Microgrids of Commercial Buildings: Strategies to Manage Mode Transfer From Grid Connected to Islanded Mode. *IEEE Transactions on Sustainable Energy*, *5*(4), 1337–1347. doi:10.1109/TSTE.2014.2305657

Modu, B., Abdullah, M. P., Sanusi, M. A., & Hamza, M. F. (2023). DC-based microgrid: Topologies, control schemes, and implementations. *Alexandria Engineering Journal*, *70*, 61–92. doi:10.1016/j.aej.2023.02.021

Mohammadi, J., Badrkhani Ajaei, F., & Stevens, G. (2019). Grounding the DC Microgrid. *IEEE Transactions on Industry Applications*, *55*(5), 4490–4499. doi:10.1109/TIA.2019.2928278

Munawer, M. E. (2018). Human health and environmental impacts of coal combustion and post-combustion wastes. *Journal of Sustainable Mining*, *17*(2), 87–96. doi:10.1016/j.jsm.2017.12.007

Nawaz, A., Zhou, M., Wu, J., & Long, C. (2022). A comprehensive review on energy management, demand response, and coordination schemes utilization in multi-microgrids network. *Applied Energy*, *323*, 119596. doi:10.1016/j.apenergy.2022.119596

Orji, U., Schantz, C., Leeb, S. B., Kirtley, J. L., Sievenpiper, B., Gerhard, K., & McCoy, T. (2017). Adaptive Zonal Protection for Ring Microgrids. *IEEE Transactions on Smart Grid*, *8*(4), 1843–1851. doi:10.1109/TSG.2015.2509018

Pan, J., Jain, R., & Paul, S. (2014). A Survey of Energy Efficiency in Buildings and Microgrids using Networking Technologies. *IEEE Communications Surveys and Tutorials*, *16*(3), 1709–1731. doi:10.1109/SURV.2014.060914.00089

Planas, E., Andreu, J., Gárate, J. I., Martínez de Alegría, I., & Ibarra, E. (2015). AC and DC technology in microgrids: A review. *Renewable & Sustainable Energy Reviews*, *43*, 726–749. doi:10.1016/j.rser.2014.11.067

Priyadarshini Padhi, P., & Deepa, K. (2021). DC Microgrids: Architecture and Challenges. *IOP Conference Series. Materials Science and Engineering*, *1070*(1), 012098. doi:10.1088/1757-899X/1070/1/012098

Rezkallah, M., Ibrahim, H., Chandra, A. F., Hamadi, A., & Hazzab, A. (2023). Control of Hybrid Off-Microgrid Configurations Based on Renewable and Non-renewable Energy Sources. doi:10.1007/978-981-19-7956-9_8

Rodriguez-Diaz, E., Savaghebi, M., Vasquez, J. C., & Guerrero, J. M. (2015). An overview of low voltage DC distribution systems for residential applications. *2015 IEEE 5th International Conference on Consumer Electronics - Berlin (ICCE-Berlin)*, 318–322. 10.1109/ICCE-Berlin.2015.7391268

Sabzehgar, R. (2017). Overview of Technical Challenges, Available Technologies and Ongoing Developments of AC/DC Microgrids. In *Development and Integration of Microgrids*. InTech. doi:10.5772/intechopen.69400

Saeed, M. H., Fangzong, W., Kalwar, B. A., & Iqbal, S. (2021). A Review on Microgrids' Challenges & Perspectives. *IEEE Access : Practical Innovations, Open Solutions*, *9*, 166502–166517. doi:10.1109/ACCESS.2021.3135083

Sahoo, S. K., Sinha, A. K., & Kishore, N. K. (2018). Control Techniques in AC, DC, and Hybrid AC–DC Microgrid: A Review. *IEEE Journal of Emerging and Selected Topics in Power Electronics*, *6*(2), 738–759. doi:10.1109/JESTPE.2017.2786588

Subburaj, D. V., Dash, R., Reddy, J., & CN, R. (2022). *International Congress on Renewable Energy (I CORE 2021)*. 10.13052/rp-9788770227537

Talapur, G. G., Suryawanshi, H. M., Xu, L., & Shitole, A. B. (2018). A Reliable Microgrid With Seamless Transition Between Grid Connected and Islanded Mode for Residential Community With Enhanced Power Quality. *IEEE Transactions on Industry Applications*, *54*(5), 5246–5255. doi:10.1109/TIA.2018.2808482

Tao, L., Schwaegerl, C., Narayanan, S., & Zhang, J. H. (2011). From laboratory Microgrid to real markets — Challenges and opportunities. *8th International Conference on Power Electronics - ECCE Asia*, 264–271. 10.1109/ICPE.2011.5944600

Tavakoli, M., Shokridehaki, F., Funsho Akorede, M., Marzband, M., Vechiu, I., & Pouresmaeil, E. (2018). CVaR-based energy management scheme for optimal resilience and operational cost in commercial building microgrids. *International Journal of Electrical Power & Energy Systems*, *100*, 1–9. doi:10.1016/j.ijepes.2018.02.022

Unamuno, E., & Barrena, J. A. (2015a). Hybrid ac/dc microgrids—Part I: Review and classification of topologies. *Renewable & Sustainable Energy Reviews*, *52*, 1251–1259. doi:10.1016/j.rser.2015.07.194

Unamuno, E., & Barrena, J. A. (2015b). Hybrid ac/dc microgrids—Part II: Review and classification of control strategies. *Renewable & Sustainable Energy Reviews*, *52*, 1123–1134. doi:10.1016/j.rser.2015.07.186

Vegunta, S. C., Higginson, M. J., Kenarangui, Y. E., Li, G. T., Zabel, D. W., Tasdighi, M., & Shadman, A. (2021). AC Microgrid Protection System Design Challenges—A Practical Experience. *Energies*, *14*(7), 2016. doi:10.3390/en14072016

Wang, F., Lei, Z., Xu, X., & Shu, X. (2017). Topology Deduction and Analysis of Voltage Balancers for DC Microgrid. *IEEE Journal of Emerging and Selected Topics in Power Electronics*, *5*(2), 672–680. doi:10.1109/JESTPE.2016.2638959

Winardi, B., Sinuraya, E. W., & Nugroho, A. (2021). Design and Implementation of Rooftop off Grid Solar Panel for Household Electricity Loads. *2021 8th International Conference on Information Technology, Computer and Electrical Engineering (ICITACEE)*, 258–263. doi:10.1109/ICITACEE53184.2021.9617551

World Energy Outlook 2022. (n.d.). www.iea.org/t&c/

Xia, Y., Wei, W., Yu, M., Wang, X., & Peng, Y. (2018). Power Management for a Hybrid AC/DC Microgrid With Multiple Subgrids. *IEEE Transactions on Power Electronics*, *33*(4), 3520–3533. doi:10.1109/TPEL.2017.2705133

Xu, L., Guerrero, J., Lashab, A., Wei, B., Bazmohammadi, N., Vasquez, J., & Abusorrah, A. (2022). A Review of DC Shipboard Microgrids—Part I: Power Architectures, Energy Storage, and Power Converters. *IEEE Transactions on Power Electronics*, *37*(5), 5155–5172. doi:10.1109/TPEL.2021.3128417

Xu, Z., Yang, P., Zheng, C., Zhang, Y., Peng, J., & Zeng, Z. (2018). Analysis on the organization and Development of multi-microgrids. *Renewable & Sustainable Energy Reviews*, *81*, 2204–2216. doi:10.1016/j.rser.2017.06.032

Yadav, N., & Tummuru, N. R. (2023). Application of Filter Capacitor Dynamics Based Short-Circuit Fault Detection Method in Ring-Type DC Microgrid. *IEEE Transactions on Industrial Electronics*, *70*(5), 5286–5295. doi:10.1109/TIE.2022.3183363

Yamashita, D. Y., Vechiu, I., & Gaubert, J.-P. (2020). A review of hierarchical control for building microgrids. *Renewable & Sustainable Energy Reviews*, *118*, 109523. doi:10.1016/j.rser.2019.109523

Zachar, M., & Daoutidis, P. (2018). Energy management and load shaping for commercial microgrids coupled with flexible building environment control. *Journal of Energy Storage*, *16*, 61–75. doi:10.1016/j.est.2017.12.017

Zhang, F., Meng, C., Yang, Y., Sun, C., Ji, C., Chen, Y., Wei, W., Qiu, H., & Yang, G. (2015). Advantages and challenges of DC microgrid for commercial building a case study from Xiamen university DC microgrid. *2015 IEEE First International Conference on DC Microgrids (ICDCM)*, 355–358. 10.1109/ICDCM.2015.7152068

Zolfaghari, M., Gharehpetian, G. B., Shafie-khah, M., & Catalão, J. P. S. (2022). Comprehensive review on the strategies for controlling the interconnection of AC and DC microgrids. *International Journal of Electrical Power & Energy Systems*, *136*, 107742. doi:10.1016/j.ijepes.2021.107742

Chapter 11
Sustainable Energy Planning and Management Using Metaheuristic Algorithms and Computational Modeling

Latifa Dekhici
https://orcid.org/0000-0002-9581-6488
University of Sciences and Technology of Oran Mohamed Boudiaf (USTO-MB), Algeria

Khaled Guerraiche
LDREI laboratory, Higher School of Electronic and Electrical Engineering of Oran (ESGEEO), Algeria

Karima Belmabrouk
https://orcid.org/0000-0003-1157-1274
University of Sciences and Technology of Oran Mohamed Boudiaf (USTO-MB), Algeria

ABSTRACT

Sustainable energy planning and management is crucial for achieving a reliable, affordable, and sustainable energy supply, while promoting environmental protection, social equity, and economic efficiency. Optimizing renewable energy systems for sustainability involves the development, deployment, and operation of renewable energy systems that maximize their environmental, social, and economic benefits while minimizing negative impacts. Metaheuristic algorithms and computational modeling techniques are increasingly being used to optimize renewable energy systems for sustainability. This chapter highlights the use of computational modeling in the field of suitable energy planning and management, with a focus on some fields as solar power forecasting and wind turbine design. Two case studies are presented, namely the environmental economic power dispatch and the environmental friendly microgrids redundancy problem.

DOI: 10.4018/978-1-6684-9130-0.ch011

I. INTRODUCTION

Sustainable energy planning and management is an essential process for achieving a reliable, affordable, and sustainable energy supply while promoting environmental protection, social equity, and economic efficiency. This process involves assessing the current energy system, identifying areas for improvement, and implementing policies and programs that encourage the use of renewable energy sources, increase energy efficiency, and reduce greenhouse gas emissions.

Optimizing renewable energy systems for sustainability is a critical aspect of sustainable energy planning and management, requiring the development, deployment, and operation of renewable energy systems that maximize their environmental, social, and economic benefits while minimizing negative impacts. These efforts can also address issues of energy equity and social justice by prioritizing renewable energy systems deployment in disadvantaged communities and ensuring equitable access to benefits.

In addition, optimizing renewable energy systems for sustainability can drive innovation and economic growth, as investing in renewable energy technologies and a skilled workforce can create new jobs and opportunities while reducing dependence on fossil fuels and promoting a sustainable and resilient economy.

Metaheuristic algorithms and computational modeling techniques are increasingly being used to optimize renewable energy systems for sustainability. Metaheuristic algorithms, such as evolutionary, swarm intelligence, and simulated annealing algorithms, can solve complex, non-linear optimization problems commonly encountered in renewable energy systems. Computational modeling involves computer simulations and mathematical models to analyze and predict the behavior of renewable energy systems under different conditions. By using these tools, the efficiency can be improved, reduce the environmental impact can be reduced, and the wider adoption of renewable energy systems can be promoted, thus facilitating the transition to a more sustainable and equitable energy system.

In this chapter, the use of computational modeling in the suitable energy planning and management is described citing many fields. Two cases studies: Environmental Economic Power Dispatch and Environmental Friendly Microgrids Redundancy Problem are detailed.

II. THE USE OF COMPUTATIONAL MODELING IN THE SUITABLE ENERGY PLANNING AND MANAGEMENT

Computational modeling involves the use of computer programs and simulations to model and analyze complex systems. This approach is often used in environmental science and policy to evaluate the impacts of different policy options and to support decision-making.

1. Solar Power Forecasting

Computational modeling can be used to predict the amount of solar power that will be generated by a photovoltaic (PV) system. This can help grid operators to manage the grid and balance the energy supply and demand. Computational modeling methods for solar power forecasting can include machine learning algorithms, such as neural networks or support vector regression, and physical models, such as radiative transfer models or clear sky models. These methods use historical solar irradiance data and weather forecasting data to predict the output of a PV system. Yu et al. (2023) proposed a hybrid forecasting system that integrates pattern recognition algorithms, data decomposition strategy, and multi-objective optimization algorithm. The proposed system achieved the best performance in all aspects of the four forecast scenarios and also performed well in the multistep advance forecast. Cheng et al. (2022) evaluated the performance of opaque deep-learning solar power forecast models towards power-grid applications. The study concluded that some simple evaluation procedures can aid a better understanding of the factors that influence the performances of opaque models, and these procedures can help in the design methods for model modifications. Lindberg et al. (2023) investigated the effect of aggregation on forecast accuracy and value at a co-located wind and photovoltaic (PV) park in Sweden using roughly three years of data. The results showed that aggregation improves forecast accuracy in terms of continuous ranked probability score, interval score, and quantile score when compared to wind or PV power forecasts alone. The optimal aggregation ratio is found to be 50%-60% wind power and the remainder PV power.

2. Wind Turbine Design

Computational modeling can be used to optimize the design of wind turbines to maximize their energy output. This includes simulating the aerodynamics of the turbine blades and the flow of wind through the turbine. computational modeling for wind turbine design can involve finite element analysis (FEA), computational fluid dynamics (CFD), and optimization algorithms. FEA is used to model the structural behavior of the turbine components under different loads, while CFD is used to simulate the aerodynamics of the turbine blades and the flow of wind through the turbine. Optimization algorithms can be used to determine the optimal turbine design parameters, such as blade geometry and rotor diameter, to maximize energy output. Researchers have proposed various methods to optimize the power output of wind turbines. Shu et al. (2022) propose a method to reduce the complexity and size of the power optimization problem for a large-scale offshore wind farm by establishing sparse communication architecture. Hassan et al. (2022) propose an optimized design of wind cubes placed in front of wind turbines to improve airflow and maximize energy output. Pustina et al. (2022) present an Economic Nonlinear Model Predictive Control strategy to maximize power production of wind turbines using a novel Reduced Order Model. Sedaghat et al. (2020) emphasize the importance of considering both the capacity factor and annual energy production (AEP) when designing or selecting a wind turbine, and introduce a new semi-empirical secondary capacity factor to determine a rated wind speed at which yearly energy and hydrogen production obtain a maximum value. These methods demonstrate effective ways to optimize wind turbine power output under various operating conditions.

These studies demonstrate the effectiveness of different approaches to optimize wind turbine power output under various operating conditions.

3. Battery Storage Optimization

Computational modeling can be used to optimize the charging and discharging of battery storage systems in renewable energy systems. This can help to maximize the efficiency of the battery system and extend its lifespan. Computational modeling for battery storage optimization can include mathematical programming, optimization algorithms, and simulation tools. These methods can be used to determine the optimal charging and discharging strategies for a given battery storage system, considering factors such as energy demand, renewable energy generation, and battery lifespan. Several recent studies have proposed optimization models for battery energy storage (BES) integrated with solar PV in residential settings. For example, Abedi and Kwon (2023) develop a rolling-horizon optimization model using recurrent neural network-driven forecasting, while Jiwei Yao et al. (2023) propose a two-level optimization framework that can extend battery lifespan by up to 5.1% and increase revenue by up to 9.8%. Ouedraogo (2023) proposes an energy management system for residential PV and battery connected to the grid that can reduce energy bills. Other recent studies have focused on optimizing the design of standalone PV/BES and PV/BES/DG systems in remote areas. Zhihan Zhang et al. (2022) propose an optimal design based on efficient energy management systems (EMS) for smart building electrification, while Zhang et al. (2023) propose an optimization model for hybrid solar energy structures in zero energy buildings, which takes into account sustainable energy indicators, interest rates, and depth of discharge. Dynamic modeling and multi-objective optimization based on the Taguchi approach have been used to evaluate off-grid hybrid renewable energy systems with battery or hydrogen storage options. Kilic and Altun (2022) demonstrate that hybridization of energy resources can significantly reduce CO2 emissions, with hydrogen storage having lower CO2 emissions in the long term despite higher costs. Lin et al. (2023) propose a solution to improve the airflow distribution of a battery energy-storage system (BESS) inspired by the ventilation system of data centers. They identify factors inducing unbalanced airflow distribution and propose a generalized solution of layout arrangement to improve temperature uniformity and coefficient of performance, demonstrating potential for effective and rapid design optimization.

4. Biofuel Production

Computational modeling can be used to optimize the production of biofuels from different feedstocks. This includes simulating the chemical processes involved in converting biomass to biofuels and evaluating the environmental impact of different production methods. Computational modeling methods for biofuel production can include process modeling, life cycle assessment (LCA), and optimization algorithms. Process modeling is used to simulate the chemical processes involved in converting biomass to biofuels, while LCA is used to evaluate the environmental impact of different production methods. Optimization algorithms can be used to determine the optimal feedstock selection and conversion process parameters to maximize biofuel production and minimize environmental impact. Meena et al. (2021) discuss how AI technologies can improve the bioenergy production system, addressing feedstock availability, economics, and supply chain management. AI techniques can predict energy from biomass and optimize process parameters for efficient bioconversion technologies, as well as manage biomass supply chains. Ziolkowska (2013) proposes a multi-objective decision analysis approach using fuzzy set theory and the PROMETHEE method to evaluate the effectiveness, efficiency, and sustainability of biofuel technologies and policies, finding algae and switchgrass as the most sustainable feedstocks. Banerjee et al. (2016) use a genetic algorithm to optimize the cultivation of Nannochloropsis sp., achieving the highest productivity of biomass, lipid, and eicosapentaenoic acid. Zarei et al. (2022) propose a multi-stage stochastic program for strategic and tactical planning in biofuel supply chain networks, considering uncertainties in biomass supply and biofuel demand, and minimizing the total annualized cost by selecting raw materials, positioning production facilities, and placing storage facilities. The study shows that considering risk can increase biofuel expected production cost by up to 88%.

5. Geothermal Energy Production

Computational modeling can be used to simulate the flow of water and heat in geothermal reservoirs. This can help to optimize the design of geothermal power plants and improve the efficiency of energy production. computational modeling for geothermal energy production can involve numerical simulation tools, such as finite difference methods or finite element methods, and optimization algorithms. Numerical simulation tools are used to model the flow of water and heat in geothermal reservoirs, while optimization algorithms can be used to determine the optimal design parameters for geothermal power plants, such as well placement and size, to maximize energy production. Geothermal energy is a promising renewable energy source, and researchers are investigating ways to optimize its use. Hoteit et al. (2023) propose a method to estimate thermal recovery and produced-enthalpy rates and quantify uncertainties to analyze geothermal fields' optimal well spacing. Tekkanat et al. (2023) design and evaluate a geothermal-based multigeneration system that is optimized using multi-objective optimization techniques to achieve high exergetic effectiveness and low total cost rate. Guo et al. (2023) propose a multi-generation geothermal-based energy system that produces various valuable products such as fuel gas, electricity, cooling, and freshwater. Li et al. (2023) establish a 3D realistic geothermal model to understand the natural status of temperature distribution in the Panzhuang karst geothermal reservoir and optimize well placement. Wang et al. (2022) propose an approach that uses a surrogate model based on the random forest technique to optimize the placement of geothermal wells in heterogeneous reservoirs. Yilmaz and Sen (2022) propose an optimization approach for a geothermal and solar-assisted sustainable energy system with hydrogen production using a genetic algorithm. Xue et al. (2023) develop an optimization framework that incorporates economic indicators, such as the levelized cost of electricity, and uses an Artificial Neural Network and Differential Evolution to optimize an enhanced geothermal system. These studies provide useful insights into optimizing the use of geothermal energy for sustainable energy systems.

6. Microgrids

Computational modeling can be used to optimize the design and operation of microgrids, which are small-scale electricity grids that can operate independently or in conjunction with the main power grid. This includes simulating the energy generation and consumption of different microgrid components, such as solar panels, wind turbines, and battery storage systems. Microgrids can be modeled using a combination of mathematical programming, optimization techniques, and simulation tools. For example, optimization algorithms can be used to determine the optimal sizing and placement of different microgrid components, while simulation tools can be used to evaluate the performance of the microgrid under different operating conditions.

7. Energy Efficiency

Computational modeling can be used to optimize the energy consumption of buildings, industrial processes, and transportation systems. This includes simulating the energy use of different components and evaluating the effectiveness of energy-saving measures. Energy efficiency modeling can use a range of methods, including energy audits, building energy simulation software, and data analytics. Energy audits involve analyzing energy consumption patterns and identifying areas for improvement. Building energy simulation software can be used to model the energy consumption of buildings under different scenarios. Data analytics can be used to identify energy-saving opportunities by analyzing large datasets of energy consumption data.

8. Renewable Energy Integration

Computational modeling can be used to optimize the integration of renewable energy sources into the power grid. This includes simulating the behavior of renewable energy technologies under different conditions and developing algorithms to manage the variability of renewable energy generation. Renewable energy integration can be modeled using a combination of power system analysis tools and optimization algorithms. Power system analysis tools can be used to model the behavior of the power grid under different operating conditions, while optimization algorithms can be used to determine the optimal placement and capacity of renewable energy sources to minimize carbon emissions and ensure grid stability.

9. Energy Storage

Computational modeling can be used to optimize the design and operation of energy storage systems, such as batteries and pumped hydroelectric storage. This includes simulating the charging and discharging of the storage system and evaluating the effectiveness of different storage technologies. Energy storage can be modeled using a range of methods, including electrochemical simulation tools, optimization algorithms, and life-cycle analysis. Electrochemical simulation tools can be used to model the behavior of batteries and other energy storage systems under different operating conditions. Optimization algorithms can be used to determine the optimal charging and discharging strategies to maximize energy efficiency. Life-cycle analysis can be used to evaluate the environmental impact of different energy storage technologies over their entire life cycle.

10. Green Hydrogen Production

Computational modeling can be used to optimize the production of green hydrogen, which is produced from renewable energy sources such as solar and wind power. This includes simulating the chemical processes involved in producing hydrogen from water and evaluating the efficiency of different production methods. Green hydrogen production can be modeled using a combination of chemical reaction modeling, thermodynamics, and optimization techniques. Chemical reaction modeling can be used to simulate the chemical processes involved in producing hydrogen from water, while thermodynamics can be used to determine the energy requirements for the process. Optimization techniques can be used to determine the optimal operating conditions and maximize energy efficiency.

11. Computational Modeling for Environmental Policy

This topic could focus on how computational modeling can be used to inform environmental policy and regulation. It could cover issues such as modeling the economic and environmental impacts of different policy options and using computational techniques to support decision-making.

III. THE INCLUSION OF METAHEURISTICS IN PROBLEMS RESOLUTION

Metaheuristics are a type of optimization algorithm that can be used to find high-quality solutions to complex problems. Examples of metaheuristics include genetic algorithms, particle swarm optimization, simulated annealing, and ant colony optimization. These algorithms are often used in operational research to optimize complex systems, including renewable energy systems, transportation networks, and resource management systems.

In recent years, the use of metaheuristic algorithms and computational modeling for renewable energy optimization has gained increasing attention in the research community. Numerous studies have evaluated the effectiveness of different metaheuristic algorithms, including genetic algorithms, particle swarm optimization, and ant colony optimization, for solving optimization problems related to renewable energy systems.

Metaheuristic algorithms can be used in various stages of the computational modeling process for the cited problems resolution, such as:

Solar power forecasting: Metaheuristic algorithms can be used to optimize the parameters of the forecasting models, such as the number of hidden layers and neurons in a neural network model. For example, a particle swarm optimization algorithm can be used to find the optimal values of these parameters to improve the accuracy of the forecasting model.

Wind turbine design: Metaheuristic algorithms can be used to optimize the turbine design parameters, such as blade geometry and rotor diameter, to maximize energy output. For example, a genetic algorithm can be used to search the design space and find the optimal values of these parameters that satisfy design constraints and maximize energy output.

Battery storage optimization: Metaheuristic algorithms can be used to find the optimal charging and discharging strategies for a given battery storage system. For example, a simulated annealing algorithm can be used to search for the optimal solution in a large solution space, considering factors such as energy demand and battery lifespan.

Biofuel production: Metaheuristic algorithms can be used to optimize the feedstock selection and conversion process parameters to maximize biofuel production and minimize environmental impact. For example, a particle swarm optimization algorithm can be used to search for the optimal values of the conversion process parameters, such as temperature and catalyst concentration, to maximize biofuel production and minimize environmental impact.

Geothermal energy production: Metaheuristic algorithms can be used to optimize the design parameters of the geothermal power plants, such as well placement and size, to maximize energy production. For example, a genetic algorithm can be used to search for the optimal solution in a large solution space, considering factors such as reservoir characteristics and energy demand.

Microgrids: Metaheuristic algorithms can be used to optimize the design and operation of microgrids by determining the optimal sizing and placement of different components, such as solar panels, wind turbines, and battery storage systems. For example, particle swarm optimization and genetic algorithms can be used to find the optimal configuration of microgrid components that minimizes the cost of energy while satisfying technical constraints.

Energy efficiency: Metaheuristic algorithms can be used to identify energy-saving opportunities in buildings, industrial processes, and transportation systems. For example, ant colony optimization can be used to determine the optimal scheduling of energy-consuming devices to minimize energy consumption while maintaining user comfort. Other metaheuristic algorithms, such as simulated annealing and tabu search, can be used to optimize the configuration of energy-saving measures in buildings and industrial processes.

Renewable energy integration: Metaheuristic algorithms can be used to optimize the integration of renewable energy sources into the power grid. For example, genetic algorithms and particle swarm optimization can be used to determine the optimal placement and capacity of renewable energy sources to minimize carbon emissions and ensure grid stability. Other metaheuristic algorithms, such as ant colony optimization and simulated annealing, can be used to manage the variability of renewable energy generation by optimizing energy storage and demand response.

Energy storage: Metaheuristic algorithms can be used to optimize the design and operation of energy storage systems, such as batteries and pumped hydroelectric storage. For example, genetic algorithms can be used to determine the optimal sizing and placement of energy storage systems to maximize energy efficiency and minimize cost. Other metaheuristic algorithms, such as particle swarm optimization and ant colony optimization, can be used to optimize the charging and discharging strategies of energy storage systems.

Green hydrogen production: Metaheuristic algorithms can be used to optimize the production of green hydrogen from renewable energy sources. For example, genetic algorithms can be used to determine the optimal operating conditions and maximize energy efficiency while minimizing the environmental impact of hydrogen production. Other metaheuristic algorithms, such as ant colony optimization and particle swarm optimization, can be used to optimize the design and operation of hydrogen production systems.

IV. CASE STUDIES AND METHODOLOGIES

1. Environmental Economic Power Dispatch Using Metaheuristics

Environmental economic power dispatch (EEPD) is an important problem in the energy sector, as it involves finding the optimal combination of power generation sources to minimize the cost of electricity generation while also reducing the environmental impact. Metaheuristic algorithms have been widely used in EEPD due to their ability to efficiently search for optimal solutions in large solution spaces. In this study, we propose a metaheuristic approach to solve the EEPD problem, which involves modeling the power generation and distribution system, as well as the environmental impacts associated with different power sources. the effectiveness is evaluated using IEEE30 and compare it with other state-of-the-art optimization methods (Dekhici et al., 2020).

1.1. Economic Dispatch Formulation

The economic dispatch problem can be represented as a quadratic fuel cost objective function, with the power plant required to provide power that satisfies demand and covers real power losses. Generator capacity and cost coefficients are taken into consideration.

Having a network with N generators nodes where:

p_{g_k} : Power of generator k.

p_L: Lost power.

p_D: Requested power.

Economic dispatch problem can be represented as a quadratic fuel cost objective function as described in equation 1.

$$f\left(p_g\right) = \sum_{k=1}^{n} f_k\left(p_{g_k}\right) \tag{1}$$

f: Total cost
f_k: Cost of node k

The power plant must provide a power which satisfies the demand p_D requested by consumers as houses, factories and energy storage facilities and cover the real power losses in transmission lines p_L.

$$\sum_{k=1}^{n} p_{g_k} - p_L = p_D \tag{2}$$

With considering this balance equality constraint (formula 2) of demand and the generator capacity equality constraint (formula 3), the cost of a power generator p_{g_k} can be formulated by (4).

$$P_{g,\min} \leq p_{g_k} \leq P_{g,\max} \tag{3}$$

$$f_k\left(p_{g_k}\right) = c_k + b_k p_{g_k} + a_k p_{g_k}^2 \tag{4}$$

So that:

a_k, b_k, c_k: are power cost coefficients

The generating units with the multi-valve steam turbines exhibit a greater variationby the fuel-cost functions, where the valve point results in theripple form of the heat-rate curve and the cost function-contains higher order nonlinearity due to the valve-point effects.

The more general fuel cost function of each thermal generator, considering the valvepoint effect, is expressed as the sum of a quadratic and a sinusoidal function. The total non-smooth fuel cost function in \$ in terms of real power output can be formulated as in formula 5.

$$f_k\left(p_{g_k}\right) = c_k + b_k p_{g_k} + a_k p_{g_k}^2 + d_k \sin e_k\left(p_{g,min} - p_{gk}\right) \tag{5}$$

Where d_k, e_k: are thermal production cost coefficients .

1.2. Environmental Power Dispatch

For thermal power, atmospheric pollutants can be modeled separately and total emissions can be expressed as a sum of quadratic and exponential functions. The aim is to minimize total fuel cost while limiting total pollutant emissions to a maximal emission threshold set by the EPA.

The total emission of these pollutants which is the sum of a quadratic and an exponential function can be expressed in Tons as in formula 6.

$$E(p) = \sum_{k=1}^{n} 10^{-2} (\alpha_k p_{gk}^2 + \beta_k p_{gk} +_k + \eta_k e^{\delta_k p_{gk}}) \tag{6}$$

where αk, βk, ck, δk, ηk are emission coefficients.

If the problem concerns economic dispatch i.e., the aim is to minimize the total fuel cost ($), an environmental constraint limiting the total pollutant emission(ton) can be added. The E value is limited to a Maximal Emission threshold (ME) (formula 7).

$$E(p) < ME \tag{7}$$

Where: M*E* is the maximum allowable amount of pollutant during the dispatch period which is the EPA's hourly emission target [9] otherwise (6) is an objective function to minimize.

1.3. Results

To optimize the bi-objective EEPD, a modified bat algorithm with generalized flight (BAG) was used.

The proposed BAG algorithm's results on the IEEE30 dataset were compared to several other optimization techniques. The BAG algorithm with generalized fly achieved the best individually optimized fuel cost of 600.1160 $/hour, which is better than the best known individually optimized cost of 600.155 $/hour achieved by Non-dominated Sorted Genetic Algorithm II(NSGAII). In addition, the BAG algorithm also achieved lower pollutant emissions than NSGAII, with a maximum emission of 0.205177 compared to 0.22188 from NSGAII.

The Pareto coefficient was set to 3000 to make both criteria in the same scale. The results were satisfactory and results shows the best BAG compromise solution, the best cost, and the best emission when selecting a multipurpose simulation. Comparisons were made with other methods (NSGA, NPGA, SPEA, and FCPSO) using a set of IEEE30 system investigated methods, and it was found that the proposed version of bat algorithm (BAG) reduces emission to 0.188538 ton/h in its best compromise solution with a satisfactory fuel cost of 614.15 $/h. It is apparent that the BAG method is better than all other methods in terms of compromise solutions, best cost, and best emission records.

2. Environmentally Friendly Microgrids Redundancy Problem Using Metaheuristics and Modelling

Microgrids are small-scale electricity grids that can operate independently or in conjunction with the main power grid. In this study, we focus on the redundancy problem in microgrids, which involves ensuring that there is enough backup power available in case of an outage. We propose a metaheuristic-based approach that uses modeling techniques to optimize the design and operation of environmentally friendly microgrids with redundant power sources. We evaluate the performance of our proposed approach using simulation tools and compare it with other optimization methods.

2.1. Problem Description

The task at hand involves the analysis of a power microgrid network system consisting of n power subsystems, including generators, substations, and lines that are interconnected in a series configuration. Each subsystem, denoted as i, comprises several device versions linked in parallel, with device j being described by its availability or reliability (Aij or Rij), its total annual cost (TACij), and its load capacity (Gij). The configuration of subsystem i is characterized by the number of parallel identical devices, kij, where j ranges from 1 to Vi, representing the available versions of type i. The main challenge is to select the optimal set of component types, which are the decision variables, that meet the performance, reliability, space, and other requirements while maximizing reliability or minimizing costs within the given system constraints, and while achieving specified maximum or minimum values for other system aspects.

2.2. Environmental and Economic Objective Function

The Total Annual Cost (TAC) of a system comprises all the expenses incurred during its lifetime, including the installation, replacement, operation, and maintenance costs of each component that constitutes the M Microgrids system, as well as the cost of fuel consumption and penalties for emissions. The objective is to decrease pollution, enhance system reliability, and reduce the overall yearly cost. Equation is utilized to compute the minimized cost:

Min {TAC = (Fuel & Capital cost + Operation & Maintenance cost + Emission cost)}

$$TAC = (F_{MC}(t) + O\&M_{MC}(t) + PL_{MC}(t)) \quad (8)$$

Where FMC(t) is a fuel and capital cost, The O&MMC(t) is the operation and maintenance, and the PLMC(t) is the emission (pollution) gas as shown in Equation (9)

Pollution Emission

$$PL_{MC}(t) = \sum_{j}^{1} P_{1j}(t) \times EC_{1j}(t) \times k_{1j} \quad (9)$$

The power and the emission depend on the kind of element micro-turbine MT, diesel generator DG and fuel cell FC:

The emission is related to coefficient of three pollutants NOx, SO2 and CO2 and using their emission cost β.

2.3. Constraints

The total power generated by DG, MT and FC should satisfy the total MG demand. Therefore, the optimization problem constraints include reliability, pollution, performance, and output power of each type of power generation units.

2.4. Results

The aim is to minimize both the Total Annual Cost (TAC) and emissions while adhering to performance and reliability constraints. The outcomes reveal that the most cost-effective and eco-friendly system for supplying the load demand using ISA and FA methods while maintaining a reliability constraint of 99% is the production by MT. The TAC of the MG system calculated using ISA is 1.8952 (M€) with a pollution rate of 555,997 (Kg), while that of FA is 1.9093 (M€) with the same pollution rate. On the other hand, the TAC obtained by BA is 2.2392 (M€) with a pollution rate of 741,3292 (Kg). It is worth noting that the BA algorithm provides a maximum reliability of 99.8%. For a reliability constraint of 97%, ISA and FA generate more cost-effective and environmentally friendly outcomes than the BA algorithm, which is better than ISA and FA at a reliability level of 99.5%.

V. CONCLUSION

In conclusion, sustainable energy planning and management is essential for achieving a reliable, affordable, and sustainable energy supply while promoting environmental protection, social equity, and economic efficiency. The optimization of renewable energy systems for sustainability is a critical aspect of this process, which requires the use of metaheuristic algorithms and computational modeling techniques. The application of these tools can enhance efficiency, reduce environmental impact, and promote the wider adoption of renewable energy systems, thereby facilitating the transition to a more sustainable and equitable energy system. The chapter highlights two case studies, the Environmental Economic Power Dispatch and the Environmentally Friendly Microgrids Redundancy Problem.

In all of the case studies, metaheuristic algorithms are investigated to solve complex optimization problems related to environmental sustainability. The methodology involves modeling the problem at hand, defining the optimization objective, selecting and implementing appropriate metaheuristic algorithms, and evaluating the results using simulation and real-world data. The proposed approaches are compared with other state-of-the-art optimization methods to demonstrate the effectiveness of our methodology in addressing important environmental sustainability challenges.

ACKNOWLEDGMENT

The authors would like to acknowledge The General Direction of Scientific Research and Technological Development (DGRST) of Algeria for its financial support and LDREI laboratory.

REFERENCES

Abedi, S., & Kwon, S. (2023). Rolling-horizon optimization integrated with recurrent neural network-driven forecasting for residential battery energy storage operations. *International Journal of Electrical Power & Energy Systems, 145*. doi:10.1016/j.ijepes.2022.108589

Banerjee, A., Guria, C., & Maiti, S. K. (2016). Fertilizer assisted optimal cultivation of microalgae using response surface method and genetic algorithm for biofuel feedstock. *Energy, 115*(1), 1272-1290. doi:10.1016/j.energy.2016.09.066

Bett, P. E., Thornton, H. E., Troccoli, A., De Felice, M., Suckling, E., Dubus, L., Saint-Drenan, Y.-M., & Brayshaw, D. J. (2022). A simplified seasonal forecasting strategy, applied to wind and solar power in Europe. *Climate Services, 27*. doi:10.1016/j.cliser.2022.100318

Bui, L. D., Nguyen, N. Q., Van Doan, B., & Sanseverino, E. R. (2022). Forecasting energy output of a solar power plant in curtailment condition based on LSTM using P/GHI coefficient and validation in training process, a case study in Vietnam. *Electric Power Systems Research, 213*. doi:10.1016/j.epsr.2022.108706

Cheng, L., Zang, H., Wei, Z., Zhang, F., & Sun, G. (2022). *Evaluation of opaque deep-learning solar power forecast models towards power-grid applications* (Vol. 198). Renewable Energy. doi:10.1016/j.renene.2022.08.054

Dekhici, L., Guerraiche, K., & Belkadi, K. (2020). Environmental economic power dispatch using bat algorithm with generalized fly and evolutionary boundary constraint handling scheme. *International Journal of Applied Metaheuristic Computing, 11*(2), 171–191. doi:10.4018/IJAMC.2020040109

Guerraiche, K., Dekhici, L., Chatelet, E., & Zeblah, A. (2023). Techno-Economic Green Optimization of Electrical Microgrid Using Swarm Metaheuristics. *Energies, 16*(4), 1803. doi:10.3390/en16041803

Guo, P., & Bharath Kumar, N. (2023). CO2 hydrogenation for geothermal energy storage through synthetic natural gas production and byproduct of refrigeration and freshwater using solid oxide electrolyzer cell (SOEC) and methanation reactor; Techno-economic evaluation and multi-objective optimization. *Journal of CO2 Utilization, 69*. doi:10.1016/j.jcou.2023.102395

Hassan, A. Y. (2022). Wind cube optimum design for wind turbine using meta-heuristic algorithms. *Alexandria Engineering Journal, 61*(6). doi:10.1016/j.aej.2021.09.059

Hoteit, H., He, X., Yan, B., & Vahrenkamp, V. (2023). Uncertainty quantification and optimization method applied to time-continuous geothermal energy extraction. *Geothermics, 110*. doi:10.1016/j.geothermics.2023.102675

Kilic, M., & Altun, A. F. (2022). Dynamic modelling and multi-objective optimization of off-grid hybrid energy systems by using battery or hydrogen storage for different climates. *International Journal of Hydrogen Energy*. doi:10.1016/j.ijhydene.2022.12.103

Li, S., Wen, D., Feng, B., Li, F., Yue, D., Zhang, Q., Wang, J., & Feng, Z. (2023). Numerical optimization of geothermal energy extraction from deep karst reservoir in North China. *Renewable Energy, 202*, 1071-1085. doi:10.1016/j.renene.2022.12.016

Lin, Y., Chen, Y.-W., & Yang, J.-T. (2023). Optimized thermal management of a battery energy-storage system (BESS) inspired by air-cooling inefficiency factor of data centers. *International Journal of Heat and Mass Transfer, 200*. https://www.sciencedirect.com/science/article/pii/S0017931022008572 doi:10.1016/j.ijheatmasstransfer.2022.123388

Lindberg, O., Lingfors, D., Arnqvist, J., van der Meer, D., & Munkhammar, J. (2023). Day-ahead probabilistic forecasting at a co-located wind and solar power park in Sweden: Trading and forecast verification. *Advances in Applied Energy, 9*. https://www.sciencedirect.com/science/article/pii/S2666792422000385 doi:10.1016/j.adapen.2022.100120

Ma, H., Ge, M., Wu, G., Du, B., & Liu, Y. (2021). Formulas of the optimized yaw angles for cooperative control of wind farms with aligned turbines to maximize the power production. *Applied Energy, 303*. doi:10.1016/j.apenergy.2021.117691

Meena, M., Shubham, S., Paritosh, K., Pareek, N., & Vivekanand, V. (2021). Production of biofuels from biomass: Predicting the energy employing artificial intelligence modelling. *Bioresource Technology, 340*. doi:10.1016/j.biortech.2021.125642

Mohammad, Shams, Niaz, Won, Lee, & Liu. (2022). Risk-based multistage stochastic mixed-integer optimization for biofuel supply chain management under multiple uncertainties. *Renewable Energy, 200*, 694-705. doi:10.1016/j.renene.2022.10.003

Ouedraogo, K. E., Ekim, P. O., & Demirok, E. (2023). Feasibility of low-cost energy management system using embedded optimization for PV and battery storage assisted residential buildings. *Energy, 271*. doi:10.1016/j.energy.2023.126922

Pombo, D. V., Rincón, M. J., Bacher, P., Bindner, H. W., Spataru, S. V., & Sørensen, P. E. (2022). Assessing stacked physics-informed machine learning models for co-located wind–solar power forecasting. *Sustainable Energy, Grids and Networks, 32*. https://www.sciencedirect.com/science/article/pii/S2352467722001886 doi:10.1016/j.segan.2022.100943

Pustina, L., Biral, F., & Serafini, J. (2022). A novel Economic Nonlinear Model Predictive Controller for power maximisation on wind turbines. *Renewable and Sustainable Energy Reviews, 170*. doi:10.1016/j.rser.2022.112964

Sedaghat, A., Mostafaeipour, A., Rezaei, M., Jahangiri, M., & Mehrabi, A. (2020). A new semi-empirical wind turbine capacity factor for maximizing annual electricity and hydrogen production. *International Journal of Hydrogen Energy, 45*(32), 15888-15903.

Shu, T., Song, D., & Joo, Y. H. (2022). Non-centralised coordinated optimisation for maximising offshore wind farm power via a sparse communication architecture. *Applied Energy, 324*. doi:10.1016/j.apenergy.2022.119705

Silva, J. G., Ferrari, R., & van Wingerden, J.-W. (2023). Wind farm control for wake-loss compensation, thrust balancing and load-limiting of turbines. *Renewable Energy, 203*, 421-433. doi:10.1016/j.renene.2022.11.113

Tekkanat, B., Yuksel, Y. E., & Ozturk, M. (2023). The evaluation of hydrogen production via a geothermal-based multigeneration system with 3E analysis and multi-objective optimization. *International Journal of Hydrogen Energy, 48*(22), 8002-8021. doi:10.1016/j.ijhydene.2022.11.185

Visser, L., AlSkaif, T., Hu, J., Louwen, A., & van Sark, W. (2023). On the value of expert knowledge in estimation and forecasting of solar photovoltaic power generation. *Solar Energy, 251*, 86-105. doi:10.1016/j.solener.2023.01.019

Wang, J., Zhao, Z., Liu, G., & Xu, H. (2022). A robust optimization approach of well placement for doublet in heterogeneous geothermal reservoirs using random forest technique and genetic algorithm. *Energy, 254*(C). doi:10.1016/j.energy.2022.124427

Wang, W., Yang, D., Hong, T., & Kleissl, J. (2022). *An archived dataset from the ECMWF Ensemble Prediction System for probabilistic solar power forecasting* (Vol. 248). Solar Energy. doi:10.1016/j.solener.2022.10.062

Xue, Z., Yao, S., Ma, H., Zhang, C., Zhang, K., & Chen, Z. (2023). Thermo-economic optimization of an enhanced geothermal system (EGS) based on machine learning and differential evolution algorithms. *Fuel, 340*. doi:10.1016/j.fuel.2023.127569

Yao, J., Hedengren, J. D., Gao, T., & Powell, K. M. (2023). A two-level optimization framework for battery energy storage systems to enhance economics and minimize long-term capacity fading. *Journal of Energy Storage, 63*. doi:10.1016/j.est.2023.106943

Yilmaz, C., & Sen, O. (2022). Thermoeconomic analysis and artificial neural network based genetic algorithm optimization of geothermal and solar energy assisted hydrogen and power generation. *International Journal of Hydrogen Energy, 47*(37), 16424-16439. doi:10.1016/j.ijhydene.2022.03.140

Yu, Y., Niu, T., Wang, J., & Jiang, H. (2023). Intermittent solar power hybrid forecasting system based on pattern recognition and feature extraction. *Energy Conversion and Management, 277*. https://www.sciencedirect.com/science/article/pii/S0196890422013577 doi:10.1016/j.enconman.2022.116579

Zhang, Z., Wen, K., & Sun, W. (2023). Optimization and sustainability analysis of a hybrid diesel-solar-battery energy storage structure for zero energy buildings at various reliability conditions. *Sustainable Energy Technologies and Assessments, 55*. doi:10.1016/j.seta.2022.102913

Ziolkowska, J. R. (2013). Evaluating sustainability of biofuels feedstocks: A multi-objective framework for supporting decision making. *Biomass and Bioenergy, 59*, 425-440. doi:10.1016/j.biombioe.2013.09.008

Chapter 12
Sustainable Technology Accessing the Software Product Line (SPL) via Model-Based Mutation

Muskan Singh
University of Petroleum and Energy Studies, India

Sudhanshu Srivastava
University of Petroleum and Energy Studies, India

Shelly Garg
University of Petroleum and Energy Studies, India

ABSTRACT

Due to frequently changing demands and for the sake of customer's satisfaction, engineers were forced to develop SPL also known as "software product line." SPL is used for the product of similar families or having something in common. FMs also known as "feature models" are used for generating the altered versions of the products through linking features with the products. Generation of SPL sometimes leads to the production of billions of products. So, it becomes very challenging to test the quality of test suites. The techniques that exist focus more towards selecting only a limited set of products. If we talk about FMs then they impact test suites in some way. In this work, the authors have paid more attention to the capabilities of the test suites to detect the existing errors. So FMs are prone to errors. Two experiments that they have conducted for calculating mutation score clearly demonstrate that dissimilar test suites are able to detect errors present in FMs very easily as compared to the similar test suites.

DOI: 10.4018/978-1-6684-9130-0.ch012

1 INTRODUCTION

The frequently changing consumer's/customer's requirements are the deriving forces that are pressurizing the production of good quality products at the low/affordable cost. The word like "SOFTWARE PRODUCT LINE (SPL)" comes into play when we refer to the products of similar families and these SPL techniques are very useful in terms of reusing and then integrating the software assets in a sequence or systematic order (Pohl et al., 2005). "FMs" which stands for "FEATURE MODELS" were introduced to represent SPL precisely (Clements & Northrop, 2001). In FMs each and every product is developed with some common and variable/different features and all these features are integrated/joined with each other in software assets (Kang et al., 1990). Model Based Testing of SPL and steps like VISUALIZATION, REASONING, CONFIGURATION all these are backbone of FMs (Benavides et al., 2010).

For example: LINUX KERNEL FMs consist of 6000 features as a result of which it becomes very complex and hectic task to develop SPL and its corresponding FM (Mendonca, Wasowski, & Czarnecki, 2009). For meeting the test budget constraints one's test suite must be realistic enough (Perrouin et al., 2010). Now here comes the term "MUTATION ANALYSIS" into picture which is nothing but the technique through which we test the quality of our testing processes (Berger, She, Lotufo, Wasowski, & Czarnecki, 2010). This consist of modifying or updating the software artifacts into altered versions known as "MUTANTS" (McGregor, 2010). Now if we talk in context of SPL then these MUTANTS are helpful in VALIDATING and EVALUATING the test cases (DeMillo & Offutt, 1991). Due to frequent changing demands and for the sake of customer's satisfaction, engineers were forced to develop SPL also known as "software product line" (Papadakis & Malevris, 2012). SPL is used for the product of similar families or having something in common (Andrews, Briand, Labiche, & Namin, 2006). FMs also known as "Feature Model" is used for generating the altered versions of the products through linking features with the products. Generation of SPL sometimes leads to the production of billions of products (Andrews, Briand, & Labiche, 2005). So, it becomes very challenging to test the quality of test suites (Hemmati, Arcuri, & Briand, 2012). The techniques that exists more towards selecting only a limited set of products (Henard, Papadakis, Perrouin, Klein, Heymans, & Traon, 2012). If we talk about FMs then they impact test suites in some or the other way (Mendonca, Branco, & Cowan, 2009). In this work we have paid more attention towards the capabilities of the test suites to detect the existing errors (Offutt, 2011). So FMs are prone to errors (Li, Dai, & Li, 2009). Two experiments that we have conducted for calculating mutation score clearly demonstrates that dissimilar test suites are able to detect errors present in FMs very easily as compared to the similar test suites (Fabbri, Maldonado, Masiero, Delamaro, & Wong, 1996). Experimentation on real feature models demonstrate that dissimilar tests suites have a higher mutant detection ability than similar ones, thus validating the relevance of similarity-driven product line testing (Jia & Harman, 2011).

Now let's address our first research question:

Q1) How can we apply mutation analysis technique on model-based product lines?

- Now here comes the interesting fact that "MUTATION ANALYSIS TECHNIQUES" till now has been applied to various models EXCEPT FMs. Two most important processes involved in mutation are:
  1. Firstly, it is used to generate test.
  2. Secondly, it is used to validate and verify the testing techniques and approaches.

Our primary goal is to focus on the second goal (Papadakis & Traon, 2012). Set of software products are represented by these "TEST SUITES" and "MUTANTS" represents the faults (Cartaxo, Machado, & Neto, 2011). Now let's address our second research question:

Q2) In context of SPL and FM do dissimilar products have higher mutant detection rate?

Introducing the technique of "MUTATION ANALYSIS" on FMs (SPL based) we generate inexact FMs by willingly introducing all the possible defects (Le Berre & Parrain, 2010).

We use "SIMILARITY HEURISTIC" to answer our second question (Yoo & Harman, 2012). Experiments conducted on similar as well as dissimilar "TEST SUITES" have the higher capability of detecting the defects that are incorporated in the modified FMs (Google, n.d.).

Main contributions of my research paper are as follows:

- Applying MUTATION ANALYSIS on SPL based on FMs.
- Performing experiments on FMs which
    1. Confirms that TEST SUITES with dissimilar features have higher ability of detecting defects than similar ones.
    2. Access a SIMILARITY-DRIVER PRIORTIZATION TECHNIQUE.

2 LITERATURE REVIEW

2.1 Feature Models

Feature model is nothing but the simple way of portraying the features along with the constraints that are present (Arcuri & Briand, 2011). Attuned construction of the software products by finalizing all the features that you want in your final product is the main function of the FMs (Do & Rothermel, 2006).

FMs in simple words is the abstraction of software assets. FMs is represented mainly in two forms:

1. Diagram: Diagram is the graphical representation of the hierarchy and the constraints that link features (Mouelhi, Fleurey, & Baudry, 2008).
2. Propositional Formula: It is just the translation of graphical representation into propositional logic which consist of all the linking constraints present between the features (Kaminski, Praphamontripong, Ammann, & Offutt, 2011).

Example: The figure shown below is a FM containing 10 features. Nevertheless, the formula of FM fulfill the constraints of the FM (Kaminski, Ammann, & Offutt, 2012).

Conjunctive Normal Form (CNF) of the FM formula is shown below:

$$FM = \bigwedge_{i=1}^{n} C_i \quad (1)$$

Figure 1. A simple feature model of a mobile phone product line (Benavides et al., 2010)

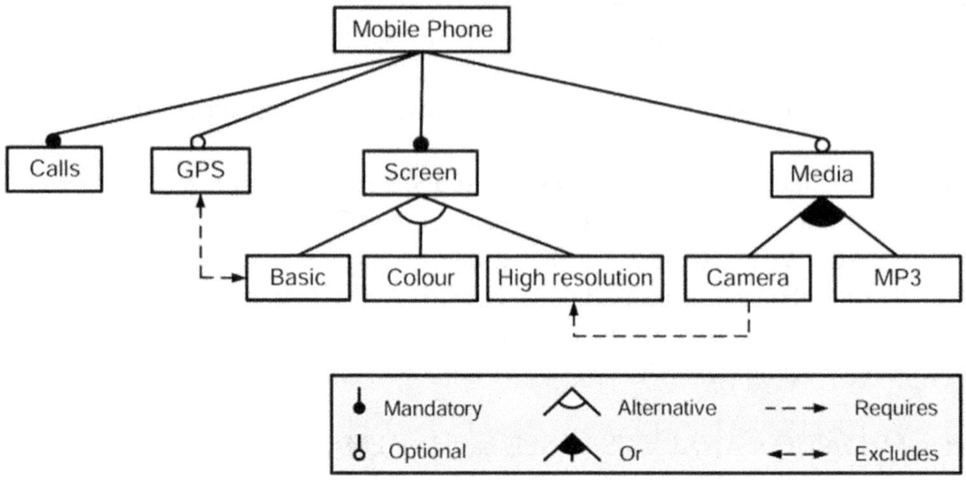

2.2 Mutation Testing and Analysis

Software testing and debugging are the techniques that are the part of mutation analysis. Attuned versions of various several program artifacts are created, some of which are source code, specification models etc (Gargantini & Fraser, 2011). And these altered versions are also known as mutants (Cartaxo, Neto, & Machado, 2007). The underlying idea behind this approach which forms its backbone is to differentiate between the original and attuned versions (Hemmati & Briand, 2010). Mutant operators which are nothing but the set of well-defined rules are used as a tool to find out all the possible defects that are present in the attuned versions (mutants) (Steffens, Oster, Lochau, & Fogdal, 2012). There is a term known as "Syntactic Descriptions" and this is used to make the changes in the objects that are developed from the syntax (Nielsen et al., 2012). And it is this "syntactic description" on which "mutant operators" are defined (Ensan et al., 2012).

Mutation Analysis is the process in which mutants are introduced (Berger, She, Lotufo, Wasowski, & Czarnecki, 2010). Capability of these utilized test-cases is to bring forward the willingly introduced mutants which are then examined for the purpose of "MUTATION TESTING" (McGregor, 2010). "KILLED MUTANTS" are referred to the mutants that are detected during the test else they are referred as "LIVE" (DeMillo & Offutt, 1991). This process is known as "MUTATION SCORE" (Papadakis & Malevris, 2012). When we apply mutant operators on the original FM it results in the production of mutants (Andrews, Briand, Labiche, & Namin, 2006). Just by speculating the fact that test satisfies the Boolean formula of modified FM you can easily perform the test evaluation (Andrews, Briand, & Labiche, 2005). Examined test always satisfy their respective Boolean formulas because they are produced on the basis of original FM. Moreover, if the formula of any mutant is not satisfied then it is referred to as "killed" (Hemmati, Arcuri, & Briand, 2012).

3 METHODOLOGY (DATA SET AND TECHNIQUES)

3.1 Approach

3.1.1. Mutation Analysis on Feature Model (Based on SPL)

In this work, we have tried introducing the approach of "mutation testing" for SPL based on FMs (Henard, Papadakis, Perrouin, Klein, Heymans, & Traon, 2012). Mutant operators are applied on the clauses of the formula in order to produce several erroneous versions (Mendonca, Branco, & Cowan, 2009). After this a valid product is produced using SAT solver and the validity of this product is examined towards the mutants (Offutt, 2011). Validity of these products towards the mutants are examined by checking whether Boolean formula is satisfied by the product or not (Li, Dai, & Li, 2009).

3.1.2 Test Suites Generation

Here in the context of this paper, one valid product is represented by one test (Fabbri, Maldonado, Masiero, Delamaro, & Wong, 1996). Collection of all the products to be tested is known as "Test Suite" (Jia & Harman, 2011). Number of features help in exploding products generated with the aid of FM (Papadakis & Traon, 2012). The example is shown in the table below (Cartaxo, Machado, & Neto, 2011). Now it is quite evident that since the size of the search space is larger hence it results in the generation of dissimilar products (Le Berre & Parrain, 2010). Thus, we have the task of producing the similar products along with the selection of random products from the search space (Yoo & Harman, 2012). After this we retrieve the adjacent products from the randomly selected ones (Google, n.d.).

Table 1. Various types of feature models (Benavides et al., 2010)

	Number of Features	Number of Valid Products
Counter strike simple Fm	24	18,176
DS sample	41	6,912
Electric Drum	52	331,776
Smart Home	60	3.87E9
Video Player	71	4.5E13
Model Transformation	88	1.65E13
Coche Ecologico	94	2.32E7
Printers	172	1.14E49
Electronic Shopping	290	4.52E49

Figure 2. Mutation analysis approach (Le Berre & Parrain, 2010)

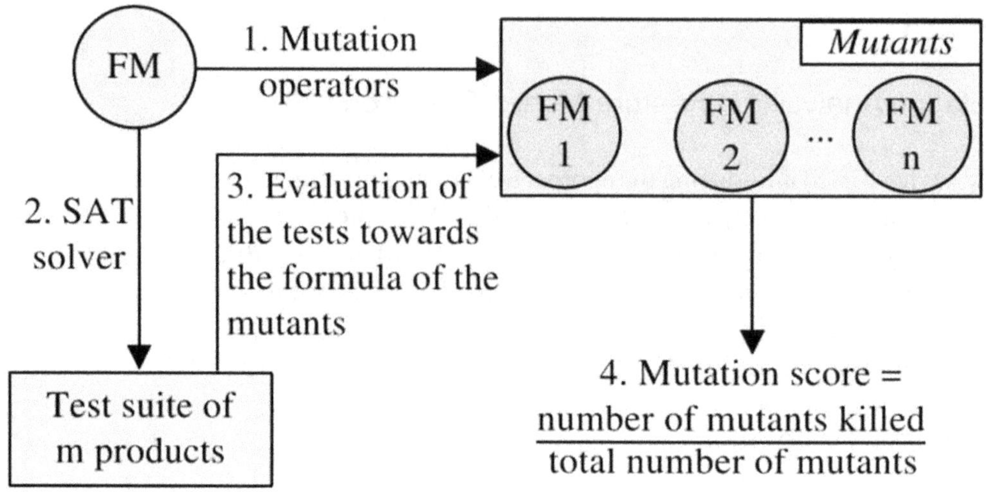

3.1.3 Evaluation of the Quality of Test Suite

Similarity-Based Distance:

Products are simply the set of selected and unselected features,

P = {+F1 and -F1, ...+Fn and -Fn}................. (2)

Where, +fi= selected features and -fi= unselected features (Arcuri & Briand, 2011)

Similarity-Driven Prioritization:

Parameter used: Distance denoted with the letter 'd' (Do & Rothermel, 2006)

Given: S which denotes the set of products along with all of its permutations denoted with Ps. The function f on the real numbers from Ps is shown below:

F: Ps = All positive real numbers............................ (3)

Here in this reference f signifies the variation accomplished by the set of all valid products S (Mouelhi, Fleurey, & Baudry, 2008). In this work we have conducted few experiments that I will mention below to establish Prioritization approach is a simple technique of finalizing the products at each step and then segregating it from the those finalized earlier (Kaminski, Praphamontripong, Ammann, & Offutt, 2011). In the Algorithm the two products belonging to S and still containing dissimilarities are appended (Kaminski, Ammann, & Offutt, 2012). Likewise, products are added in L and removed from S based upon the highest number of dissimilarities that they contain (Gargantini & Fraser, 2011). These steps are repeated continuously until S is empty (Cartaxo, Neto, & Machado, 2007).

4.1 Experiments

Aim of this particular experiment is to evaluate the quality of test suites based on mutation score (Hemmati & Briand, 2010). Compering whether the dissimilar test suites are capable of killing more mutants than the similar ones is the main objective of this experiment (Steffens, Oster, Lochau, & Fogdal, 2012).

SETUP: Corresponding to all the 12 FMs mentioned in Table II we have generated 100 mutants. Set the mutation operator to 0.5 which indicates the chances of producing the mutants by using any one among the two mutation operators. Three test suites generated by me are mentioned below:

- Test suites consisting of only dissimilar products.
- Test suites consisting of equal number of similar and dissimilar products.
- Test suite consisting of only similar products.

Sizes of test suites generated for these three conditions are: 2, 10 and 50 products. Now we calculated mutation score simply by evaluating these test suites towards the 100 mutants. This step of producing, computing was repeated 100 times. Figure shown below represents the 100 p-values resulting from the Mann-Whitney U Test between mutation score of similar and dissimilar test suites of 100 execution.

4.1.1 Results and Discussions

Table III represents all the result set that is recorded by me. This table contains the average, maximum and minimum mutation score for each FM. But there are some exceptional cases as well, for example: LINUX KERNEL has 26,286 FMs with 50 test suites. Here the mutation score that we have got with dissimilar test suites is approximately three times larger in size than that achieved with similar test suites. Specified set of guidelines were followed by me to find out the statistical significance of these differences. These guidelines were proposed by none other than Arcuri and Briand. In this we check one of the two samples just to check that which one has the larger value than the other. The point to keep in mind is that both of these samples have their independent observations. Total number of executions performed by me is 100 and hence in total we have 100 p-values. From the results recorded in Table III, Observations drawn are:

- Differences are significant for 10 and 50 products test suites because for all of them the p-values recorded is lower than 5%.

Table 2. Mutation score achieved by different types of products (Mendonca, Wasowski, & Czarnecki, 2009)

Types of Products Dissim. Sim. Dissim. Half Sim./Dissim. Sim.
Counter Strike avg 69.50 53.02 82.29 80.09 64.40
Min 54 31 77 75 45
Max 77 71 85 84 75
DS sample avg 50.64 35.91 69.26 64.71 52.72
Min 46 22 60 56 39
Max 58 50 81 75 63
Electronic Drum avg 69.50 53.02 82.29 80.09 64.40
Min 54 31 77 75 45
Max 77 71 85 84 75
Smart Home avg 50.64 35.91 69.26 64.71 52.72
Min 46 22 60 56 39
Max 58 50 81 75 63
Video Player avg 69.16 53.64 82.50 80.70 58.36
Min 31 25 77 57 29
Max 81 72 85 84 73
Model Tra. avg 68.69 52.33 83.17 80.24 54.21
Min 60 39 80 66 41
Max 74 65 84 84 66
Coche Eco. avg 69.16 53.64 82.50 80.70 58.36
Min 31 25 77 57 29
Max 81 72 85 84 73
Printers avg 68.69 52.33 83.17 80.24 54.21
Min 60 39 80 66 41
Max 74 65 84 84 66

4.2 Mutation Score Evaluation Towards the Similarity Driven Prioritization

Aim of this experiment is to find out that whether the k-products selected according to the prioritization technique are able to kill more mutants or not as compared to the k-products that are randomly prioritized.

SETUP: 100 test suites are produced for each FM consisting of both similar and dissimilar products. Now task is to prioritize each test suite based on the prioritization technique. In order to randomly order them I applied 100 times random prioritization. In order to calculate the prioritization approaches one need to calculate the area under the curve using the TRAPEZOIDAL RULE: More effective prioritization is expressed by higher area under the curve.

4.2.1 Result for This Experiment

Prioritization table is shown below. Observations made from the result set recorded:

- Similarity driven prioritization show higher area under the curve as compared to the random ones.

Table 3. Area under curve observed for two prioritization techniques (Cartaxo, Machado, & Neto, 2011)

Counter Strike DS Sample Electronic Drum Smart Home
Similarity Driven 8346 8567.5 8,177.5 9,276.5
Random (min) 7,983 7.995 9,117.5 8142.5
Random (avg) 8,187 8,473 331,776 9175
Random(max) 8,133.5 8,118.5 331,776 9268.5

In simple words, if 80% of the similarity driven prioritization can be achieved just from 5 products that are selected using prioritization technique then randomly selected ones will require 20 products. Mann-Whitney U Test is conducted in order to check whether the difference achieved from similarity driven prioritization and random are significant or not. Figure shown below represents the 100 p-values per FM.

5 CONCLUSION AND FUTURE SCOPE

We did the evaluation of FAULT DETECTION POWER of various types of test suites that were generated from correct FM and also generated 100 erroneous FMs.There is a involvement of mutation operator for altering the clauses which works as the Boolean formula of FM. The solution that I reached is that evaluation of mutation score depending on two factors:

- Different types of tests
- Similarity driven prioritization

Clearly indicates that dissimilar test suites are capable of killing more mutants than the similar ones.In this research paper I have applied mutation analysis approach for software product line based on FM.And let me mention over here that to the best of my knowledge this is the first mutation analysis approach applied on SPL. Not only this but in this research paper I have also evaluated mutation analysis approach towards the similar as well as dissimilar test suites. In this work the second experiment that I have performed on 12 real FMs clearly showcase the effectiveness of this approach. Two parameters that are:

- Mutation score
- Prioritization evaluation

Clearly prove the benefits of dissimilar test suites. Till now mutation testing has been widely used on:

- Security policies metamodels and finite state machines.

Also, more attention has been towards checking the efficiency of test suites which formed the basis of my research paper. In this work we have focused on FMs, the Boolean model of SPLs. Also, rather than paying more attention towards the efficiency I have focused on the bug detection of different type of test suites. Also, the most important factor of my research paper is that here I have used mutation score as a factor for testing the quality of test suites.

REFERENCES

Andrews, J. H., Briand, L. C., & Labiche, Y. (2005). *Is mutation an appropriate tool for testing experiments? In Proceedings of the 27th international conference on Software engineering, ser. ICSE '05.* ACM. https://doi.acm.org/10.1145/1062455.1062530

Andrews, J. H., Briand, L. C., Labiche, Y., & Namin, A. S. (2006, August). Using mutation analysis for assessing and comparing testing coverage criteria. *IEEE Trans. Softw. Eng., 32*(8), 608–624. Advance online publication. doi:10.1109/TSE.2006.83

Arcuri, A., & Briand, L. (2011). *A practical guide for using statistical tests to assess randomized algorithms in software engineering. In Proceedings of the 33rd International Conference on Software Engineering, ser. ICSE '11*. ACM. https://doi.acm.org/10.1145/1985793.1985795

Benavides, D., Segura, S., & Ruiz-Cortes, A. (2010, September). Automated analysis of feature models 20 years later: A literature review. *Inf. Syst., 35*(6), 615–636. Advance online publication. doi:10.1016/j.is.2010.01.001

Berger, She, Lotufo, Wasowski, & Czarnecki. (2010). Variability modeling in the real: a perspective from the operating systems domain. *ASE*, 73–82.

Cartaxo, Neto, & Machado. (2007). Automated test case selection based on a similarity function. *GI Jahrestagung,* (2), 399–404.

Cartaxo, E. G., Machado, P. D. L., & Neto, F. G. O. (2011). On the use of a similarity function for test case selection in the context of model-based testing. Software Testing, Verification and Reliability, 21(2), 75–100. doi:10.1002/stvr.413

Clements, P., & Northrop, L. (2001). *Software Product Lines: Practices and Patterns*. Addison Wesley.

DeMillo & Offutt. (1991). Constraint-based automatic test data generation. *IEEE Trans. Softw. Eng., 17*(9), 900–910. doi:10.1109/32.92910

Do, H., & Rothermel, G. (2006, September). On the use of mutation faults in empirical assessments of test case prioritization techniques. *IEEE Trans. Softw. Eng., 32*(9), 733–752. Advance online publication. doi:10.1109/TSE.2006.92

Ensan, F., Bagheri, E., & Gasevic, D. (2012). *Evolutionary searchbased test generation for software product line feature models. In CAISE*. Springer.

Fabbri, S. C. P. F., Maldonado, J. C., Masiero, P. C., Delamaro, M. E., & Wong, E. (1996). *Mutation testing applied to validate specifications based on petri nets. In Proceedings of the IFIP TC6 Eighth International Conference on Formal Description Techniques VIII*. Chapman & Hall, Ltd. https://dl.acm.org/citation.cfm?id=646214.681539

Gargantini, A., & Fraser, G. (2011). Generating minimal fault detecting test suites for general boolean specifications. *Information and Software Technology, 53*(11), 1263–1273. doi:10.1016/j.infsof.2011.06.008

Google. (n.d.). http://code.google.com/p/linux-variability-analysis-tools/ source/browse/?repo=formulas

Hemmati, H., Arcuri, A., & Briand, L. (2012). Achieving scalable model-based testing through test case diversity. *ACM Transactions on Software Engineering and Methodology, 22*(1), 1–42. doi:10.1145/2430536.2430540

Hemmati, H., & Briand, L. (2010). An industrial investigation of similarity measures for model-based test case selection. In *Proceedings of the 2010 IEEE 21st International Symposium on Software Reliability Engineering, ser. ISSRE '10*. IEEE Computer Society. 10.1109/ISSRE.2010.9

HenardC.PapadakisM.PerrouinG.KleinJ.HeymansP.TraonY. L. (2012). *Bypassing the combinatorial explosion: Using similarity to generate and prioritize t-wise test suites for large software product lines.* Tech. Rep. http://arxiv.org/abs/1211.5451

Jia & Harman. (2011). An analysis and survey of the development of mutation testing. *IEEE Transactions on Software Engineering, 37*(5), 649–678.

Johansen, M. F., Haugen, Ø., Fleurey, F., Carlson, E., Endresen, J., & Wien, T. (2012). *A technique for agile and automatic interaction testing for product lines.* In B. Nielsen & C. Weise (Eds.), *ICTSS* (Vol. 7641, pp. 39–54). Springer.

Kaminski, G., Ammann, P., & Offutt, J. (2012). Improving logicbased testing. *Journal of Systems and Software.*

Kaminski, G. K., Praphamontripong, U., Ammann, P., & Offutt, J. (2011). A logic mutation approach to selective mutation for programs and queries. *Information and Software Technology, 53*(10), 1137–1152. doi:10.1016/j.infsof.2011.03.009

Kang, Cohen, Hess, Novak, & Peterson. (1990). *Feature-Oriented Domain Analysis (FODA) Feasibility Study.* Carnegie-Mellon University Software Engineering Institute, Tech. Rep.

Le Berre, D., & Parrain, A. (2010). The sat4j library, release 2.2, system description. *Journal on Satisfiability, Boolean Modeling and Computation, 7*(2-3), 59–64. doi:10.3233/SAT190075

Li, J.-h., Dai, G.-x., & Li, H.-h. (2009). Mutation analysis for testing finite state machines. *Proceedings of the 2009 Second International Symposium on Electronic Commerce and Security*, 620–624. 10.1109/ISECS.2009.158

McGregor, J. (2010). *Testing a software product line. In Testing Techniques in Software Engineering* (Vol. 6153). Springer.

Mendonca, M., Branco, M., & Cowan, D. (2009). *S.p.l.o.t.: software product lines online tools.* http://doi.acm.org/ 10.1145/1639950.1640002

Mendonca, M., Wasowski, A., & Czarnecki, K. (2009). *Sat-based analysis of feature models is easy. In Proceedings of the 13th International Software Product Line Conference, ser. SPLC '09.* Carnegie Mellon University. http: //dl.acm.org/citation.cfm?id=1753235.1753267

Mouelhi, T., Fleurey, F., & Baudry, B. (2008). A generic metamodel for security policies mutation. *Software Testing Verification and Validation Workshop, ICSTW '08. IEEE International Conference on*, 278–286.

Offutt, J. (2011). A mutation carol: Past, present and future. *Information and Software Technology, 53*(10), 1098–1107. doi:10.1016/j.infsof.2011.03.007

Papadakis, M., & Malevris, N. (2012, September). Mutation based test case generation via a path selection strategy. *Inf. Softw. Technol., 54*(9), 915–932. Advance online publication. doi:10.1016/j.infsof.2012.02.004

Papadakis, M., & Traon, Y. L. (2012). Using mutants to locate "unknown" faults. *Software Testing, Verification, and Validation, 2008 International Conference on,* 691–700. 10.1109/ICST.2012.159

Perrouin, G., Sen, S., Klein, J., Baudry, B., & Traon, Y. L. (2010). *Automated and scalable t-wise test case generation strategies for software product lines. In ICST.* IEEE Computer Society.

Pohl, K., Bockle, G., & van der Linden, F. J. (2005). *Software Product Line Engineering: Foundations, Principles and Techniques.* Springer-Verlag. doi:10.1007/3-540-28901-1

Steffens, Oster, Lochau, & Fogdal. (2012). Industrial evaluation of pairwise spl testing with moso-polite. *VaMoS*, 55–62.

Yoo, S., & Harman, M. (2012, March). Regression testing minimization, selection and prioritization: A survey. *Softw. Test. Verif. Reliab., 22*(2), 67–120. Advance online publication. doi:10.1002/stv.430

Chapter 13
System Modeling and Functional Analysis of Renewable Energies for Management of Smart Grids

Hamza Wertani
National Engineering School of Carthage, Carthage University, Tunisia

ABSTRACT

Smart grids studies focus on improving their reliability and reducing the cost of energy, particularly when they are operated onshore. A smart grid can integrate electric vehicles that will greatly modify the urban infrastructure as well as the electrical network. This chapter presents an overview of system modeling for management of smart grids based on renewable energies and functional analysis techniques. In fact, this work allows a discussion of a hybrid system on the one hand and the description of the functional and operational control-command applications method SADT (structured analysis, design technique) and SA-RT (structured analysis real time), on the other hand.

1. INTRODUCTION

The context of consideration by large global players and investors of Smart Grids (SG) is justified by key energy goals such as increasing the share of energy from renewable sources in total consumption as well as reducing greenhouse gas emissions without forgetting the even more ambitious objectives currently developed for 2050. The most significant contribution that the electricity sector can make to the reduction of greenhouse gas emissions will be to replace energy production from fossil fuels through low or zero carbon technologies (Kopsidas & Abogaleela, 2019; Liu et al., 2019). However, other key components of the supply chain, networks and demand management will also have vital roles to play. SG are believed to be a key enabler in the transition to a low carbon energy sector (Song et al., 2019).

It was also recognized that the challenges will be different for transport and distribution, with potential changes, at the operational and planning level, more significant for the distribution networks. The need to

DOI: 10.4018/978-1-6684-9130-0.ch013

introduce SG solutions will strongly depend on the characteristics of the local systems, in particular the current and future composition of all the different means of production (Boukettaya & Krichen, 2014).

In recent decades, SG has an active research field, because almost all the technologies required to build a SG are mature enough. Thus, the SG can reduce electricity consumption, it can also provide a more reliable and versatile service than the traditional power grid (Aghaei & Alizadeh, 2013).

Although the world's SG infrastructure is far from being studied, there is no doubt that our daily lives will benefit from the SG. Therefore, a lot of research has been done recently to highlight the challenges and needs of the future SG (Mahmoud et al., 2020).

The question is how to use the massive data captured by smart meters to provide such "smart" services that automatically read the meter information.

One of the biggest issues in power grids is the analysis and modeling of a SG, which is a major challenge for security, automation, energy efficiency and power consumption. The full implementation of a SG grid will evolve over time. However, before a new set of infrastructure is invested to build the SG, proper analysis and modeling is required to avoid wasting resources (Neureiter et al., 2016). Modeling also helps to identify and prioritize appropriate systems.

This chapter begins with a presentation of a hybrid system, consisting of a wind turbine and photovoltaic sources connected to grid, its domains as well as its advantage, on the other hand brought by the word intelligence over the conventional grid. Then, the challenges, potentials, outstanding issues and future trends of a SG are discussed.

Then, a complete SG model has been proposed using a systemic method. The model is used to analyze the constraints of a future SG. Then, to achieve this objective, we start with a functional analysis using the SADT and the SA-RT methods.

2. WIND SOURCE

2.1. History

For centuries, man has used the energy of the wind to move boats, grind grain or pump water. This energy was used to provide mechanical work. Subsequently, for several decades, wind energy was used to generate electrical power in remote areas not connected to an electrical grid.

2.2. Operating Principle

The wind turbine works, as its name indicates, with the wind, since Aeolus is the god of the winds in Greek mythology. Wind turbines are composed of blades (usually 3) carried by a rotor and installed at the top of a vertical mast. This assembly is attached to a nacelle that houses a generator. An electric motor is used to orient the nacelle so that the rotor is always facing the wind. The wind makes the blades turn between 10 and 25 turns per minute. The generator transforms the mechanical energy created into electrical energy that is fed into the grid in compliance with applicable electrical standards.

Figure 1. Elements constituting a wind turbine (Li, 2010)

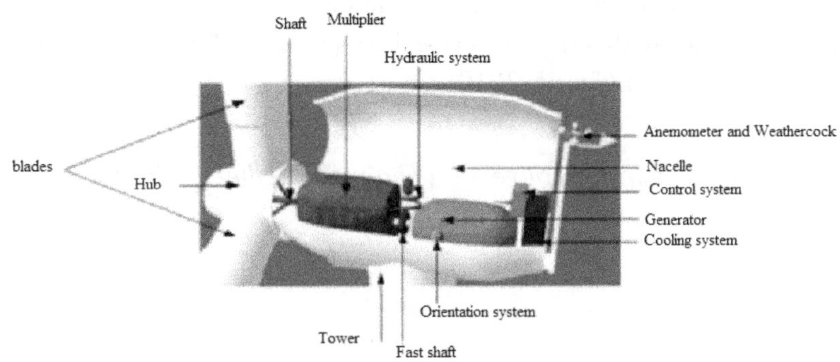

2.3. Different Types of Wind Turbines

There are two types of wind turbines: vertical axis wind turbine and horizontal axis wind turbine (the most common).

- Vertical axis wind turbine

Vertical axis wind turbines were the first structures developed to produce electricity paradoxically at odds with the traditional horizontal axis windmill. They have the advantage of having the control elements and the generator at ground level, therefore easily accessible. Many variations have been tested since the 1920s with some success. These types of wind turbines are hardly used any more, since the mechanical stresses at ground level take up more space.

- Horizontal axis wind turbine

The horizontal axis wind turbines presented in figure 26 are the most used at present compared to the vertical axis ones since they are less expensive. In addition, they are less exposed to mechanical stresses and are made up of one or more blades to generate a driving torque for rotation. The three-bladed rotor is the most used because it is a compromise between the power coefficient, the price and the speed of rotation of the wind sensor. In the literature, the main wind turbines studied are three-bladed horizontal axis.

2.4. Physical Modeling of the Wind Turbine

The reliable and cost-effective wind turbine is the ideal source of electricity for many applications. Wind turbines come in many sizes, from small, single-watt turbines to megawatt turbines.

The widely dominant technology today is horizontal axis, three-bladed and sometimes two-bladed turbines with upwind rotors. These wind turbines have a nominal power between 5KW and 5MW, they can operate at fixed or variable speed (Muselli et al., 1999; Thabor, 1988).

- Squirrel cage asynchronous generators

Asynchronous generators with cage are used for high power (over 100 kW). They only require a rather basic installation. They are often associated with a capacitor bank to compensate for reactive power and an automatic progressive starter with a dimmer or resistors to limit the transient inrush of current at the time of connection to the grid (Ben Cherif, 2010).

Figure 2. Asynchronous cage generator conversion chain (Ben Cherif, 2010)

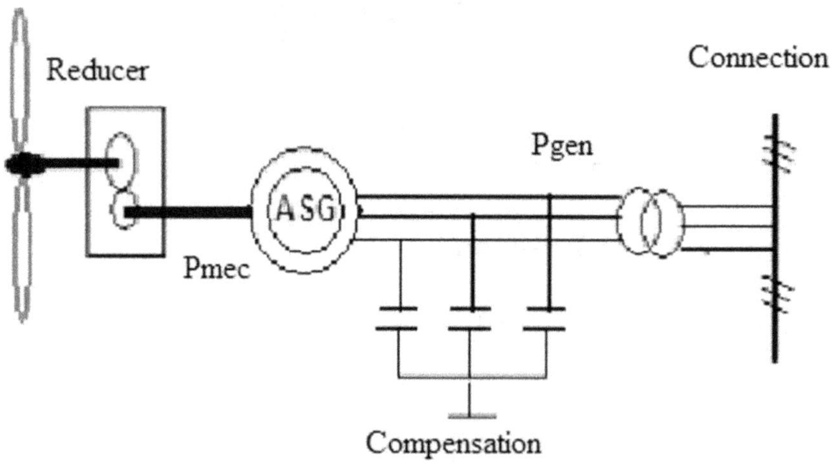

If the electric generator is of the synchronous or asynchronous type directly connected to the network, the speed is substantially constant and the aerodynamic efficiency can only be maximum for a single wind speed. In order to obtain a variable speed, there are three families of systems: asynchronous machine with variable number of poles that can operate in two speeds, asynchronous cage machine with frequency converter and thus synchronous machines with frequency converter (Chikh, 1994).

The type of generator with variable number of poles operates at two speeds according to the two-in-one principle, which allows them to operate alternately at two different speeds. These systems consist of a small power generator for low wind speeds and another for fairly high wind speeds.

Figure 3. Conversion chain with a two-speed cage asynchronous generator (Ben Cherif, 2010)

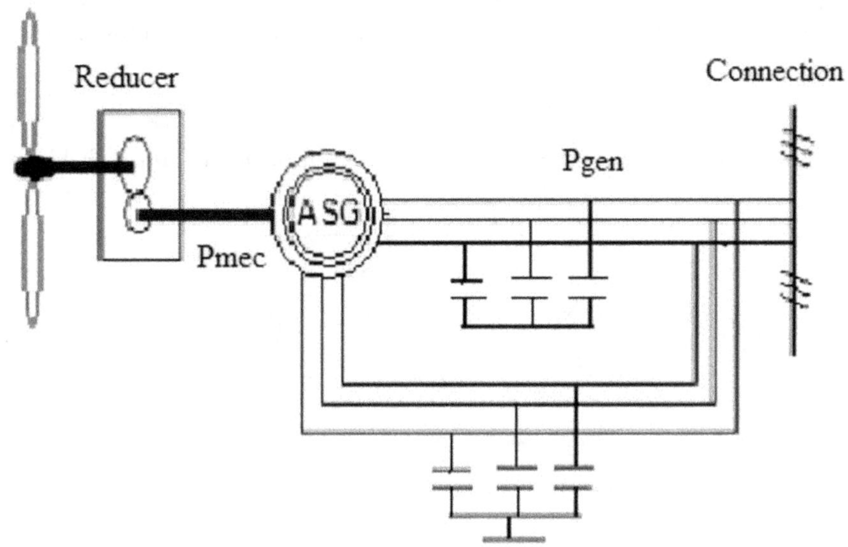

The introduction of a power converter between the machine and the grid allows the frequency of the grid to be decoupled from the rotational speed of the machine and thus to operate the machine at variable speed, the rotational speed follows the wind speed to optimize the aerodynamic efficiency of the wind turbine.

Figure 4. Asynchronous generator converter chain with frequency converter (Ben Cherif, 2010)

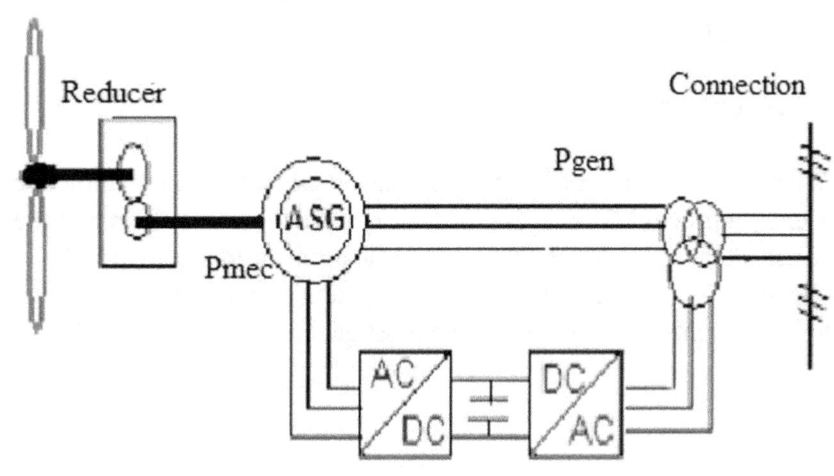

- Wind turbine based on a synchronous machine

The design of small wind systems is quite different from that of wind turbines connected to large grids. These wind power systems use synchronous generators with permanent magnets and operating at variable speed. These wind turbines are coupled to a conversion system to obtain an alternating voltage at constant frequency (Beyer et al., 1995).

Figure 5. Wind turbine based on direct-drive synchronous generator (Mladjao & Al Anfaf, 2016)

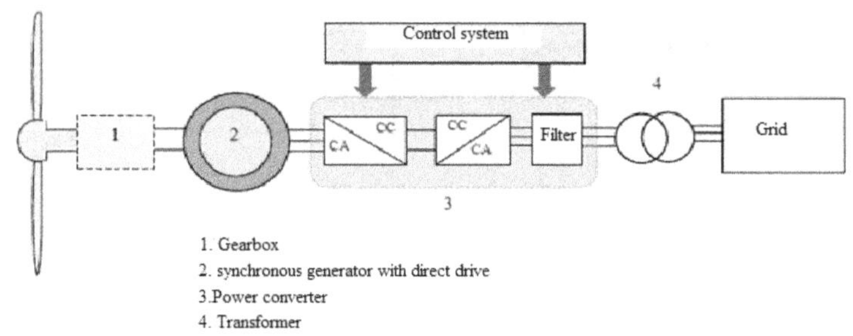

3. PHOTOVOLTAIC SOURCE

3.1. Generalities

The photovoltaic effect allows the transformation of light energy into electricity. This principle is based on semiconductor technology. It consists in using the photons to release the electrons and create a potential difference between the terminals of the cell which generates a direct electric current (Astier & Dakyo, 1999; Wertani, Ben Salem, & Lakhoua, 2020).

The transformation of photons into electrons is generally provided by silicon: it is the most important element in photovoltaic cells.

3.2. Photovoltaic Systems

A photovoltaic system consists of a set of distinct organs or subsystems in relation to each other, representing a complex energy process. The basic element is the collection system and then the set of conditioning of the electrical energy produced to the specifications of the receivers. This assembly groups all the equipment between the solar energy conversion system and the final load (Ben Cherif, 2010).

The figure below represents the elements of the most general system.

Figure 6. Block diagram deriving the most general photovoltaic system (Ben Cherif, 2010)

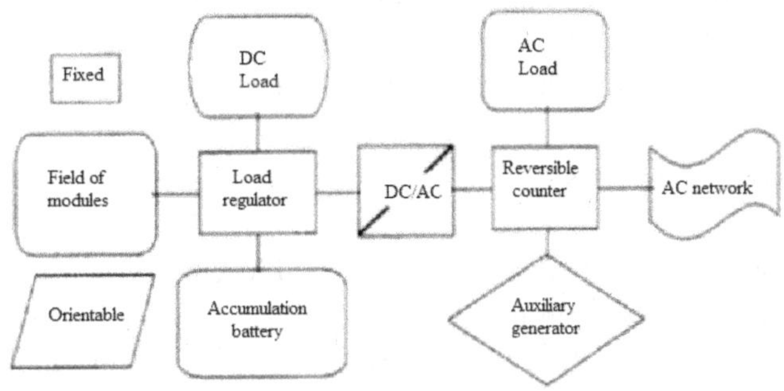

A photovoltaic system will therefore consist of a generator generally associated with one or more of the following elements (Delfosse, 1998):

- A guidance or follow-up system.
- Electronic management (storage, current shaping, energy transfer).
- Storage that compensates for the random nature of the solar source.
- A DC / AC converter.
- A load in low voltage direct current or in standard alternating current.

3.3. Different Types of Photovoltaic Systems

The most widely used photovoltaic systems are of three types: autonomous systems, hybrid systems and systems connected to a grid.

- Autonomous photovoltaic systems

These photovoltaic systems depend solely on solar energy to meet the demand for electricity. They are installed for autonomous operation without the need for other energy sources.

The energy produced can be used immediately or stored in accumulators.

Typically, these systems are used in isolated and remote areas of the network.

Figure 7. Autonomous system (Ben Cherif, 2010)

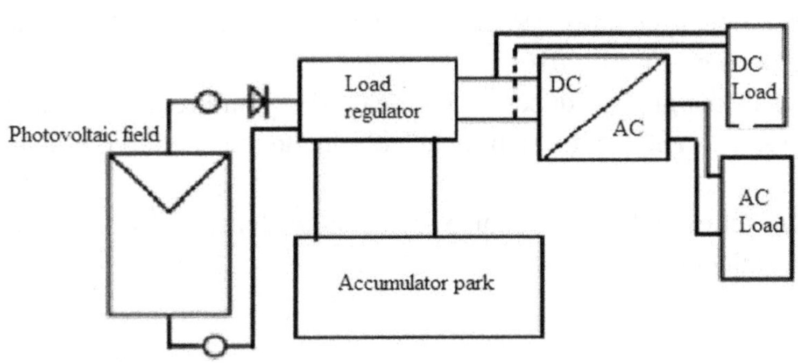

- Photovoltaic panels produce a continuous electric current.
- The regulator optimizes the charge and discharge of the battery.
- The inverter converts direct current into alternating current to supply the AC loads.
- The batteries are charged during the day to be able to supply power at night or on bad weather days.
    - Hybrid photovoltaic systems

Hybrid energy systems combine two or more renewable energy sources as well as one or more conventional energy sources. Renewable energy sources, such as photovoltaics and wind turbines do not deliver constant power, but given their complementarities, their combination allows for continuous electricity production. Such a system is a good choice for applications that require a continuous supply of fairly high power.

Figure 8. Block diagram describing the most general a hybrid system (Ben Cherif, 2010)

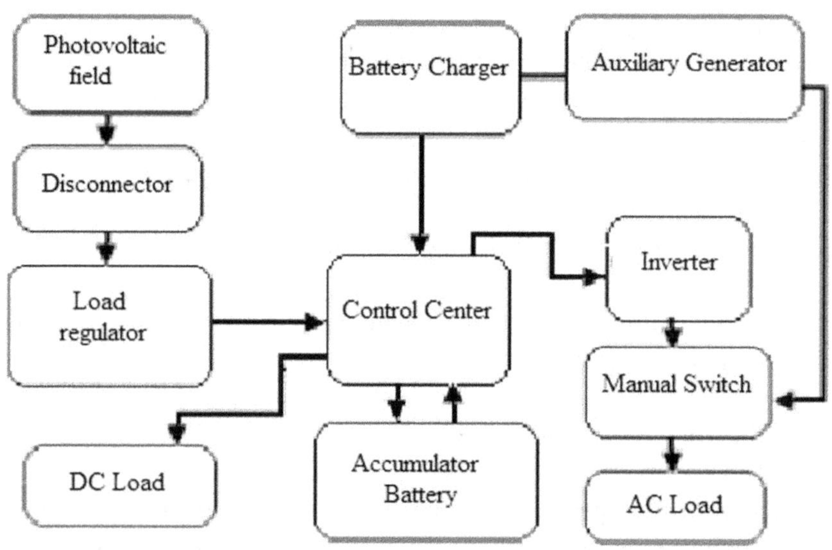

- Grid-connected photovoltaic systems

Grid-connected photovoltaic power generation systems are a result of the trend towards decentralization of the power grid. Energy is produced closer to where it is consumed and not just by large thermal or hydroelectric plants (Ben Cherif, 2010).

A system connected to a grid produces its own electricity and routes its excess energy to the grid, from which it is supplied as needed; these transfers eliminate the need to purchase and maintain a storage battery.

Two electrical connection architectures can be considered, depending on whether the energy produced is consumed on the production site or not.

- Photovoltaic system connected to the grid with total injection

This solution is achieved with connection to the public network at one point: the customer consumes part of his production and injects the remainder into the network. When consumption exceeds production, the customer withdraws the balance from the network. Only one additional counter is added to the existing counter.

Figure 9. Schematic diagram of the surplus energy injection configuration (Essakhi & Farhat, 2019)

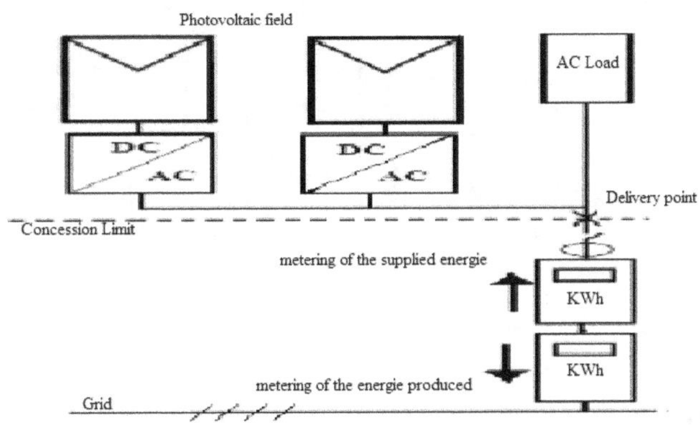

- Photovoltaic system connected to the grid with surplus injection

All the electrical energy produced by the photovoltaic collectors is sent to be sold on the distribution network. The separation between the two consumption and injection delivery points therefore leads to starring the connection in two parts:

- The connection of the consumer which remains identical with its consumption (you cannot use your own production).
- The new connection allowing all of the production to be injected into the network, has two meters: one for production and the other for non-consumption.

Figure 10. Schematic diagram of the total energy injection configuration (Meinhardt & Wimmer, 2001)

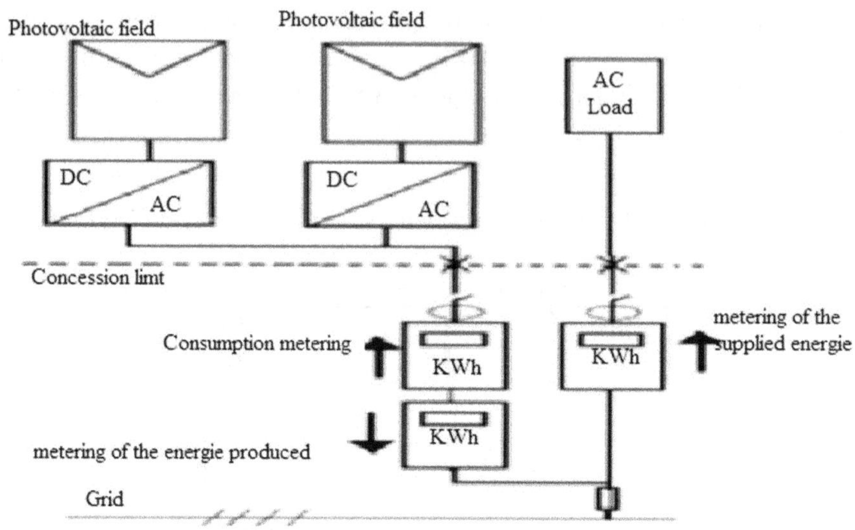

4. HYBRID SYSTEM

4.1. Definitions

A hybrid system with renewable energy sources is an electrical system, comprising more than one source of energy, of which at least one is renewable. The hybrid system may include a storage device. From a more global point of view, the energy system of a given country can be considered as a hybrid system (Saheb-Koussa, 2007; Wertani, Ben Salem, & Lakhoua, 2020).

4.2. Main Components of a Hybrid Energy System

Hybrid systems typically include:

- A generating system.
- A regulation system.
- A storage system.
- Power equipment.
- A loads.

Figure 11. The components of a hybrid system (El Khadimi, 1998)

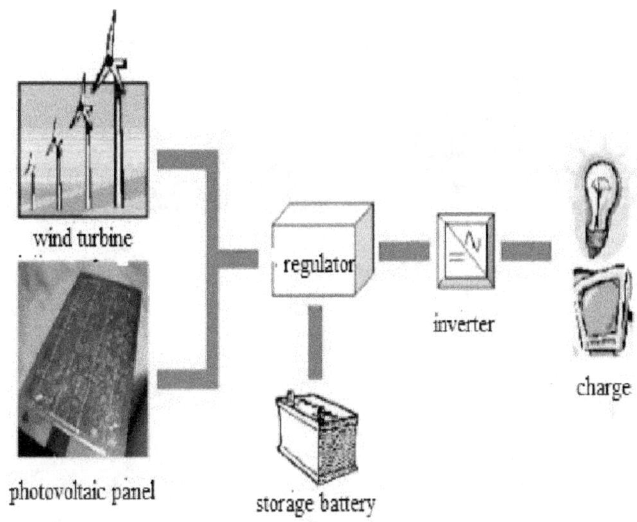

4.3. Main Combinations of a Hybrid Energy System

To achieve a hybrid system, several combinations are possible:

- Photovoltaic - electric generator.
- Wind - diesel.
- Wind - photovoltaic - diesel.
- Photovoltaic - wind, battery storage.
- Photovoltaic - wind, hydrogen storage.

Figure 12. Hybrid system containing a photovoltaic field and a wind park (Stoyanov, 2011)

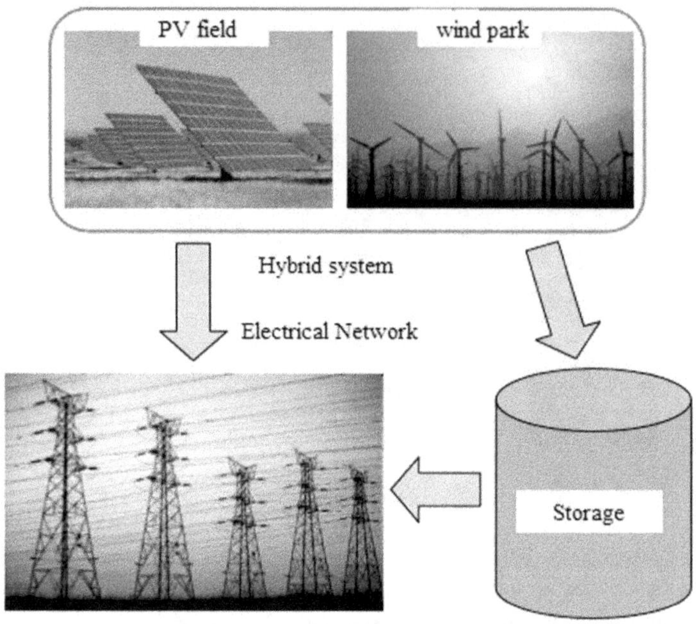

Although the electric generator and the diesel engine are not renewable sources of energy, they are typically used in hybrid systems as standby power.

4.4. The Benefits of a Hybrid System

The realization of a hybrid system represents different advantages (Wertani & Lakhoua, 2018):

- Smoothing production,
- Increasing reliability,
- Reducing storage.
- Reducing production costs.

4.5. The Disadvantages of a Hybrid System

The effectiveness of a hybrid system depends on many fundamental factors. The form and type of the loads, the cost, the availability of energy, the storage system.

For these reasons, we find some inconveniences, which we quote:

- High capital cost compared to diesel generators.
- Increasing the complexity of the system.
- The difficulty of optimizing the various organs of the system.

To overcome these disadvantages, it is necessary to know the processes that take place within this system, to model the different elements and to simulate their behavior.

5. RESEARCH WORKS ON A SG

In the literature, we find many researches that have been used to develop the modeling of electrical systems. Here we present some studies that have been presented in various researches:

Woroghi (2015) proposed to use UML (Unified Modeling Language) as an object-oriented tool to describe the concept and reduce the complexity of the system. In order to develop a simple model, they are considered the SG as an implementation of several micro-grids which are defined by small electrical systems, these systems are made up of energy sources including renewable energy sources. To implement and evaluate the proposed model, they proposed to use a simulation tool, it avoids the deployment of real and expensive infrastructures and reduces the experimental risks. Their goal is to model SG with optimal energy flow, including energy storage systems. The model is used to analyze and study the performance of the system, including its operational and management parts.

Nasraoui et al. (2016) used SysML (Systems Modeling Language) to provide a simple yet powerful description for modeling a micro network system. Indeed, they presented a functional modeling with the use case diagram. Then, a complement to the functional modeling with the requirements diagram. Then, they presented a behavioral modeling with the sequence diagram and a structural modeling with the block definition diagram (BDD) and the internal block diagram (IBD). Finally, they presented modeling with the state machine diagram.

Dey (2011) proposed a SG model using Generalized Stochastic Petri Nets (GSPN) petri nets. The model was used to analyze the constraints and deliverables of the future SG. The GSPN model of the SG, designed in this paper, introduced the basics of functional analysis and SG performance. The simulation of the model analyzed how close the behavior of the model is to the behavior of the real SG. The average distribution of tokens was studied and compared for different numbers of initial random shots. Since then, the designed model has proven to support all SG functionalities and incorporate all technologies used in the SG; the model can be used by SG engineers and designers for hypothetical analyses and all kinds of experiments. The model could have been broken down into its four phases and the stochastic analysis of the phases could have been done separately. However, for a proper GSPN analysis, valid rates must be assigned to the timed transitions and accurate weights to the immediate transitions. This could only have been done if a collaboration with an electric utility had been possible.

Glaa et al. (2017) presented a system modeling based on the OOPP method (Objective Oriented Projects Planning), applied on a photovoltaic generation system. The system analysis is presented to create a communication interface between operators, to analyze and plan the photovoltaic system at different stages, to study the installation site and to make the right choice of components. On the other hand, a photovoltaic production chain is described with its different phases in both modes. Thus, the simulation with Matlab software allowed studying the characteristics of a photovoltaic (PV) module in optimal conditions and under different climatic conditions such as the influence of solar radiation and temperature values on the characteristics of the photovoltaic system. With the SG, the field of production becomes very complex, first of all, the consumer participates in the production of electricity. He can use his own production and if there is a production higher than his consumption, it will be sold or stored in a battery. The unidirectional direction of the power grid and the data information between the generation and the consumer domain is changed to a bidirectional direction.

Seghiri (2016) has presented methods and tools allowing to model, then to simulate an enterprise architecture in order to validate it. In this context, he proposed a multi-view framework, named Execute EA, to facilitate the modeling of enterprise architectures by automating the analysis of their structures and behaviors through simulation. Execute EA treats each of the business, functional and application views according to three aspects: information, process and objectives. To meet the need for business / IT alignment, he introduced an additional view: the integration view. In this view he presented to model the inter and intra views coherence links. He also used techniques from model-driven engineering as support techniques for modeling and simulating enterprise architecture. Finally, he validates the proposal through a SG business case relating to the management of a fleet of electric vehicles.

Chabaud (2014) has proposed to use a strategy in order to develop and validate, in simulation, an intelligent approach (predictive or not) and multi-criteria of energy management of a micro-grid, at the scale of an individual home equipped with production and storage systems. This management strategy takes into account the state of the electricity network in order to minimize the impact of local production and electricity consumption. The individual habitat was modeled, as well as the production systems (photovoltaic solar panels and a vertical axis wind turbine) and storage systems (batteries). In addition, several energy and economic criteria are proposed to assess the sizing of these systems and the relevance of the proposed strategy.

6. FUNCTIONAL ANALYSIS OF SG

In this part, we present a functional analysis study of a SG.

6.1. External Functional Analysis

This analysis consists of highlighting the purpose of the study and specifies the relationship between the SG and its external environment. The basic functions are deduced after a study of the external environments

Figure 13 represents the materialization of the need for our system. On this figure, we can see that the SG provides a service to the producer (answer to the first question to whom?), it acts on the best available consumption information (answer to the second question about what?) and it allows to optimize the system operation by a better control and management of the production and consumption (answer to the last question why ?).

Figure 13. Formalization of a SG need

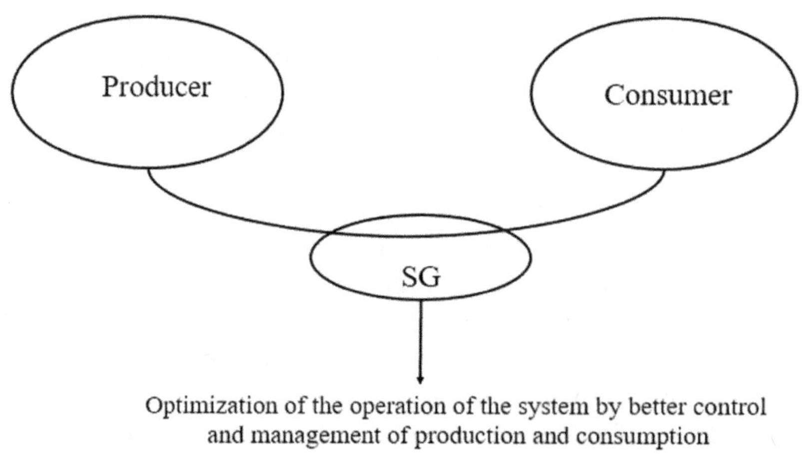

After formalizing the need for the SG, it is important to establish the various relationships between the system and the elements of its external environment (Figure 14).

Figure 14. Octopus diagram of a SG

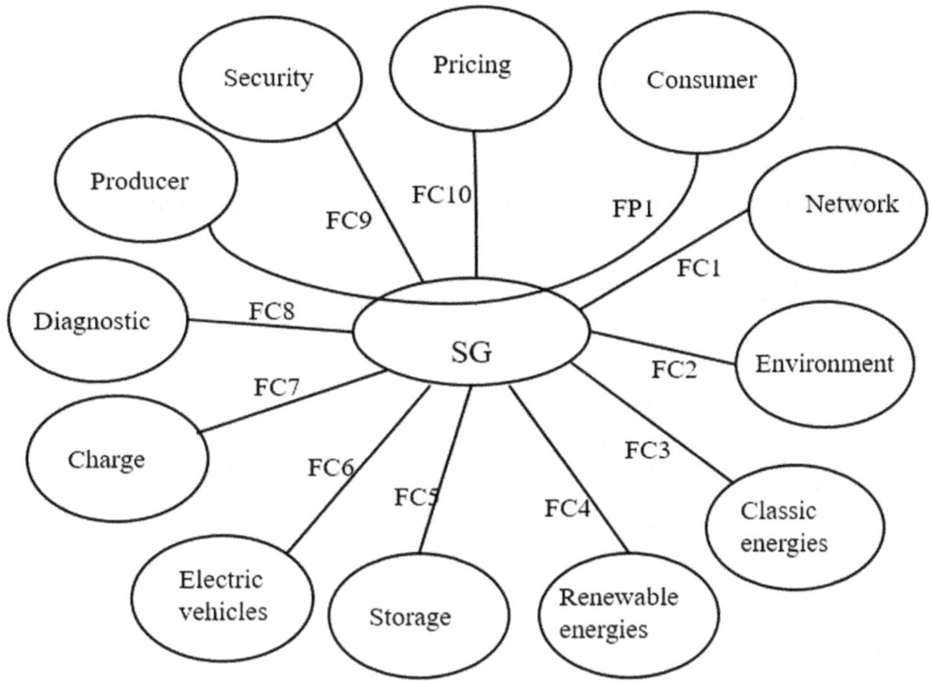

For this we use the Octopus diagram which allows us to highlight the FP (Principal Function) of a SG which consists in optimizing the operation of the system by a better control and management of production and consumption as well as the FC (Constraint Function) undergone by this system.

These functions are summarized as follows:

FP1: Optimize the system's operation through better control and management of production and consumption.
FC1: Optimize the control and operation of the network.
FC2: Reduce the environmental impact of the electrical network and increase the energy efficiency.
FC3: Make better use of conventional energy sources.
FC4: Integrate renewable energies that promote energy independence and smooth the load curve.
FC5: Limit the effects of fluctuating energy production introduced by the intermittency of decentralized production, sufficient production and distribution capacity for the electrical network.
FC6: Integrate electric vehicles that will greatly modify the urban infrastructure as well as the electrical network.
FC7: To know in real time the load at any point of the network.
FC8: Immediately detect outages and reduce response times.
FC9: Maintain and improve system reliability, quality and safety.

FC10: Choose a flexible pricing system according to usage (valuing good consumer behavior).

6.2. Internal Functional Analysis

6.2.1. SADT Method of a SG

The proposed functional analysis for a SG provides self-healing, reliable, cost-effective and sustainable power services to users, giving them the freedom to choose and prioritize according to their needs and budgets.

The following figure shows the A-0 node of a SG using SADT method. This is how each module of the SG will be taken separately. The first SG action program is the production of electricity

Figure 15. Level A-0 of the SADT model of a SG

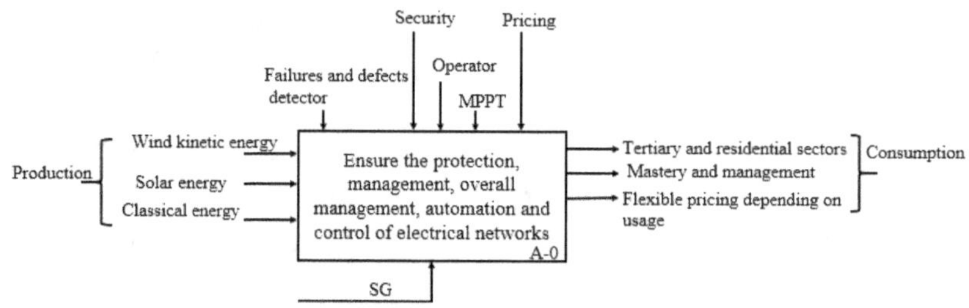

The second diagram is the A0 level which is shown in Figure 16.

Figure 16. Level A0 of the SADT model of a SG

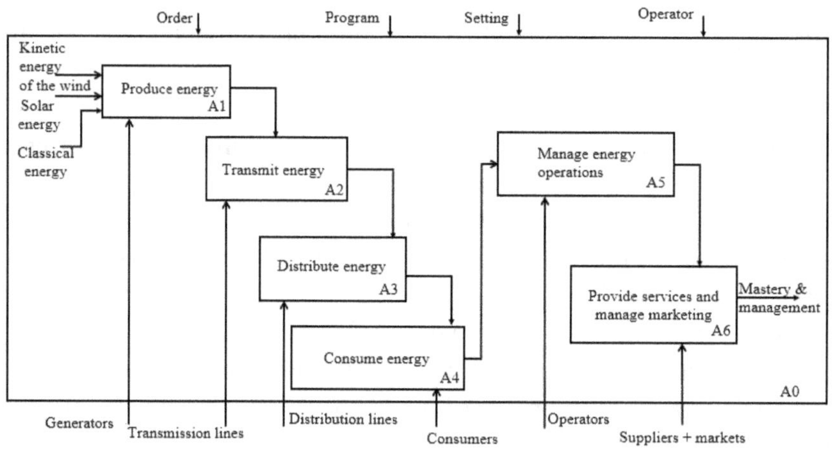

This diagram contains 6 boxes that describe one of the seven areas of a SG.

The following figures explain the A0 node of the SG of the SADT model as well as the interaction between its modules.

The diagram shown in Figure 17 is the level A1 and it is a decomposition of the box that allows to produce energy which corresponds to the field of production which is based on renewable sources and non-renewable sources. This diagram contains 4 boxes.

Figure 17. Level A1 of the SADT model of a SG

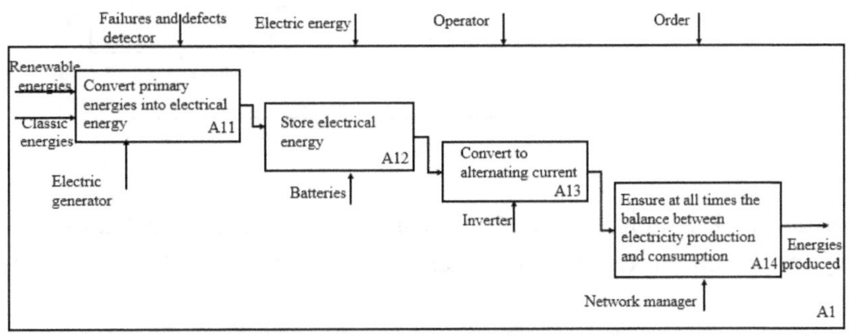

Box A2 represents the energy transmission with the lines. It consists of three stages describing insured activities which are illustrated in Figure 18.

Figure 18. Level A2 of the SADT model of a SG

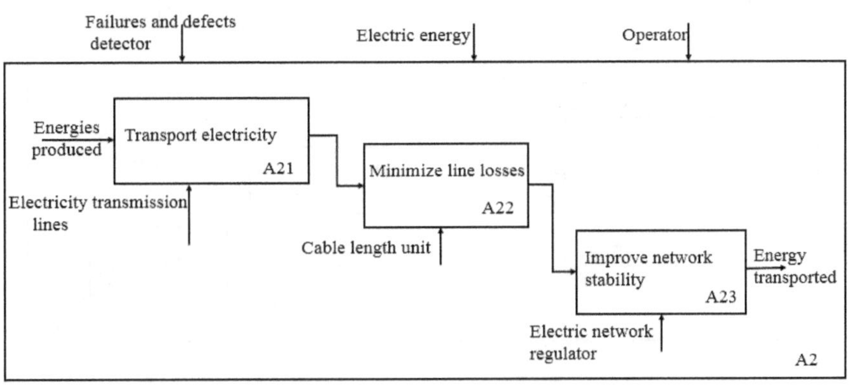

After the electricity transmission, this step must be distributed to consumers.

Thus, it is necessary to present the components of this domain, box A3 represents the functions performed by this domain as well as the communication provided by the other domains of electrical network.

Figure 19. Level A3 of the SADT model of a SG

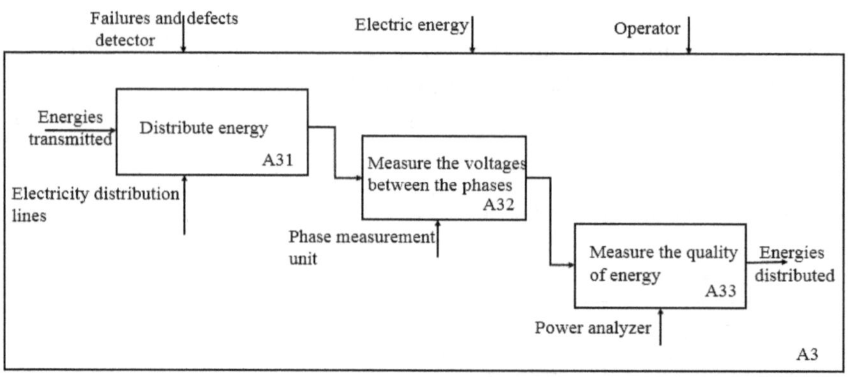

The distribution network supplies in medium voltage (MV) the industrial consumers and in low voltage (LV) a very important number of consumers such as commercial and residential (home, house, university, etc.)

The SG allows the consumer to be a producer with the installation of small parks on the ground of buildings and houses, and to be a seller of electricity and have a direct relationship with the distribution area. This domain allows communication with almost all domains of the SG.

The A4 level is a decomposition of this domain, it contains 4 boxes.

Figure 20. Level A4 of the SADT model of a SG

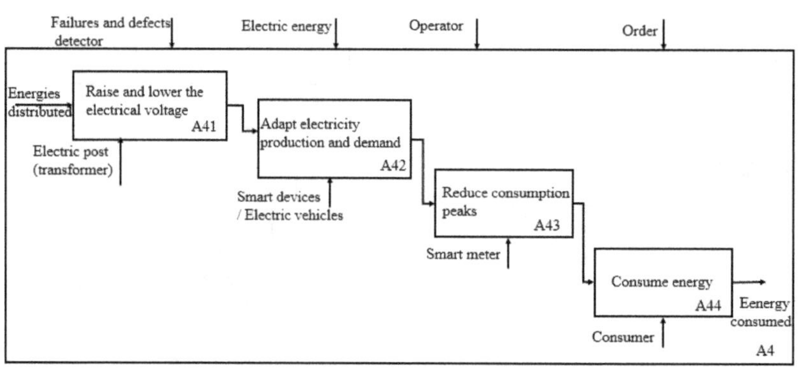

The SG system analysis based on the SADT method offers several pieces of information allowing the system to be modeled. However, the static behavior of this method does not specify the sequences and the order of the activities nor the data flows circulating in the system.

In the following, we will rely on the dynamic behavior of the SA-RT method in order to present the order of the processing, control and information data between the elements of the model.

6.2.2. SA-RT Model of a SG

This part presents a modeling of a SG based on the SA-RT systemic analysis method to lead to an operational and functional hierarchical breakdown of the system studied.

- **Data Context Diagram (DCD)**

The DCD is the first and extremely important analysis phase in the functional analysis by the SA-RT method, since it represents a realization relationship between the external environment and our system. It identifies the external entities with which the SG must communicate.

The edges of this model or terminations will appear only in this diagram. This step allows us to develop the environment modeling, more precisely the application and the list of events (Figure 21).

Figure 21. DCD of the SA-RT model of a SG

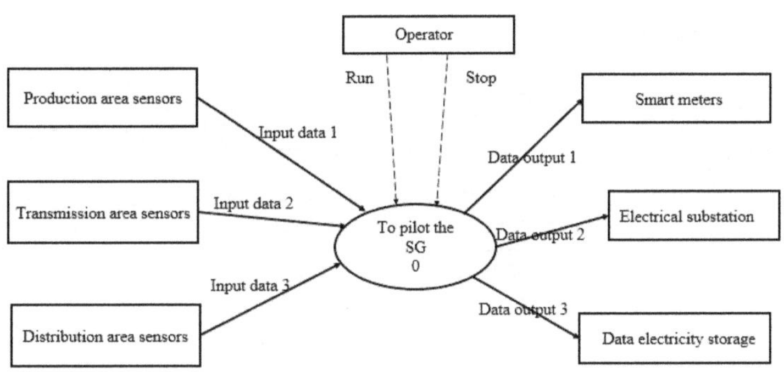

Figure 21 shows the DCD of a SG which integrates the initial main process 0 "to pilot the SG" which constitutes the control - command application to be studied, the 7 model boxes corresponding to the 3 inputs (sensors) and 3 outputs (synoptic and actuators) and one edge corresponds to the operator console. In addition, we have 2 events which are "on" and "off".

The table below shows the meanings of the data shown on the DCD.

Table 1. Context diagram data flows

Inputs (Data)	Outputs (Commands)
- Input data 1: Manage temperature and lighting data and orient the panels according to the irradiation; regulate the blades and orient the nacelle in the direction of the wind; display the storage level; manage data from conventional energies. - Input data 2: Report the conditions of the advanced instrument to analyze the transmission lines. - Input data 3: Detect faults and overloads in the distribution area. - Operation of the control application. - Stopping the control - command application.	- Data output 1: Track in detail, and often in real time, the price of electricity and the consumer's electricity consumption. - Output data 2: Transform the voltage level by reducing or increasing it. - Data output 3: Store electricity.

All inputs and outputs of the studied system are represented as data streams which are interfaces between the terminations and the main process: taking into account that all the streams flowing to the main process are inputs, and the outputs are all the outgoing streams.

The second step of the functional analysis by the SA-RT method is the DFD, the DFC and the DET.

- **Data flow diagram (DFD)**

The passage of data between all functional processes can be directly or through storage units. The DFD, shown in Figure 22, presents a functional analysis or decomposition of the initial functional process 0 to have the six functional processes 1.0, 2.0, 3.0, 4.0, 5.0, 6.0.

Figure 22. DFD of the SA-RT model of a SG

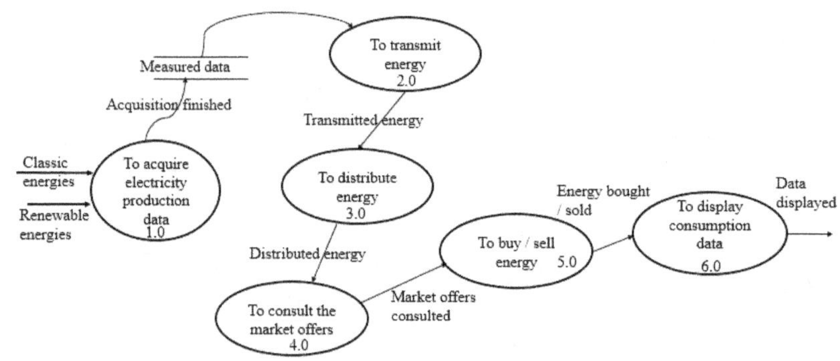

- **Control Flow Diagram (DFC)**

The DFC, presented in figure 23 presents a functional analysis or decomposition of the initial main process to have the six functional processes of the DFD and a control process. It therefore represents the additional step to the DFD.

Indeed, the six basic functional processes identified are: acquiring electricity production data; transmit energy; distribute energy; buy / sell energy; Consult the market offers; Buy / sell energy; display consumption data.

Figure 23. DFC of the SA-RT model of a SG

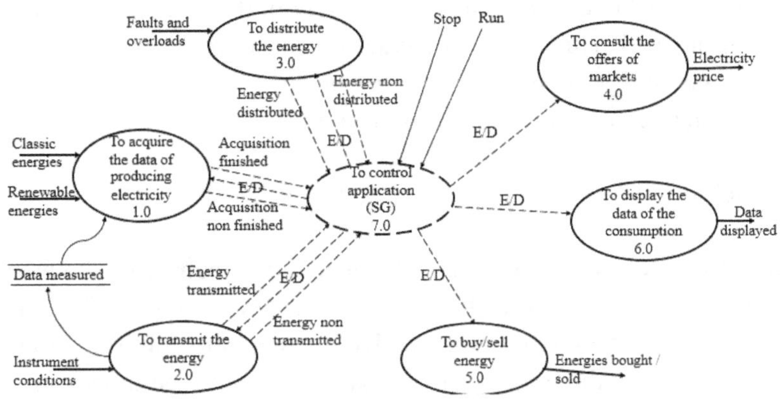

The E/D events are used together to drive an endless loop process.

The purpose of the control process is to express the execution or chaining of the six basic functional processes.

The objective is not completely achieved by the DFC. It is therefore necessary to add additional information describing the operation of the control process, this is usually done by the DET to present the operating dynamics and explain the operation of the control process.

- **State Transition Diagram (DET)**

In view of the complexities of the DET diagram, the SG comprises four main areas which are generation, transmission, distribution and finally consumption as shown in the following paragraphs.

- **Production Field**

The field of production also becomes very complex. First, the consumer participates in the production of electricity. He can use his own production and if there is more production than his consumption, it will be sold or stored in batteries. Secondly, the one-way direction of the power grid production and information data between the production domain and the operator domain is changed to a two-way direction. For this reason, a system analysis is very important to properly structure the activities of the generation domain.

When energy sources are very intense and more electricity than usual is needed, the grid will be activated to harness the energy and meet the electricity needs. The availability of these finite energy sources depends on the load and the amount of additional usage required to balance the SG system.

In any power system, electricity is generated from non-renewable energy sources, such as coal, oil and natural gas. The event is triggered only when the amount of energy is above a certain minimum level and is activated with the availability of coal, oil and natural gas.

The event will be activated with the availability of energy but is only triggered when the amount of energy in the centralized energy storage falls below a certain minimum. In the centralized energy storage, all generated electricity is stored. The energy is generated according to the requirements because energy storage can be expensive.

The centralized power generator converts this energy into electricity. The power generation is represented by E1.x is activated whenever the energy is present in the storage, but the power generation is based on the demand of the grid. This enables demand-response management as expected from a SG.

The distributed storage energy is used to generate electricity on-site using the distributed power generator. The energy generated on-site must be used as it is.

- **Transmission Field**

The next module is the power transmission of the generated electricity. The electricity produced at the centralized power generator is then transferred to the transmission grid. Several transmission lines are activated in the transmission system, superconducting wires to minimize the power loss during transmission and also to increase the transmission capacity. To keep the model simple, only two transmission lines are shown.

The E2.x timed transitions place the generated electricity on the transmission lines. The generated power is transmitted across the line as required by the associated consumer(s). The electricity on the line is transmitted to its utility. At the utility level, advanced sensors are randomly placed on the transmission line.

The sensors for the transmission lines are at the locations of the advanced sensors. The real-time situation can be monitored using sensors and visualization tools to generate reports and use the results for mapping and analysis. If the sensors detect a fault, the relevant unit is removed from the line. This removed power unit is gathered at a location, to temporarily block the faulty line.

E2.x can only be triggered if there is no event at the advanced sensors. Thus, if the sensors detect an overload or fault in the line, that line cannot be used to transmit power until the fault is repaired or the load is balanced.

The sensors report the line conditions in real time to the advanced instrument to analyze the event, i.e., the power transmission unit, from the network if a fault is identified. The sensors are always active and are represented as immediate transitions. The advanced instruments and technologies provide large-scale power system responsiveness and the fastest source of information in situations such as a power outage or inability to take cascading actions and improve the quality of power supply for the transmission line in the SG system.

- **Distribution Field**

After the transmission stage, the electricity must be distributed to the consumers. Thus, it is necessary to present the behavior of this area.

Each transmission line splits into several distribution lines before the electricity reaches the end users (E3.x). Again, to keep the model simple, only one of the multiple transmission lines has been split into two distribution lines.

From the utility, the transmission line is displayed. The power distribution event to the distribution lines was shown by the timed transitions (Power Distribution Line). As soon as at least one unit of electricity reaches the utility, the transition will be activated. On the other hand, the timing of the activated transition depends on many factors, including customer requirements. Before the electricity reaches the substation, the distribution line must be selected, which depends on availability and capacity. It will go through the phase measurement unit (PMU) for analysis and measurement of supply according to the needs of consumers. The PMU measures power quality to control congestion. The power then reaches the substation.

- **Consumer Field**

The next module is the energy consumption module (E6.x). A SG consumer is represented by a set of smart appliances and/or plug-in electric vehicles. The electricity consumption pattern is monitored using an advanced metering infrastructure (AMI). The AMI records the amount of electricity used and the time the electricity is used so that this information can be used to charge a premium to customers using electricity during peak hours. The consumer's bill is prepared from this information. The higher the number of events, the higher the number of billable units used by the consumer. The electricity was generated at the consumer's site using the distributed electrical generator. The electricity then passes through the microgrids based on the distributed energy resources that reduce the load before transferring it to the smart consumer devices, etc.

- **Full DET of a SG**

The complete DET for the entire SG is shown in Figure 24. All SG features and technologies have been successfully integrated into this model. To control the flow of electricity in the model. Consumers have different electricity needs at different times. This controlled system also gives consumers the freedom to use electricity according to their needs and gives them the ability to choose and customize their smart appliances according to the electricity rates during peak and low consumption periods.

Figure 24. DET of the SA-RT model of a SG

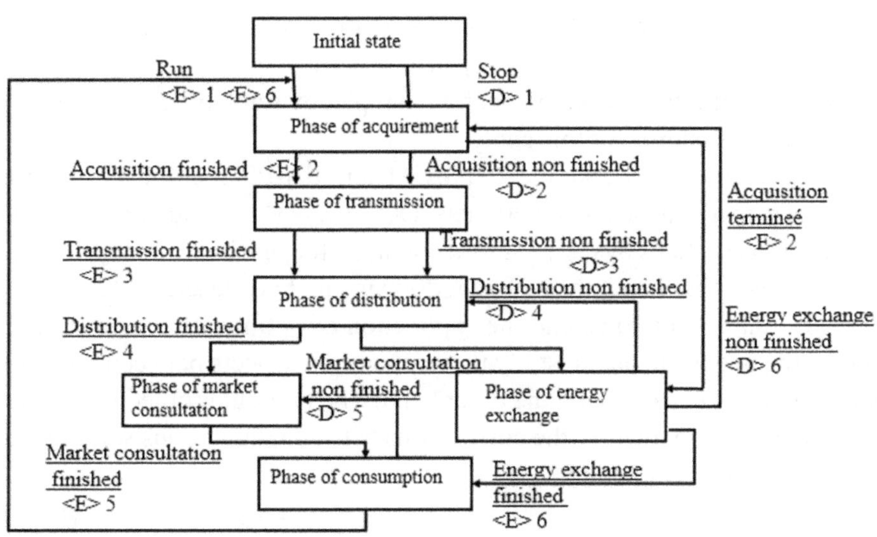

The model also satisfies the alternative power supply in case of transmission line failure and provides the power supply from another renewable energies sources.

7. CONCLUSION

In this chapter, we have presented the problematic of a SG systems and we have proposed a methodology to face these problems generated by these systems.

In order to estimate the reliability of a complex power system, it is important, first of all, to model it. For this reason, the first step of the methodology that we propose is the functional analysis that will provide us with all the necessary information about the operation of an electrical network.

In order to obtain this information, we have used the functional analysis with these two types external and internal. These two functional analyses only provide information about the functioning of a SG. It is important to have information about the different activities of this system. It is for this reason that we carried out a systemic modeling based on the two methods SADT and SA-RT making it possible to apprehend the various sources of renewable energies of a SG, to work out a hierarchical approach and to organize the information in order to determine the data, the controls as well as the various states of a system.

Also, the modeling of electrical systems, with a high integration of components of different technologies and from several disciplines, requires from the beginning of the study a collaborative work between the different actors of the development, in order to achieve a competitive and quality product. It is, therefore, necessary to have a unique approach to build reliability.

The contribution of this chapter is both at the systemic methodological level as external functional analysis allows illustrating the relationships between a system and its external environment. However, we do not have any information concerning these internal functions, hence the need to perform an internal functional analysis. The internal functional analysis allows realizing a hierarchical and tree-like decomposition of the system in elements. It also describes the technical functions of a SG. We chose to use the SADT method because it applies to the whole electrical system, i.e. it is well adapted to electronic and mechanical components as well as to software. We are also interested in the SA-RT method in order to take into account the dynamic aspect, which is missing in the SADT method.

To facilitate the future use of our methodology in an electrical context, it is necessary to specify the information system in its entirety. This could be, for example, by implementing other research methods to develop maintenance procedures for this analysis,

In the case of functional analysis, we use the SADT and SA-RT methods. For some time, several improvements have been made to the tools: Petri nets, OOPP, SysML, Bond Graph and UML that allows treating complex electrical systems by taking into account, on the one hand, the dynamic, reconfigurable and hybrid aspects, on the other hand, the different domains that constitute a SG.

Moreover, it is also interesting to extend the application of optimization algorithms to all the other domains of a SG, on the one hand, and to study the cooperation aspects between the various techniques and methods of modeling and analysis, on the other hand. The particular objective of such a connection is to obtain a global approach to identify the correlations between the various solutions obtained for the global resolution of problems concerning a SG.

REFERENCES

Aghaei, A., & Alizadeh, M. I. (2013, February). Demand response in smart electricity grids equipped with renewable Energy sources: A review. *Renewable & Sustainable Energy Reviews, 18*, 64–72. doi:10.1016/j.rser.2012.09.019

Astier & Dakyo. (1999). Les systèmes photovoltaïques autonomes non embarqués pour la production d'énergie électrique. *Journées EEA*.

Ben Cherif, M. (2010). *Modélisation de systèmes énergétiques photovoltaïques et éoliens intégration dans un système hybride basse tension* [Thèse de doctorat]. l'Université Abou-Bekr Belkeid Tlemcen.

Beyer, H. G., Degner, T., & Gabler, H. (1995). Operational behavior of wind diesel systems incorporating short-term storage: An analysis via simulation calculations. *Solar Energy, 54*(6), 429–439. doi:10.1016/0038-092X(95)00015-J

Boukettaya, G., & Krichen, L. (2014). A dynamic power management strategy of a grid connected hybrid generation system using wind, photovoltaic and Flywheel Energy Storage System in residential applications. *Energy, 71*, 148–159. doi:10.1016/j.energy.2014.04.039

Chabaud, A. (2014). *Micro-réseau intelligent pour la gestion des ressources énergétiques*. University de Perpignan.

Chikh, M. (1994). *Etude analytique conceptuelle des systèmes photovoltaïque de puissance adapté aux sites isolés* [Thèse de magister]. CDER.

Delfosse, F. (1998). Détermination de l'état de charge des batteries d'un véhicule électrique. Université de Liège.

Dey, A. (2011). Modeling Smart Grid using Generalized Stochastic Petri Net. *Renewable & Sustainable Energy Reviews, 43*, 51–64.

El Khadimi, A. (1998). *Modélisation de l'Irradiation Solaire Globale d'un Plan Horizontal et de la Vitesse du Vent en Vue de leur Utilisation dans le Dimensionnement d'un Système Hybride* [Thèse de Doctorat]. Université Cadi Ayyad.

Essakhi & Farhat. (2019). Modélisation et simulation d'un module photovoltaïque. *Journées internationale matériaux et environnement*.

Glaa, Lakhoua, & El Amraoui. (2017). Application of system modeling and simulation of the photovoltaic production. *Advances in Science and Technology Research Journal, 11*(3).

Kopsidas, K., & Abogaleela, M. (2019, May). Utilizing Demand Response to Improve Network Reliability and Ageing Resilience. *IEEE Transactions on Power Systems, 34*(3), 2216–2227. doi:10.1109/TPWRS.2018.2883612

Li, W. (2010). Real-Time Simulation of a Wind Turbine Generator Coupled With a Battery Supercapacitor Energy Storage System. *IEEE Transactions on Industrial Electronics, 57*(4), 1137–1145. doi:10.1109/TIE.2009.2037103

Liu, Y., Lei, S., & Hou, Y. (2019, September). Restoration of Power Distribution Systems With Multiple Data Centers as Critical Loads. *IEEE Transactions on Smart Grid, 10*(5), 5294–5307. doi:10.1109/TSG.2018.2880255

Mahmoud, M. A., Tang, Y. C., Maseleno, A., Lim, F.-C., Kasim, H., & Yong, C. (2020). *Towards the Development of a Smart Energy Grid.* Springer, LLC. doi:10.1007/978-3-030-33582-3_63

Meinhardt, M., & Wimmer, D. (2001). Multistring-converter: The next step in evolution of string converter. *Conference: 9th European Conference on Power Electronics and Applications.*

Mladjao, M., & Al Anfaf, M. (2016). *Contribution à la modélisation et à l'optimisation de systèmes énergétiques multi-sources et multi charges* [Thèse de doctorat]. l'Université de Lorraine.

Muselli, M., Notton, G., & Louche, A. (1999). Design of hybrid-photovoltaic power generator, with optimization of energy management. Solar Energy, 65(3), 143–157. doi:10.1016/S0038-092X(98)00139-X

Nasraoui, K., Lakhoua, M. N., & El Amraoui, L. (2016). Analysis and Modeling of a Wind Power System based on SysML. *Proceedings of the International Conference on Recent Advances in Electrical Systems.*

Neureiter, C., Uslar, M., Engel, D., & Lastro, G. (2016). A standards-based approach for domain specific modelling of Smart Grid system architectures. *11th System of Systems Engineering Conference.* 10.1109/SYSOSE.2016.7542888

Saheb-Koussa, D. (2007). Production d'électricité sans interruption moyennant un système hybride (éolien – photovoltaïque – diesel). Revue des Energies Renouvelables, 121 – 128.

Seghiri. (2016). *Modélisation et simulation d'une architecture d'entreprise - application aux Smart Grids.* Université Paris-Saclay préparée à centrale Supelec.

Song, M., Gao, C., Shahidehpour, M., Li, Z., Yang, J., & Yan, H. (2019, July). State Space Modeling and Control of Aggregated TCLs for Regulation Services in Power Grids. *IEEE Transactions on Smart Grid, 10*(4), 4095–4106. doi:10.1109/TSG.2018.2849321

Stoyanov, L. (2011). *Etude de différentes structures de systèmes hybrides à sources d'énergie renouvelables.* Université Pascal Paoli.

Thabor, M. Z. (1988). Small diesel power. *International Conference on Small Power Supplies, Tasmanie.*

Wertani, H., Ben Salem, J., & Lakhoua, M. N. (2020a, July). Analysis and supervision of a smart grid system with a systemic tool. *The Electricity Journal, 33*(6), 106784. doi:10.1016/j.tej.2020.106784

Wertani, H., Ben Salem, J., & Lakhoua, M. N. (2020b, May). Application of Bond Graph of Renewable Energy Sources in Smart Grids. *Journal of Electrical and Electronics Engineering (Oradea), 13.*

Wertani, H., & Lakhoua, M. N. (2018). *Overview of Smart Grids Architecture and Design.* SmartNets2018, Hammamet, Tunisia. 10.1109/SMARTNETS.2018.8707404

Woroghi, I. (2015). Modeling a Smart Grid using Objects Interaction. *Proceedings of the IEEE power and energy society general meeting.*

Compilation of References

Abbey, C., Brissette, Y., & Venne, P. (2013). An autoground system for anti-islanding protection of distributed generation. *IEEE Transactions on Power Systems*, *29*(2), 873–880. doi:10.1109/TPWRS.2013.2284670

Abdolmaleki, B., & Bergna-Diaz, G. (2022). Distributed Control and Optimization of DC Microgrids: A Port-Hamiltonian Approach. *IEEE Access : Practical Innovations, Open Solutions*, *10*, 64222–64233. doi:10.1109/ACCESS.2022.3183209

Abedi, S., & Kwon, S. (2023). Rolling-horizon optimization integrated with recurrent neural network-driven forecasting for residential battery energy storage operations. *International Journal of Electrical Power & Energy Systems, 145*. doi:10.1016/j.ijepes.2022.108589

Abhishek, A., Ranjan, A., Devassy, S., Kumar Verma, B., Ram, S. K., & Dhakar, A. K. (2020). Review of hierarchical control strategies for DC microgrid. *IET Renewable Power Generation*, *14*(10), 1631–1640. doi:10.1049/iet-rpg.2019.1136

Abhishek, B., & Tyagi, A. K. (2022). An Useful Survey on Supervised Machine Learning Algorithms: Comparisons and Classifications. In T. Sengodan, M. Murugappan, & S. Misra (Eds.), *Advances in Electrical and Computer Technologies. Lecture Notes in Electrical Engineering* (Vol. 881). Springer. doi:10.1007/978-981-19-1111-8_24

Abu-Elzait, S., & Parkin, R. (2019). Economic and Environmental Advantages of Renewable-based Microgrids over Conventional Microgrids. *2019 IEEE Green Technologies Conference(GreenTech)*, 1–4. 10.1109/GreenTech.2019.8767146

Ackora-Prah, J., Acheson, V., Owusu-Ansah, E., & Nkrumah, S. K. (2023). A proposed method for finding initial solutions to transportation problems. *Pakistan Journal of Statistics and Operation Research*, *19*(1), 63–75. doi:10.18187/pjsor.v19i1.4196

Adams, R., Jeanrenaud, S., Bessant, J., Denyer, D., & Overy, P. (2016). Sustainability-oriented Innovation: A Systematic Review. *International Journal of Management Reviews*, *18*(2), 180–205. doi:10.1111/ijmr.12068

Adefarati, T., & Bansal, R. C. (2019). Reliability, economic and environmental analysis of a microgrid system in the presence of renewable energy resources. *Applied Energy*, *236*, 1089–1114. doi:10.1016/j.apenergy.2018.12.050

Agarwal, D., & Tripathi, K. (2022). A Framework for Structural Damage detection system in automobiles for flexible Insurance claim using IOT and Machine Learning. *2022 International Mobile and Embedded Technology Conference (MECON),* 5-8. 10.1109/MECON53876.2022.9751889

Aghaei, A., & Alizadeh, M. I. (2013, February). Demand response in smart electricity grids equipped with renewable Energy sources: A review. *Renewable & Sustainable Energy Reviews*, *18*, 64–72. doi:10.1016/j.rser.2012.09.019

Aghaei, J., Akbari, M. A., Roosta, A., Gitizadeh, M., & Niknam, T. (2012). Integrated renewable–conventional generation expansion planning using multiobjective framework. *IET Generation, Transmission & Distribution*, *6*(8), 773–784. doi:10.1049/iet-gtd.2011.0816

Agrell, P. J., & Bogetoft, P. (2005). Economic and environmental efficiency of district heating plants. *Energy Policy*, *33*(10), 1351–1362. doi:10.1016/j.enpol.2003.12.011

Ahmad, F., Alam, M. S., & Alsaidan, I. S. (2020). Campus Microgrid: A Case Study. *2020 IEEE/PES Transmission and Distribution Conference and Exposition (T&D)*, 1–6. 10.1109/TD39804.2020.9299998

Ahmed, A. A., & Ahmed, A. A. (2020). A comparative study between new proposed Technique (LPAM) and the other existing techniques to Find IBFS for transportation problem. *Journal of Pure & Applied Sciences*, *19*(5), 40–45.

Akbar, Y. R., Zain, I., Septianingsih, R., & Nuraini, P. (2023). The transportation method for efficient cost of shipping goods. *International Journal of Economics, Management, Business, and Social Science*, *3*(1), 67–76.

Akinci, O., & Olmstead-Rumsey, J. (2018). How effective are macroprudential policies? An empirical investigation. *Journal of Financial Intermediation*, *33*, 33–57. doi:10.1016/j.jfi.2017.04.001

Akpofure, N., & Jude, O. (2022). An explicit and efficient self-developed algorithm for IBFS of a transportation problem. *Asian Journal of Multidimensional Research*, *11*(11), 58–72.

Al Hasibi, R. A., Hadi, S. P., & Sarjiya, S. (2018). Integrated and simultaneous model of power expansion planning with distributed generation. *International Review of Electrical Engineering*, *13*(2), 116–127. doi:10.15866/iree.v13i2.14748

Alam, M., Al-Ismail, F. S., Al-Sulaiman, F. A., & Abido, M. A. (2023). Energy management in DC microgrid with an efficient voltage compensation mechanism. *Electric Power Systems Research*, *214*, 108842. doi:10.1016/j.epsr.2022.108842

Alfen, H. W., & Ten Brink, P. (Eds.). (2018). *Nature-based solutions for urban resilience*. Springer.

Ali, A., Ahmadini, H., & Ahmad, F. (2021). Solving intuitionistic fuzzy multiobjective linear programming problem under neutrosophic environment. *AIMS Mathematics*, *6*(5), 4556–4580. doi:10.3934/math.2021269

Ali, S., Zheng, Z., Aillerie, M., Sawicki, J.-P., Péra, M.-C., & Hissel, D. (2021). A Review of DC Microgrid Energy Management Systems Dedicated to Residential Applications. *Energies*, *14*(14), 4308. doi:10.3390/en14144308

Al-Ismail, F. S. (2021). DC Microgrid Planning, Operation, and Control: A Comprehensive Review. *IEEE Access : Practical Innovations, Open Solutions*, *9*, 36154–36172. doi:10.1109/ACCESS.2021.3062840

Alizadeh, B., & Jadid, S. (2014). Uncertainty handling in power system expansion planning under a robust multi-objective framework. *IET Generation, Transmission & Distribution*, *8*(12), 2012–2026. doi:10.1049/iet-gtd.2013.0674

Almeshaiei, E., & Soltan, H. (2011). A methodology for electric power load forecasting. *Alexandria Engineering Journal*, *50*(2), 137–144. doi:10.1016/j.aej.2011.01.015

Altaf, M. W., Arif, M. T., Islam, S. N., & Haque, M. (2022). Microgrid Protection Challenges and Mitigation Approaches–A Comprehensive Review. *IEEE Access : Practical Innovations, Open Solutions*, *10*, 38895–38922. doi:10.1109/ACCESS.2022.3165011

Amado, M. P., Lucas, V., & Ribeiro, M. (2013) Sustainable Construction: Value of Certification. *Proceedings of the ACSEE - International Conference on Advances in Civil, Structural and Environmental Engineering, Institute of Research Engineers and Doctors*, 180-187. https://doi.org/10.3850/ 978-981-07-7965-8_83

Amirteimoori, A., Kordrostami, S., & Sarparast, M. (2006). Modeling undesirable factors in data envelopment analysis. *Applied Mathematics and Computation*, *180*(2), 444–452. doi:10.1016/j.amc.2005.12.029

Andeobu, L., Wibowo, S., & Grandhi, S. (2022). Artificial intelligence applications for sustainable solid waste management practices in Australia: A systematic review. *The Science of the Total Environment*, *155389*. Advance online publication. doi:10.1016/j.scitotenv.2022.155389 PMID:35460765

Andrews, J. H., Briand, L. C., & Labiche, Y. (2005). *Is mutation an appropriate tool for testing experiments? In Proceedings of the 27th international conference on Software engineering, ser. ICSE '05*. ACM. https://doi.acm.org/10.1145/1062455.1062530

Andrews, J. H., Briand, L. C., Labiche, Y., & Namin, A. S. (2006, August). Using mutation analysis for assessing and comparing testing coverage criteria. *IEEE Trans. Softw. Eng., 32*(8), 608–624. Advance online publication. doi:10.1109/TSE.2006.83

Angelov, P. P. (1997). Optimization in an intuitionistic fuzzy environment. *Fuzzy Sets and Systems, 86*(3), 299–306.

Angel, S., Sheppard, S., & Civco, D. L. (Eds.). (2011). *The dynamics of global urban expansion*. World Bank Publications.

An, Q., Pang, Z., Chen, H., & Liang, L. (2015). Closest targets in environmental efficiency evaluation based on enhanced Russell measure. *Ecological Indicators, 51*, 59–66. doi:10.1016/j.ecolind.2014.09.008

Arabi, B., Munisamy, S., Emrouznejad, A., & Shadman, F. (2014). Power industry restructuring and eco-efficiency changes: A new slacks-based model in Malmquist–Luenberger Index measurement. *Energy Policy, 68*, 132–145. doi:10.1016/j.enpol.2014.01.016

Arcuri, A., & Briand, L. (2011). *A practical guide for using statistical tests to assess randomized algorithms in software engineering. In Proceedings of the 33rd International Conference on Software Engineering, ser. ICSE '11*. ACM. https://doi.acm.org/10.1145/1985793.1985795

Ashok Kumar, A., & Amutha Prabha, N. (2022). A comprehensive review of DC microgrid in market segments and control technique. *Heliyon, 8*(11), e11694. doi:10.1016/j.heliyon.2022.e11694 PMID:36425418

Astier & Dakyo. (1999). Les systèmes photovoltaïques autonomes non embarqués pour la production d'énergie électrique. *Journées EEA*.

Atanassov, K.T. (1983). *Intuitionistic fuzzy sets, VII ITKR's Session, Sofia, June 1983*. Deposed in Central Sci. - Techn. Library of Bulg. Acad. of Sci., 1697/84.

Atanassov, K.T. (1999). Intuitionistic Fuzzy Sets. *Studies in Fuzziness and Soft Computing*. doi:10.1007/978-3-7908-1870-3

Atanassov, K. T. (1995). Ideas for intuitionistic fuzzy equations, inequalities and optimization. *Notes on Intuitionistic Fuzzy Sets, 1*(1), 17–24.

Atkinson, R. D. (2013). *Competitiveness, Innovation and Productivity: Clearing up the Confusion*. The Information Technology & Innovation Foundation.

Augustine, S., Quiroz, J. E., Reno, M. J., & Brahma, S. (2018). *DC Microgrid Protection: Review and Challenges*. doi:10.2172/1465634

Azadeh, A., Gaeini, Z., & Moradi, B. (2014). Optimization of HSE in maintenance activities by integration of continuous improvement cycle and fuzzy multivariate approach: A gas refinery. *Journal of Loss Prevention in the Process Industries, 32*, 415–427. doi:10.1016/j.jlp.2014.10.006

Azadeh, A., Mokhtari, Z., Sharahi, Z. J., & Zarrin, M. (2015). An integrated experiment for identification of best decision styles and teamworks with respect to HSE and ergonomics program: The case of a large oil refinery. *Accident; Analysis and Prevention, 85*, 30–44. doi:10.1016/j.aap.2015.08.016 PMID:26397195

Azadeh, A., & Mousavi Ahranjani, P. (2014). The impact of job security, satisfaction and stress on performance assessment and optimization of generation companies. *Journal of Loss Prevention in the Process Industries, 32*, 343–348. doi:10.1016/j.jlp.2014.10.002

Azadeh, A., Rahimi-Golkhandan, A., & Moghaddam, M. (2014). Location optimization of wind power generation–transmission systems under uncertainty using hierarchical fuzzy DEA: A case study. *Renewable & Sustainable Energy Reviews*, *30*, 877–885. doi:10.1016/j.rser.2013.10.020

Babazadeh, R., Razmi, J., Pishvaee, M. S., & Rabbani, M. (2015). A non-radial DEA model for location optimization of Jatropha curcas L. cultivation. *Industrial Crops and Products*, *69*, 197–203. doi:10.1016/j.indcrop.2015.02.006

Bagdadioglu, N., Waddams Price, C. M., & Weyman-Jones, T. G. (1996). Efficiency and ownership in electricity distribution: A non-parametric model of the Turkish experience. *Energy Economics*, *18*(1), 1–23. doi:10.1016/0140-9883(95)00042-9

Baidya, S., & Nandi, C. (2022). A comprehensive review on DC Microgrid protection schemes. *Electric Power Systems Research*, *210*, 108051. doi:10.1016/j.epsr.2022.108051

Bai, X., & Imura, H. (Eds.). (2016). *Urbanization and sustainability: Linking urban ecology, environmental justice and global environmental change*. Springer.

Baležentis, T., Li, T., Streimikiene, D., & Baležentis, A. (2016). Is the Lithuanian economy approaching the goals of sustainable energy and climate change mitigation? Evidence from DEA-based environmental performance index. *Journal of Cleaner Production*, *116*, 23–31. doi:10.1016/j.jclepro.2015.12.088

Bama, C. S., & Sumathi, P. (2022, November). A new proposal for getting transportation cost using redundancy. In. AIP Conference Proceedings: Vol. 2516. *No. 1* (p. 320011). AIP Publishing LLC. doi:10.1063/5.0108708

Ban, A. (2008). Trapezoidal approximations of intuitionistic fuzzy numbers expressed by value, ambiguity, width and weighted expected value. *Notes on Intuitionistic Fuzzy Sets*, *14*(1), 38–47.

Banerjee, A., Guria, C., & Maiti, S. K. (2016). Fertilizer assisted optimal cultivation of microalgae using response surface method and genetic algorithm for biofuel feedstock. *Energy, 115*(1), 1272-1290. doi:10.1016/j.energy.2016.09.066

Banu, I. V., Barkat, F., Istrate, M., Guerrero, J. M., Culea, G., Livinti, P., Motas, J. G., Neagu, B., & Andrioaia, D. (2023). Passive anti-Islanding protection for Three-Phase Grid-Connected photovoltaic power systems. *International Journal of Electrical Power & Energy Systems*, *148*, 108946. doi:10.1016/j.ijepes.2023.108946

Batlle, C., & Rodilla, P. (2013). An enhanced screening curves method for considering thermal cycling operation costs in generation expansion planning. *IEEE Transactions on Power Systems*, *28*(4), 3683–3691. doi:10.1109/TPWRS.2013.2249540

Batty, M. (2018). *The new science of cities*. MIT Press.

Bayrak, G., & Kabalci, E. (2016). Implementation of a new remote islanding detection method for wind–solar hybrid power plants. *Renewable & Sustainable Energy Reviews*, *58*, 1–15. doi:10.1016/j.rser.2015.12.227

Bayraktar, M. E., & Owens, C. R. (2010). LEED Implementation Guide for Construction Practitioners. *Journal of Architectural Engineering*, *16*(3), 85–93. doi:10.1061/(ASCE)AE.1943-5568.0000013

Beise, M., & Rennings, K. (2005). Lead markets and regulation: A framework for analyzing the international diffusion of environmental innovations. *Ecological Economics*, *52*(1), 5–17. doi:10.1016/j.ecolecon.2004.06.007

Beltrán-Esteve, M., & Picazo-Tadeo, A. J. (2015). Assessing environmental performance trends in the transport industry: Eco-innovation or catching-up? *Energy Economics*, *51*, 570–580. doi:10.1016/j.eneco.2015.08.018

Ben Cherif, M. (2010). *Modélisation de systèmes énergétiques photovoltaïques et éoliens intégration dans un système hybride basse tension* [Thèse de doctorat]. l'Université Abou-Bekr Belkeid Tlemcen.

Benavides, D., Segura, S., & Ruiz-Cortes, A. (2010, September). Automated analysis of feature models 20 years later: A literature review. *Inf. Syst., 35*(6), 615–636. Advance online publication. doi:10.1016/j.is.2010.01.001

Berardi, U. (2012). Sustainability Assessment in the Construction Sector: Rating Systems and Rated Buildings. *Sustainable Development (Bradford), 20*(6), 411–424. doi:10.1002/sd.532

Berger, She, Lotufo, Wasowski, & Czarnecki. (2010). Variability modeling in the real: a perspective from the operating systems domain. *ASE*, 73–82.

Bett, P. E., Thornton, H. E., Troccoli, A., De Felice, M., Suckling, E., Dubus, L., Saint-Drenan, Y.-M., & Brayshaw, D. J. (2022). A simplified seasonal forecasting strategy, applied to wind and solar power in Europe. *Climate Services, 27*. doi:10.1016/j.cliser.2022.100318

Beyer, H. G., Degner, T., & Gabler, H. (1995). Operational behavior of wind diesel systems incorporating short-term storage: An analysis via simulation calculations. *Solar Energy*, *54*(6), 429–439. doi:10.1016/0038-092X(95)00015-J

Bhagat, R., Chauhan, V., & Bhagat, P. (2022). Investigating the impact of artificial intelligence on consumer's purchase intention in e-retailing. Foresight. doi:10.1108/FS-10-2021-0218

Bharati, S. K. (2019). Detailed study of uncertainty and hesitation in transportation problem. *Life Cycle Reliability and Safety Engineering*, *8*(4), 357–364. doi:10.1007/s41872-019-00095-y

Blancard, S., & Hoarau, J.-F. (2013). A new sustainable human development indicator for small island developing states: A reappraisal from data envelopment analysis. *Economic Modelling*, *30*, 623–635. doi:10.1016/j.econmod.2012.10.016

Blokhuis, E., Advokaat, B., & Schaefer, W. (2012). Assessing the performance of Dutch local energy companies. *Energy Policy*, *45*, 680–690. doi:10.1016/j.enpol.2012.03.021

Bossink, B. A. (2007a). Leadership for sustainable innovation. *International Journal of Technology Management & Sustainable Development*, *6*(2), 135–149. doi:10.1386/ijtm.6.2.135_1

Bossink, B. A. (2007b). The interorganizational innovation processes of sustainable building: A Dutch case of joint building innovation in sustainability. *Building and Environment*, *42*(12), 4086–4092. doi:10.1016/j.buildenv.2006.11.020

Boukettaya, G., & Krichen, L. (2014). A dynamic power management strategy of a grid connected hybrid generation system using wind, photovoltaic and Flywheel Energy Storage System in residential applications. *Energy*, *71*, 148–159. doi:10.1016/j.energy.2014.04.039

Bowlin, W. F. (1998). Measuring Performance: An Introduction to Data Envelopment Analysis (DEA). *The Journal of Cost Analysis*, *15*(2), 3–27. doi:10.1080/08823871.1998.10462318

Bragança, L. (2005). *Princípios de Desenho e Metodologias de Avaliação da Sustentabilidade das Construções*. Revista Espaços. http://repositorium.sdum.uminho.pt/handle/1822/4943%0A

Brandenburger, A., & Nalebuff, B. (1998). Co-opetition: 1. A Revolutionary Mindset that Combines Competition and Co-operation. 2. In The Game Theory Strategy that's Changing the Game of Business. Currency Doubleday.

Bui, L. D., Nguyen, N. Q., Van Doan, B., & Sanseverino, E. R. (2022). Forecasting energy output of a solar power plant in curtailment condition based on LSTM using P/GHI coefficient and validation in training process, a case study in Vietnam. *Electric Power Systems Research, 213*. doi:10.1016/j.epsr.2022.108706

Buonocore, J. J., Salimifard, P., Michanowicz, D. R., & Allen, J. G. (2021). A decade of the U.S. energy mix transitioning away from coal: Historical reconstruction of the reductions in the public health burden of energy. *Environmental Research Letters*, *16*(5), 054030. doi:10.1088/1748-9326/abe74c

Burillo, P., Bustince, H., & Mohedano, V. (1994). Some definitions of intuitionistic fuzzy number first properties. *Proceedings of the First Workshop on Fuzzy Based Expert System*, 53-55.

Callens, I., & Tyteca, D. (1999). Towards indicators of sustainable development for firms: A productive efficiency perspective. *Ecological Economics*, *28*(1), 41–53. doi:10.1016/S0921-8009(98)00035-4

Camaréna, S. (2020). Artificial intelligence in the design of the transitions to sustainable food systems. *Journal of Cleaner Production*, *271*, 122574. doi:10.1016/j.jclepro.2020.122574

Camioto, F. de C., Mariano, E. B., & Rebelatto, D. A. (2014). Efficiency in Brazil's industrial sectors in terms of energy and sustainable development. *Environmental Science & Policy*, *37*, 50–60. doi:10.1016/j.envsci.2013.08.007

Campiglio, E. (2016). Beyond carbon pricing: The role of banking and monetary policy in financing the transition to a low-carbon economy. Ecological Economics, 121, 220–230. http://dx.doi.org/.ecolecon.2015.03.020 doi:10.1016/j

Careri, F., Genesi, C., Marannino, P., Montagna, M., Rossi, S., & Siviero, I. (2011). Generation expansion planning in the age of green economy. *IEEE Transactions on Power Systems*, *26*(4), 2214–2223. doi:10.1109/TPWRS.2011.2107753

Cartaxo, E. G., Machado, P. D. L., & Neto, F. G. O. (2011). On the use of a similarity function for test case selection in the context of model-based testing. Software Testing, Verification and Reliability, 21(2), 75–100. doi:10.1002/stvr.413

Cartaxo, Neto, & Machado. (2007). Automated test case selection based on a similarity function. *GI Jahrestagung*, (2), 399–404.

Cerutti, E., Claessens, S., & Laeven, L. (2017). The use and effectiveness of macroprudential policies: New evidence. *Journal of Financial Stability*, *28*, 203–224. doi:10.1016/j.jfs.2015.10.004

Chabaud, A. (2014). *Micro-réseau intelligent pour la gestion des ressources énergétiques*. University de Perpignan.

Chang, D.-S., & Yang, F.-C. (2011). Assessing the power generation, pollution control, and overall efficiencies of municipal solid waste incinerators in Taiwan. *Energy Policy*, *39*(2), 651–663. doi:10.1016/j.enpol.2010.10.038

Chang, Y. T., Zhang, N., Danao, D., & Zhang, N. (2013). Environmental efficiency analysis of transportation system in China: A non-radial DEA approach. *Energy Policy*, *58*, 277–283. doi:10.1016/j.enpol.2013.03.011

Chang, Y.-T., Park, H., Jeong, J., & Lee, J. (2014). Evaluating economic and environmental efficiency of global airlines: A SBM-DEA approach. *Transportation Research Part D, Transport and Environment*, *27*, 46–50. doi:10.1016/j.trd.2013.12.013

Charles, V., Aparicio, J., & Zhu, J. (2019). The curse of dimensionality of decision-making units: A simple approach to increase the discriminatory power of data envelopment analysis. *European Journal of Operational Research*, *279*(3), 929–940. doi:10.1016/j.ejor.2019.06.025

Charnes, A., & Cooper, W. W. (1954). The stepping- stone method for explaining linear programming calculation in transportation problem. *Management Science*, *1*(1), 49–69.

Charnes, A., & Cooper, W. W. (1962). Programming with linear fractional functionals. *Naval Research Logistics Quarterly*, *9*(3–4), 181–186. doi:10.1002/nav.3800090303

Charnes, A., Cooper, W. W., & Rhodes, E. (1978). Measuring the efficiency of decision making units. *European Journal of Operational Research*, *2*(6), 429–444. doi:10.1016/0377-2217(78)90138-8

Chen, C., Zhu, J., Yu, J.-Y., & Noori, H. (2012). A new methodology for evaluating sustainable product design performance with two-stage network data envelopment analysis. *European Journal of Operational Research*, *221*(2), 348–359. doi:10.1016/j.ejor.2012.03.043

Cheng, L., Zang, H., Wei, Z., Zhang, F., & Sun, G. (2022). *Evaluation of opaque deep-learning solar power forecast models towards power-grid applications* (Vol. 198). Renewable Energy. doi:10.1016/j.renene.2022.08.054

Chen, J., Song, M., & Xu, L. (2015). Evaluation of environmental efficiency in China using data envelopment analysis. *Ecological Indicators*, *52*, 577–583. doi:10.1016/j.ecolind.2014.05.008

Chen, L., Wang, Y.-M., & Wang, L. (2016). Congestion measurement under different policy objectives: An analysis of Chinese industry. *Journal of Cleaner Production*, *112*, 2943–2952. doi:10.1016/j.jclepro.2015.09.064

Chhibber, D., Srivastava, P. K., & Bisht, D. C. (2019). Average duo triangle ranking technique to solve fully and type-2 intuitionistic fuzzy transportation problem. *Nonlinear Studies*, *26*(3), 487–504.

Chien, T., & Hu, J.-L. (2007). Renewable energy and macroeconomic efficiency of OECD and non-OECD economies. *Energy Policy*, *35*(7), 3606–3615. doi:10.1016/j.enpol.2006.12.033

Chikh, M. (1994). *Etude analytique conceptuelle des systèmes photovoltaïque de puissance adapté aux sites isolés* [Thèse de magister]. CDER.

Choi, D. G., & Thomas, V. M. (2012). An electricity generation planning model incorporating demand response. *Energy Policy*, *42*, 429–441. doi:10.1016/j.enpol.2011.12.008

Chopra, R., & Saxena, R. R. (2019). Linear programming problems with coefficients as trapezoidal intuitionistic fuzzy numbers. *Advances in Fuzzy Mathematics*, *14*(1), 41–52.

Chowdhury, S. P., Chowdhury, S., & Crossley, P. A. (2009). Islanding protection of active distribution networks with renewable distributed generators: A comprehensive survey. *Electric Power Systems Research*, *79*(6), 984–992. doi:10.1016/j.epsr.2008.12.012

Chujai, P., Kerdprasop, N., & Kerdprasop, K. (2013, March). Time series analysis of household electric consumption with ARIMA and ARMA models. In *Proceedings of the international multiconference of engineers and computer scientists* (Vol. 1, pp. 295-300). Hong Kong: IAENG.

Chutia, R. (2021). Ordering intuitionistic fuzzy numbers by a convex combination of values and multiple of ambiguity inclusion functions with ambiguities of membership and nonmembership functions. *International Journal of Intelligent Systems*, *36*(10), 5785–5815. doi:10.1002/int.22530

Chutia, R., Saikia, S., & Gogoi, M. K. (2021). A theoretical approach to ranking of parametric fuzzy numbers using value and left–right ambiguity. *Mathematical Sciences*, 1–17. doi:10.1007/s40096-021-00422-4

Clements, P., & Northrop, L. (2001). *Software Product Lines: Practices and Patterns*. Addison Wesley.

Coady, D., Parry, I. W. H., Sears, L., & Shang, B. (2015). *How Large Are Global Energy Subsidies?* IMF Working Paper.

Coaffee, J., Healey, P., & O'Brien, P. (2018). *Resilient cities: Rethinking urban protection*. Springer.

Cole, R. J. (2005). Building environmental assessment methods: Redefining intentions and roles. *Building Research and Information*, *35*(5), 455–467. doi:10.1080/09613210500219063

Commission Brundtland. (1987). Our Common Future. Oxford University Press. doi:10.1080/07488008808408783

Conchado, A., Linares, P., Lago, O., & Santamaría, A. (2016). An estimation of the economic and environmental benefits of a demand-response electricity program for Spain. *Sustainable Production and Consumption*, *8*, 108–119. doi:10.1016/j.spc.2016.09.004

Cook, W. D., & Green, R. H. (2005). Evaluating power plant efficiency: A hierarchical model. *Computers & Operations Research, 32*(4), 813–823. doi:10.1016/j.cor.2003.08.019

Cooper, W. W., Seiford, L. M., & Tone, K. (2007). Data envelopment analysis: A comprehensive text with models, applications, references and DEA-solver software. In *Data Envelopment Analysis: A Comprehensive Text with Models* (2nd ed.). Applications, References and DEA-Solver Software. doi:10.1007/978-0-387-45283-8

Cooper, W. W., Seiford, L. M., & Zhu, J. (2004). Data envelopment analysis: History, models and interpretations. In L. M. Cooper, L. M. Seiford, & J. Zhu (Eds.), *Handbook on Data Envelopment Analysis* (pp. 1–39). Kluwer Academic Publishers. doi:10.1007/1-4020-7798-X_1

Crama, Y., & Pironet, T. (2019). Vehicle allocation problem with uncertain transportation requests over a multi-period rolling horizon. *Logistics Research, 12*(1), 1–22. doi:10.23773/2019_1

Cubric, M. (2020). Drivers, barriers and social considerations for AI adoption in business and management: A tertiary study. *Technology in Society, 62*, 101257. doi:10.1016/j.techsoc.2020.101257

Cui, Q. (2019). Investigating the airlines emission reduction through carbon trading under CNG2020 strategy via a Network Weak Disposability DEA. *Energy, 180*, 763–771. doi:10.1016/j.energy.2019.05.159

D'silva, S., Shadmand, M., Bayhan, S., & Abu-Rub, H. (2020). Towards Grid of Microgrids: Seamless Transition between Grid-Connected and Islanded Modes of Operation. *IEEE Open Journal of the Industrial Electronics Society, 1*, 66–81. doi:10.1109/OJIES.2020.2988618

Dantzig, G. B. (1963). *Linear programming and extensions*. Princeton Univ. Press.

Darehmiraki, M. (2019). A novel parametric ranking method for intuitionistic fuzzy numbers. *Iranian Journal of Fuzzy Systems, 16*(1), 129–143. doi:10.22111/IJFS.2019.4489

Das, S. K., Mandal, T., & Behera, D. (2019). A new approach for solving fully fuzzy linear programming problem. *International Journal of Mathematics in Operational Research, 15*(3), 296–309. doi:10.1504/ijmor.2019.102074

Datta, A., Saha, D., Ray, A., & Das, P. (2014). Anti-islanding selection for grid-connected solar photovoltaic system applications: A MCDM based distance approach. *Solar Energy, 110*, 519–532. doi:10.1016/j.solener.2014.09.042

Dehghan, S., Amjady, N., & Kazemi, A. (2013). Two-stage robust generation expansion planning: A mixed integer linear programming model. *IEEE Transactions on Power Systems, 29*(2), 584–597. doi:10.1109/TPWRS.2013.2287457

Dekhici, L., Guerraiche, K., & Belkadi, K. (2020). Environmental economic power dispatch using bat algorithm with generalized fly and evolutionary boundary constraint handling scheme. *International Journal of Applied Metaheuristic Computing, 11*(2), 171–191. doi:10.4018/IJAMC.2020040109

Delfosse, F. (1998). Détermination de l'état de charge des batteries d'un véhicule électrique. Université de Liège.

Delima, A. J. P. (2019). Application of Time Series Analysis in Projecting Philippines' Electric Consumption. *International Journal of Machine Learning and Computing, 9*(5), 694–699. doi:10.18178/ijmlc.2019.9.5.860

DeMillo & Offutt. (1991). Constraint-based automatic test data generation. *IEEE Trans. Softw. Eng., 17*(9), 900–910. doi:10.1109/32.92910

Deshmukh, A., Mhaske, A. S., Jadhav, A., & Bondar, K. L. (2022). New methods for ranking of trapezoidal fuzzy to obtained optimum solution to fuzzy transportation problem. *Jisuanji Jicheng Zhizao Xitong, 28*(11), 257–264.

Deshmukh, A., Patil, D. S., Soni, G., & Tyagi, A. K. (2023). Cyber Security: New Realities for Industry 4.0 and Society 5.0. In A. Tyagi (Ed.), *Handbook of Research on Quantum Computing for Smart Environments* (pp. 299–325). IGI Global. doi:10.4018/978-1-6684-6697-1.ch017

Dey, A. (2011). Modeling Smart Grid using Generalized Stochastic Petri Net. *Renewable & Sustainable Energy Reviews, 43*, 51–64.

Dey, S., Chakraborty, A., & Basak, P. (2020). Forecasting electricity consumption in India: An ARDL approach. *Energy Reports, 6*, 300–308.

Dias, A. S. (2018). *Inovação Orientada para a Sustentabilidade : Uma análise ao caso da Corticeira Amorim*. Universidade de Aveiro, Departamento de Economia, Gestão, Engenharia Industrial e Turismo.

Dijkstra-Silva, S., Schaltegger, S., & Beske-Janssen, P. (2022). Understanding positive contributions to sustainability. A systematic review. *Journal of Environmental Management, 320*, 115802. doi:10.1016/j.jenvman.2022.115802 PMID:35952560

Dinagar, D. S., & Keerthivasan, R. (2020). Solving transportation problem with modern zero suffix method under fuzzy environment. In AIP Conference Proceedings (Vol. 2261, No. 1, p. 030060). AIP Publishing LLC. doi:10.1063/5.0016998

Ding, G. (2005). Developing a multicriteria approach for the measurement of sustainable performance. *Building Research and Information, 33*(1), 3–16. doi:10.1080/0961321042000322618

Ding, G. (2008). Sustainable construction-The role of environmental assessment tools. *Journal of Environmental Management, 86*(3), 451–464. doi:10.1016/j.jenvman.2006.12.025 PMID:17289255

Dobiás, J., & Macek, D. (2014). Leadership in Energy and Environmental Design (LEED) and its impact on building operational expenditures. *Procedia Engineering, 85*, 132–139. doi:10.1016/j.proeng.2014.10.537

Do, H., & Rothermel, G. (2006, September). On the use of mutation faults in empirical assessments of test case prioritization techniques. *IEEE Trans. Softw. Eng., 32*(9), 733–752. Advance online publication. doi:10.1109/TSE.2006.92

Durmaz, A., Demir, H., & Sezen, B. (2021). The role of negative entropy within supply chain sustainability. *Sustainable Production and Consumption, 28*, 218–230. doi:10.1016/j.spc.2021.04.014 PMID:34722845

Dutta, P., & Saikia, B. (2019). Arithmetic operations on normal semi elliptic intuitionistic fuzzy numbers and their application in decision-making. *Granular Computing*, 1-17. doi:10.1007/s41066-019-00175-5

Dyllick, T., & Hockerts, K. (2002). Beyond the business case for corporate sustainability. *Business Strategy and the Environment, 11*(2), 130–141. doi:10.1002/bse.323

Dymova, L. (2020). A numerical solution of fully fuzzy distribution problem. In *International Conference on Artificial Intelligence and Soft Computing* (pp. 305-314). Springer. doi:10.1007/978-3-030-61401-0_29

Dyson, R. G., Allen, R., Camanho, A. S., Podinovski, V. V., Sarrico, C. S., & Shale, E. A. (2001). Pitfalls and protocols in DEA. *European Journal of Operational Research, 132*(2), 245–259. doi:10.1016/S0377-2217(00)00149-1

Egilmez, G., Kucukvar, M., Tatari, O., & Bhutta, M. K. S. (2014). Supply chain sustainability assessment of the U.S. food manufacturing sectors: A life cycle-based frontier approach. *Resources, Conservation and Recycling, 82*, 8–20. doi:10.1016/j.resconrec.2013.10.008

El-Latif, A. A. AZobaa, A. FElsonbaty, A. H. (2021). Short-term load forecasting using SARIMA and artificial neural network models. *Electric Power Systems Research, 190*, 106823.

El Alaoui, M. (2020). Intuitionistic fully fuzzy balanced transportation problem. *Notes on Intuitionistic Fuzzy Sets*, *26*(1), 69–80. doi:10.7546/nifs.2020.26.1.69-80

El Alaoui, M. (2021). Type 2 fully fuzzy linear programming. *International Journal of Fuzzy System Applications*, *10*(4), 37–56. doi:10.4018/ijfsa.2021100103

El Azzaoui, M. (2021). Islanding detection method for distributed generation with wavelet based nuisance tripping suppression. *Electric Power Systems Research*, *199*, 107366. doi:10.1016/j.epsr.2021.107366

El Khadimi, A. (1998). *Modélisation de l'Irradiation Solaire Globale d'un Plan Horizontal et de la Vitesse du Vent en Vue de leur Utilisation dans le Dimensionnement d'un Système Hybride* [Thèse de Doctorat]. Université Cadi Ayyad.

Ela, E., Milligan, M., Parsons, B., Lew, D., & Corbus, D. (2009, July). *The evolution of wind power integration studies: Past, present, and future. In 2009 IEEE Power & Energy Society General Meeting*. IEEE.

Elaibi, W. M., & Hasanain, W. S. (2022). Optimality criterion for three proposed methods to unbalanced transportation models. *Italian Journal of Pure and Applied Mathematics, 48*, 67-79.

Elsaraiti, M., Ali, G., Musbah, H., Merabet, A., & Little, T. (2021, April). Time series analysis of electricity consumption forecasting using ARIMA model. In 2021 IEEE Green Technologies Conference (GreenTech) (pp. 259-262). IEEE. doi:10.1109/GreenTech48523.2021.00049

Elshrief, Y. A., Helmi, D. H., Abd-Elhaleem, S., Abozalam, B. A., & Asham, A. D. (2021). Fast and accurate islanding detection technique for microgrid connected to photovoltaic system. *Journal of Radiation Research and Applied Sciences*, *14*(1), 210–221. doi:10.1080/16878507.2021.1923913

Emam, O., Haggag, R. M. Y., & Nabil, N. (2022). Using a fuzzy genetic algorithm for solving transportation logistics problems. *Journal of Business and Management Studies*, *4*(4), 195–211. doi:10.32996/jbms.2022.4.4.19

Emas, R. (2015). *Brief for GSDR 2015 The Concept of Sustainable Development: Definition and Defining Principles.* https://asset-pdf.scinapse.io/prod/2184349672/2184349672.pdf

Ensan, F., Bagheri, E., & Gasevic, D. (2012). *Evolutionary searchbased test generation for software product line feature models. In CAISE.* Springer.

Essakhi & Farhat. (2019). Modélisation et simulation d'un module photovoltaïque. *Journées internationale matériaux et environnement.*

European Union. (2008). *Energy and climate change elements of the final compromise*. Technical report, European Union.

Fabbri, S. C. P. F., Maldonado, J. C., Masiero, P. C., Delamaro, M. E., & Wong, E. (1996). *Mutation testing applied to validate specifications based on petri nets. In Proceedings of the IFIP TC6 Eighth International Conference on Formal Description Techniques VIII.* Chapman & Hall, Ltd. https://dl.acm.org/citation.cfm?id=646214.681539

Faere, R., & Grosskopf, S. (1989). *Multilateral Productivity Comparisons When Some Outputs are Undesirable : A Nonparametric Approach.* The MIT Press. https://www.jstor.org/stable/192

Fallahi, A., Ebrahimi, R., & Ghaderi, S. F. (2011). Measuring efficiency and productivity change in power electric generation management companies by using data envelopment analysis: A case study. *Energy*, *36*(11), 6398–6405. doi:10.1016/j.energy.2011.09.034

Fallahpour, A., Olugu, E. U., Musa, S. N., Khezrimotlagh, D., & Wong, K. Y. (2016). An integrated model for green supplier selection under fuzzy environment: Application of data envelopment analysis and genetic programming approach. *Neural Computing & Applications*, *27*(3), 707–725. doi:10.1007/s00521-015-1890-3

Fang, H., Wu, J., & Zeng, C. (2009). Comparative study on efficiency performance of listed coal mining companies in China and the US. *Energy Policy*, *37*(12), 5140–5148. doi:10.1016/j.enpol.2009.07.027

Fang, X., Misra, S., Xue, G., & Yang, D. (2012). Smart Grid — The New and Improved Power Grid: A Survey. *IEEE Communications Surveys and Tutorials*, *14*(4), 944–980. doi:10.1109/SURV.2011.101911.00087

Färe, R., Grosskopf, S., Noh, D.-W., & Weber, W. (2005). Characteristics of a polluting technology: Theory and practice. *Journal of Econometrics*, *126*(2), 469–492. doi:10.1016/j.jeconom.2004.05.010

Farrell, M. J. (1957). The Measurement of Productive Efficiency. *Journal of the Royal Statistical Society. Series A (General)*, *120*(3), 253–281. doi:10.2307/2343100

Fazal, S., Enamul Haque, M., Taufiqul Arif, M., Gargoom, A., & Oo, A. M. T. (2023). Grid integration impacts and control strategies for renewable based microgrid. *Sustainable Energy Technologies and Assessments*, *56*, 103069. doi:10.1016/j.seta.2023.103069

Fellows, R., & Liu, A. (2013). Use and misuse of the concept of culture. *Construction Management and Economics*, *31*(5), 401–422. doi:10.1080/01446193.2013.794296

Feng, C., Chu, F., Ding, J., Bi, G., & Liang, L. (2015). Carbon Emissions Abatement (CEA) allocation and compensation schemes based on DEA. *Omega*, *53*, 78–89. doi:10.1016/j.omega.2014.12.005

Ferreira, R. R., Colorado, P. J., Grilo, A. P., Teixeira, J. C., & Santos, R. C. (2019). Method for identification of grid operating conditions for adaptive overcurrent protection during intentional islanding operation. *International Journal of Electrical Power & Energy Systems*, *105*, 632–641. doi:10.1016/j.ijepes.2018.09.004

Ferrera, M. B., Litran, S. P., Duran, E., & Andujar, J. M. (2015). A SEPIC-Cuk converter combination for bipolar DC microgrid applications. *2015 IEEE International Conference on Industrial Technology (ICIT)*, 884–889. 10.1109/ICIT.2015.7125209

Finkelman, R. B., Wolfe, A., & Hendryx, M. S. (2021). The future environmental and health impacts of coal. *Energy Geoscience*, *2*(2), 99–112. doi:10.1016/j.engeos.2020.11.001

Frankfurt School-UNEP Centre. (2018). Global trends in renewable energy investment 2018. Frankfurt School of Finance & Management.

Fraser, E., Dougill, A., Mabee, W., Reed, M., & McAlpine, P. (2006). Bottom up and top down: Analysis of participatory processes for sustainability indicator identification as a pathway to community empowerment and sustainable environmental management. *Journal of Environmental Management*, *78*(2), 114–127. doi:10.1016/j.jenvman.2005.04.009 PMID:16095806

Fukuyama, H., & Weber, W. L. (2008). Japanese banking inefficiency and shadow pricing. *Mathematical and Computer Modelling*, *48*(11), 1854–1867. doi:10.1016/j.mcm.2008.03.004

Galaz, V., Centeno, M. A., Callahan, P. W., Causevic, A., Patterson, T., Brass, I., Baum, S., Farber, D., Fischer, J., Garcia, D., McPhearson, T., Jimenez, D., King, B., Larcey, P., & Levy, K. (2021). Artificial intelligence, systemic risks, and sustainability. *Technology in Society*, *67*, 101741. doi:10.1016/j.techsoc.2021.101741

Gallagher, K. S., Grübler, A., Kuhl, L., Nemet, G., & Wilson, C. (2012). The energy technology innovation system. *Annual Review of Environment and Resources*, *37*(1), 137–162. doi:10.1146/annurev-environ-060311-133915

Gao, F., Wang, X., Yang, P., Kou, S., & Sun, M. (2020). Research and Simulation of Hybrid AC/DC Microgrid. *2020 4th International Conference on HVDC (HVDC)*, 1276–1280. 10.1109/HVDC50696.2020.9292793

Gargantini, A., & Fraser, G. (2011). Generating minimal fault detecting test suites for general boolean specifications. *Information and Software Technology*, *53*(11), 1263–1273. doi:10.1016/j.infsof.2011.06.008

Gattiker, T., & Carter, C. R. (2010). Understanding project champions' ability to gain intra-organizational commitment for environmental projects. *Journal of Operations Management*, *28*(1), 72–85. doi:10.1016/j.jom.2009.09.001

Ghaderi, A., Moghaddam, M. P., & Sheikh-El-Eslami, M. K. (2014). Energy efficiency resource modeling in generation expansion planning. *Energy*, *68*, 529–537. doi:10.1016/j.energy.2014.02.028

Ghanbari, R., Ghorbani-Moghadam, K., Mahdavi-Amiri, N., & De Baets, B. (2019). Fuzzy linear programming problems: Models and solutions. *Soft Computing*, 1–31. doi:10.1007/s00500-019-04519-w

Giffinger, R., Fertner, C., Kramar, H., Kalasek, R., Pichler-Milanović, N., & Meijers, E. (2007). Smart cities: Ranking of European medium-sized cities. Centre of Regional Science (SRF*)*. Vienna University of Technology. 10.1145/1947940.1947974

Gil, E., Aravena, I., & Cárdenas, R. (2014). Generation capacity expansion planning under hydro uncertainty using stochastic mixed integer programming and scenario reduction. *IEEE Transactions on Power Systems*, *30*(4), 1838–1847. doi:10.1109/TPWRS.2014.2351374

Gimenez, C., Sierra, V., & Rodon, J. (2012). Sustainable operations: Their impact on the triple bottom line. *International Journal of Production Economics*, *140*(1), 149–159. doi:10.1016/j.ijpe.2012.01.035

Giri, P. K., Maiti, M. K., & Maiti, M. (2023). *Profit maximization fuzzy 4D-TP with budget constraint for breakable substitute items: a swarm based optimization approach*. OPSEARCH. doi:10.1007/s12597-023-00621-8

Glaa, Lakhoua, & El Amraoui. (2017). Application of system modeling and simulation of the photovoltaic production. *Advances in Science and Technology Research Journal, 11*(3).

Glaeser, E. L. (2011). *Triumph of the city: How our greatest invention makes us richer, smarter, greener, healthier, and happier*. Penguin.

Glavič, P., & Lukman, R. (2007). Review of sustainability terms and their definitions. *Journal of Cleaner Production*, *15*(18), 1875–1885. doi:10.1016/j.jclepro.2006.12.006

Golany, B., & Roll, Y. (1989). An application procedure for DEA. *Omega*, *17*(3), 237–250. doi:10.1016/0305-0483(89)90029-7

Golmohamadi, H. (2022). Demand-Side Flexibility in Power Systems: A Survey of Residential, Industrial, Commercial, and Agricultural Sectors. *Sustainability (Basel)*, *14*(13), 7916. doi:10.3390/su14137916

Gomathi, L., Mishra, A. K., & Tyagi, A. K. (2023). Industry 5.0 for Healthcare 5.0: Opportunities, Challenges and Future Research Possibilities. *2023 7th International Conference on Trends in Electronics and Informatics (ICOEI),* 204-213. 10.1109/ICOEI56765.2023.10125660

Gómez-Calvet, R., Conesa, D., Gómez-Calvet, A. R., & Tortosa-Ausina, E. (2016). On the dynamics of eco-efficiency performance in the European Union. *Computers & Operations Research*, *66*, 336–350. doi:10.1016/j.cor.2015.07.018

Google. (n.d.). http://code.google.com/p/linux-variability-analysis-tools/ source/browse/?repo=formulas

Goralski, M. A., & Tan, T. K. (2022). Artificial intelligence and poverty alleviation: Emerging innovations and their implications for management education and sustainable development. *International Journal of Management Education*, *20*(3), 100662. doi:10.1016/j.ijme.2022.100662

Goto, M., Otsuka, A., & Sueyoshi, T. (2014). DEA (Data Envelopment Analysis) assessment of operational and environmental efficiencies on Japanese regional industries. *Energy*, *66*, 535–549. doi:10.1016/j.energy.2013.12.020

Grzegorzewski, P. (2003a). Distance and orderings in a family of intuitionistic fuzzy numbers. *Proceedings of the Third Conference of the European Society for Fuzzy Logic and Technology*, 223-227.

Grzegorzewski, P. (2003b). Intuitionistic fuzzy numbers. *Proceedings of the IFSA 2003 World Congress.*

Guerraiche, K., Dekhici, L., Chatelet, E., & Zeblah, A. (2023). Techno-Economic Green Optimization of Electrical Microgrid Using Swarm Metaheuristics. *Energies*, *16*(4), 1803. doi:10.3390/en16041803

Guha, D., & Chakraborty, D. (2010). A theoretical development of distance measure for intuitionistic fuzzy numbers. *International Journal of Mathematics and Mathematical Sciences.*

Guo, P., & Bharath Kumar, N. (2023). CO2 hydrogenation for geothermal energy storage through synthetic natural gas production and byproduct of refrigeration and freshwater using solid oxide electrolyzer cell (SOEC) and methanation reactor; Techno-economic evaluation and multi-objective optimization. *Journal of CO2 Utilization, 69*. doi:10.1016/j.jcou.2023.102395

Guo, C., Liao, J., & Zhang, Y. (2022). Adaptive droop control of unbalanced voltage in the multi-node bipolar DC microgrid based on fuzzy control. *International Journal of Electrical Power & Energy Systems*, *142*, 108300. doi:10.1016/j.ijepes.2022.108300

Gupta, B. B., Gaurav, A., Panigrahi, P. K., & Arya, V. (2023). Analysis of artificial intelligence-based technologies and approaches on sustainable entrepreneurship. *Technological Forecasting and Social Change*, *186*, 122152. doi:10.1016/j.techfore.2022.122152

Haider, H. T., See, O. H., & Elmenreich, W. (2016). A review of residential demand response of smart grid. *Renewable & Sustainable Energy Reviews*, *59*, 166–178. doi:10.1016/j.rser.2016.01.016

Hailu, A., & Veeman, T. S. (2001). Non-parametric Productivity Analysis with Undesirable Outputs: An Application to the Canadian Pulp and Paper Industry. *American Journal of Agricultural Economics*, *83*(3), 605–616. doi:10.1111/0002-9092.00181

Hai, T., Zhou, J., & Muranaka, K. (2023). Energy management and operational planning of renewable energy resources-based microgrid with energy saving. *Electric Power Systems Research*, *214*, 108792. doi:10.1016/j.epsr.2022.108792

Halkos, G. E., & Tzeremes, N. G. (2011). Oil consumption and economic efficiency: A comparative analysis of advanced, developing and emerging economies. *Ecological Economics*, *70*(7), 1354–1362. doi:10.1016/j.ecolecon.2011.02.010

Halkos, G., & Petrou, K. N. (2019). Treating undesirable outputs in DEA: A critical review. *Economic Analysis and Policy*, *62*, 97–104. doi:10.1016/j.eap.2019.01.005

Hampf, B., & Rødseth, K. L. (2015). Carbon dioxide emission standards for U.S. power plants: An efficiency analysis perspective. *Energy Economics*, *50*, 140–153. doi:10.1016/j.eneco.2015.04.001

Han, C. H., Ko, B. K., Song, S. Y., & Jang, G. (2017). Application of PV Inverter on distribution system with high penetration of PV. *Journal of International Council on Electrical Engineering*, *7*(1), 263–268. doi:10.1080/22348972.2017.1373726

Hansen, E. G., Grosse-Dunker, F., & Reichwald, R. (2009). Sustainability innovation cube - A framework to evaluate sustainability-oriented innovations. *International Journal of Innovation Management*, *13*(4), 683–713. doi:10.1142/S1363919609002479

Haque, S., Bhurjee, A. K., & Kumar, P. (2022). Multi-objective non-linear solid transportation problem with fixed charge, budget constraints under uncertain environments. *Systems Science & Control Engineering*, *10*(1), 899–909. doi:10.1080/21642583.2022.2137707

Hartmann, A. (2008). Overcoming resistance to innovation: The integration champion in construction. In P. S. Brandon & S.-L. Lu (Eds.), *Clients Driving Innovation* (pp. 157–167). Blackwell. doi:10.1002/9781444301342.ch16

Hart, S. L., & Milstein, M. B. (2003). Creating sustainable value. *The Academy of Management Perspectives*, *17*(2), 56–69. doi:10.5465/ame.2003.10025194

Harty, C. (2005). Innovation in construction: A sociology of technology approach. *Building Research and Information*, *33*(6), 512–522. doi:10.1080/09613210500288605

Hassan, A. Y. (2022). Wind cube optimum design for wind turbine using meta-heuristic algorithms. *Alexandria Engineering Journal, 61*(6). doi:10.1016/j.aej.2021.09.059

Hathroubi, S., Peypoch, N., & Robinot, E. (2014). Technical efficiency and environmental management: The Tunisian case. *Journal of Hospitality and Tourism Management*, *21*, 27–33. doi:10.1016/j.jhtm.2014.03.002

Hatziargyriou, N., Asano, H., Iravani, R., & Marnay, C. (2007). Microgrids. *IEEE Power & Energy Magazine*, *5*(4), 78–94. doi:10.1109/MPAE.2007.376583

Heaven, D. (2019). Why deep-learning AIs are so easy to fool. *Nature*, *574*(7777), 163–166. doi:10.1038/d41586-019-03013-5 PMID:31597977

He, F., Zhang, Q., Lei, J., Fu, W., & Xu, X. (2013). Energy efficiency and productivity change of China's iron and steel industry: Accounting for undesirable outputs. *Energy Policy*, *54*, 204–213. doi:10.1016/j.enpol.2012.11.020

Heins, C. (2022). Artificial intelligence in retail – a systematic literature review. Foresight. doi:10.1108/FS-10-2021-0210

Hemmati, H., Arcuri, A., & Briand, L. (2012). Achieving scalable model-based testing through test case diversity. *ACM Transactions on Software Engineering and Methodology*, *22*(1), 1–42. doi:10.1145/2430536.2430540

Hemmati, H., & Briand, L. (2010). An industrial investigation of similarity measures for model-based test case selection. In *Proceedings of the 2010 IEEE 21st International Symposium on Software Reliability Engineering, ser. ISSRE '10.* IEEE Computer Society. 10.1109/ISSRE.2010.9

Hemmati, R., Hooshmand, R. A., & Khodabakhshian, A. (2013). Reliability constrained generation expansion planning with consideration of wind farms uncertainties in deregulated electricity market. *Energy Conversion and Management*, *76*, 517–526. doi:10.1016/j.enconman.2013.08.002

HenardC.PapadakisM.PerrouinG.KleinJ.HeymansP.TraonY. L. (2012). *Bypassing the combinatorial explosion: Using similarity to generate and prioritize t-wise test suites for large software product lines*. Tech. Rep. http://arxiv.org/abs/1211.5451

Herazo, B., & Lizarralde, G. (2015). The influence of green building certifications in collaboration and innovation processes. *Construction Management and Economics*, *33*(4), 279–298. doi:10.1080/01446193.2015.1047879

Hitchcock, F. L. (1941). The distribution of a product from several sources to numerous localities. *Journal of Mathematics and Physics*, *20*(2), 224–230. doi:10.1002/sapm1941201224

Hoteit, H., He, X., Yan, B., & Vahrenkamp, V. (2023). Uncertainty quantification and optimization method applied to time-continuous geothermal energy extraction. *Geothermics, 110*. doi:10.1016/j.geothermics.2023.102675

Howell, J. M., & Higgins, C. A. (1990). Champions of technological innovation. *Administrative Science Quarterly*, *35*(2), 317–341. doi:10.2307/2393393

Huang, J., Yang, X., Cheng, G., & Wang, S. (2014). A comprehensive eco-efficiency model and dynamics of regional eco-efficiency in China. *Journal of Cleaner Production*, *67*, 228–238. doi:10.1016/j.jclepro.2013.12.003

Huber, J. (2008). Technological environmental innovations (TEIs) in a chain-analytical and life-cycle-analytical perspective. *Journal of Cleaner Production*, *16*(18), 1980–1986. doi:10.1016/j.jclepro.2008.01.014

Hu, J.-L., & Wang, S.-C. (2006). Total-factor energy efficiency of regions in China. *Energy Policy*, *34*(17), 3206–3217. doi:10.1016/j.enpol.2005.06.015

Husein, M., & Chung, I.-Y. (2018). Optimal design and financial feasibility of a university campus microgrid considering renewable energy incentives. *Applied Energy*, *225*, 273–289. doi:10.1016/j.apenergy.2018.05.036

Hussain, R. J., & Kumar, P. S. (2012a). The transportation problem in an intuitionistic fuzzy environment. *International Journal of Mathematics Research*, *4*(4), 411–420.

Hussain, R. J., & Kumar, P. S. (2012b). Algorithmic approach for solving intuitionistic fuzzy transportation problem. *Applied Mathematical Sciences*, *6*(77-80), 3981–3989.

Hussain, R. J., & Kumar, P. S. (2012c). The transportation problem with the aid of triangular intuitionistic fuzzy numbers. *Proceedings in International Conference on Mathematical Modeling and Applied Soft Computing (MMASC-2012)*, 819-825.

Hussain, R. J., & Kumar, P. S. (2013). An optimal more-for-less solution of mixed constraints intuitionistic fuzzy transportation problems. *International Journal of Contemporary Mathematical Sciences*, *8*(12), 565–576. doi:10.12988/ijcms.2013.13056

IEA and IRENA. (2017). *Perspectives for the energy transition: Investment needs for a low-carbon energy system*. IRENA.

IEA. (2015). *World Energy Balances 2015 edition (Database)*. IEA.

IEA. (2017). *World Energy Outlook 2017*. IEA.

IEA. (2018a). *Global Energy & CO2 Status Report 2017*. IEA.

IEA. (2018b). *Perspectives for the Energy Transition: The role of Energy Efficiency*. IEA.

Iglesias, G., Castellanos, P., & Seijas, A. (2010). Measurement of productive efficiency with frontier methods: A case study for wind farms. *Energy Economics*, *32*(5), 1199–1208. doi:10.1016/j.eneco.2010.03.004

IPCC. (2018). *Global warming of 1.5°C: An IPCC special report on the impacts of global warming of 1.5°C above pre-industrial levels and related global greenhouse gas emission pathways, in the context of strengthening the global response to the threat of climate change*. Cambridge University Press.

IPCC. (2021 in press). Summary for Policymakers. In *Climate Change 2021: The Physical Science Basis. Contribution of Working Group I to the Sixth Assessment Report of the Intergovernmental Panel on Climate Change*. Cambridge University Press.

Iqbal, S., & Yaqoob, N. (2023). Ranking of linear Diophantine fuzzy numbers using circumcenter of centroids. *AIMS Mathematics*, *8*(4), 9840–9861. doi:10.3934/math.2023497

IRENA & CPI. (2018). *Global landscape of renewable energy finance 2018*. CPI.

IRENA. (2016). *Renewable Energy Benefits: Measuring the Economics*. IRENA.

IRENA. (2017a). *Accelerating the Energy Transition through Innovation*. IRENA.

IRENA. (2017b). *Stranded Assets and Renewables: How the energy transition affects the value of energy reserves, buildings and capital stock*. IRENA.

IRENA. (2017d). *Renewable Energy and Jobs- Annual Review 2017*. IRENA.

IRENA. (2017e). *Untapped potential for climate action: Renewable energy in Nationally Determined Contributions.* IRENA.

IRENA. (2017f). *Renewable Energy Benefits: Leveraging Local Capacity for Solar PV.* LBNL.

IRENA. (2017g). *Renewable Energy Benefits: Leveraging Local Capacity for Onshore Wind.* IRENA.

IRENA. (2017h). *Renewable Energy Benefits: Leveraging Local Industries, International Renewable Energy Agency.* IRENA.

IRENA. (2018a). *Renewable Capacity Statistics 2018.* IRENA.

IRENA. (2018b). *New Global Commission to Examine Geopolitics of Energy Transformation.* http://www.irena. org/newsroom/pressreleases/2018/Jan/New-GlobalCommission-to-Examine-Geopolitics-of-EnergyTransformation

IRENA. (2018c). *Renewable Power Generation Costs in 2017.* IRENA.

IRENA. (2018d). *Renewable Energy Benefits: Leveraging Local Capacity for Offshore Wind.* IRENA.

Ismail, Z., Tai, J. C., Kong, K. K., Law, K. H., Shirazi, S. M., & Karim, R. (2013). Using data envelopment analysis in comparing the environmental performance and technical efficiency of selected companies in their global petroleum operations. *Measurement, 46*(9), 3401–3413. doi:10.1016/j.measurement.2013.04.076

Jain, P. K., Quamer, W., & Pamula, R. (2018). Electricity consumption forecasting using time series analysis. In *Advances in Computing and Data Sciences: Second International Conference, ICACDS 2018, Dehradun, India, April 20-21, 2018, Revised Selected Papers, Part II 2* (pp. 327-335). Springer Singapore. 10.1007/978-981-13-1813-9_33

Jain, D., & Saxena, D. (2023). Comprehensive review on control schemes and stability investigation of hybrid AC-DC microgrid. *Electric Power Systems Research, 218*, 109182. doi:10.1016/j.epsr.2023.109182

Jain, V., Sethi, P., Verma, R., Arya, S., & Chawla, C. (2020). An analytical solution of transportation problem from modified distribution (MODI) approach. *Wesleyan Journal of Research, 13*(16), 43–55.

Jajula, S. K., Tripathi, K., & Bajaj, S. B. (2023). Review of Detection of Packets Inspection and Attacks in Network Security. In P. Dutta, S. Chakrabarti, A. Bhattacharya, S. Dutta, & V. Piuri (Eds.), *Emerging Technologies in Data Mining and Information Security. Lecture Notes in Networks and Systems* (Vol. 491). Springer. doi:10.1007/978-981-19-4193-1_58

Jansi Rani, J., Dhanasekar, S., Micheal, D. R., & Manivannan, A. (2023). On solving fully intuitionistic fuzzy transportation problem via branch and bound technique. *Journal of Intelligent & Fuzzy Systems*, 1–11. doi:10.3233/JIFS-221345

Jansi Rani, J., Manivannan, A., & Dhanasekar, S. (2023). Interval valued intuitionistic fuzzy diagonal optimal algorithm to solve transportation problems. *International Journal of Fuzzy Systems*, 1–15. doi:10.1007/s40815-022-01446-1

Javanmard, M., & Nehi, H. M. (2019). A solving method for fuzzy linear programming problem with interval type-2 fuzzy numbers. *International Journal of Fuzzy Systems, 21*(3), 882–891. doi:10.1007/s40815-018-0591-3

Jia & Harman. (2011). An analysis and survey of the development of mutation testing. *IEEE Transactions on Software Engineering, 37*(5), 649–678.

Jia, K., Bi, T., Liu, B., Thomas, D., & Goodman, A. (2014). Advanced islanding detection utilized in distribution systems with DFIG. *International Journal of Electrical Power & Energy Systems, 63*, 113–123. doi:10.1016/j.ijepes.2014.05.003

Jin, S., Botterud, A., & Ryan, S. M. (2014). Temporal versus stochastic granularity in thermal generation capacity planning with wind power. *IEEE Transactions on Power Systems, 29*(5), 2033–2041. doi:10.1109/TPWRS.2014.2299760

Jirdehi, M. A., Tabar, V. S., Ghassemzadeh, S., & Tohidi, S. (2020). Different aspects of microgrid management: A comprehensive review. *Journal of Energy Storage*, *30*, 101457. doi:10.1016/j.est.2020.101457

Johansen, M. F., Haugen, Ø., Fleurey, F., Carlson, E., Endresen, J., & Wien, T. (2012). *A technique for agile and automatic interaction testing for product lines*. In B. Nielsen & C. Weise (Eds.), *ICTSS* (Vol. 7641, pp. 39–54). Springer.

John, V., Ye, Z., & Kolwalkar, A. (2003). *Investigation of antiislanding protection of power converter based distributed generators using frequency domain analysis. In 2003 IEEE Power Engineering Society General Meeting*. IEEE.

Ju, K., Su, B., Zhou, D., Zhou, P., & Zhang, Y. (2015). Oil price crisis response: Capability assessment and key indicator identification. *Energy*, *93*, 1353–1360. doi:10.1016/j.energy.2015.09.124

Kadhem, A. A., Wahab, N. A., Aris, I., Jasni, J., Abdalla, A. N., & Matsukawa, Y. (2017). Reliability assessment of generating systems with wind power penetration via BPSO. *Journal of International Journal on Advanced Science. Engineering and Information Technology*, *7*(4), 1248–1254.

Kagawa, S., Takezono, K., Suh, S., & Kudoh, Y. (2013). Production possibility frontier analysis of biodiesel from waste cooking oil. *Energy Policy*, *55*, 362–368. doi:10.1016/j.enpol.2012.12.016

Kalimoldayev, M., Drozdenko, A., Koplyk, I., Marinich, T., Abdildayeva, A., & Zhukabayeva, T. (2020). Analysis of modern approaches for the prediction of electric energy consumption. *Open Engineering*, *10*(1), 350–361. doi:10.1515/eng-2020-0028

Kalonda, P. O., & Omekanda, A. M. (2020). *Kruskal's Algorithm, Vogel's Approximation and Modified Distribution Methods for the Design of Optimal Electrical Networks in the Democratic Republic of Congo. In 2020 IEEE PES/IAS PowerAfrica*. IEEE. doi:10.1109/PowerAfrica49420.2020.9219801

Kalyani, S., & Nagarani, S. (2020). A fully fuzzy transportation problem with hexagonal fuzzy number. In AIP Conference Proceedings (Vol. 2261, No. 1, p. 030078). AIP Publishing LLC. doi:10.1063/5.0016903

Kamalinia, S., & Shahidehpour, M. (2010). Generation expansion planning in wind-thermal power systems. *IET Generation, Transmission & Distribution*, *4*(8), 940–951. doi:10.1049/iet-gtd.2009.0695

Kaminski, G. K., Praphamontripong, U., Ammann, P., & Offutt, J. (2011). A logic mutation approach to selective mutation for programs and queries. *Information and Software Technology*, *53*(10), 1137–1152. doi:10.1016/j.infsof.2011.03.009

Kaminski, G., Ammann, P., & Offutt, J. (2012). Improving logicbased testing. *Journal of Systems and Software*.

Kang, Cohen, Hess, Novak, & Peterson. (1990). *Feature-Oriented Domain Analysis (FODA) Feasibility Study*. Carnegie-Mellon University Software Engineering Institute, Tech. Rep.

Kant, K., & Gupta, O. H. (2022). DC Microgrid: A Comprehensive Review on Protection Challenges and Schemes. *IETE Technical Review*, 1–17. doi:10.1080/02564602.2022.2136271

Kao, C. (2014). Network data envelopment analysis: A review. *European Journal of Operational Research*, *239*(1), 1–16. doi:10.1016/j.ejor.2014.02.039

Kao, C. (2020). Measuring efficiency in a general production possibility set allowing for negative data. *European Journal of Operational Research*, *282*(3), 980–988. doi:10.1016/j.ejor.2019.10.027

Kao, C. (2022). Measuring efficiency in a general production possibility set allowing for negative data: An extension and a focus on returns to scale. *European Journal of Operational Research*, *296*(1), 267–276. doi:10.1016/j.ejor.2021.03.061

Kaplan, A., & Haenlein, M. (2019). Siri, siri, in my hand: Who's the fairest in the land? On the interpretations, illustrations, and implications of artificial intelligence. *Business Horizons*, *62*(1), 15–25. doi:10.1016/j.bushor.2018.08.004

Kar, A. K., Choudhary, S. K., & Singh, V. K. (2022). How can artificial intelligence impact sustainability: A systematic literature review. *Journal of Cleaner Production, 134120*. Advance online publication. doi:10.1016/j.jclepro.2022.134120

Karunanithi, K., Saravanan, S., Prabakar, B. R., Kannan, S., & Thangaraj, C. (2017). Integration of demand and supply side management strategies in generation expansion planning. *Renewable & Sustainable Energy Reviews*, *73*, 966–982. doi:10.1016/j.rser.2017.01.017

Kaur, A., Kacprzyk, J., & Kumar, A. (2020). New methods for solving the fully fuzzy transportation problems with the LR flat fuzzy numbers. In *Fuzzy Transportation and Transshipment Problems* (pp. 81–101). Springer. doi:10.1007/978-3-030-26676-9_5

Kazancoglu, I., Sagnak, M., Kumar Mangla, S., & Kazancoglu, Y. (2021). Circular economy and the policy: A framework for improving the corporate environmental management in supply chains. *Business Strategy and the Environment*, *30*(1), 590–608. doi:10.1002/bse.2641

Khalifa, A. E. W., Kumar, P., & Alharbi, M. G. (2020). Enhancement of capacitated transportation problem in fuzzy environment. *Advances in Fuzzy Systems*, *2020*, 1–9. doi:10.1155/2020/8893976

Khalili-Damghani, K., & Shahmir, Z. (2015). Uncertain network data envelopment analysis with undesirable outputs to evaluate the efficiency of electricity power production and distribution processes. *Computers & Industrial Engineering*, *88*, 131–150. doi:10.1016/j.cie.2015.06.013

Khodakarami, M., Shabani, A., Farzipoor Saen, R., & Azadi, M. (2015). Developing distinctive two-stage data envelopment analysis models: An application in evaluating the sustainability of supply chain management. *Measurement*, *70*, 62–74. doi:10.1016/j.measurement.2015.03.024

Khoshnevisan, B., Bolandnazar, E., Shamshirband, S., Shariati, H. M., Anuar, N. B., & Mat Kiah, M. L. (2015). Decreasing environmental impacts of cropping systems using life cycle assessment (LCA) and multi-objective genetic algorithm. *Journal of Cleaner Production*, *86*, 67–77. doi:10.1016/j.jclepro.2014.08.062

Khosravi, H., Samet, H., & Tajdinian, M. (2021). Empirical mode decomposition based algorithm for islanding detection in micro-grids. *Electric Power Systems Research*, *201*, 107542. doi:10.1016/j.epsr.2021.107542

Kilic, M., & Altun, A. F. (2022). Dynamic modelling and multi-objective optimization of off-grid hybrid energy systems by using battery or hydrogen storage for different climates. *International Journal of Hydrogen Energy*. doi:10.1016/j.ijhydene.2022.12.103

Kishor, Y., Kamesh Rao, C. H., & Patel, R. N. (2022). Comprehensive Analysis on DC-Microgrid Application for Remote Electrification. In *Renewable Energy Technologies* (pp. 435–475). Wiley. doi:10.1002/9781119827634.ch13

Klewitz, J., & Hansen, E. G. (2014). Sustainability-oriented innovation of SMEs: A systematic review. *Journal of Cleaner Production*, *65*, 57–75. doi:10.1016/j.jclepro.2013.07.017

Koltsaklis, N. E., Liu, P., & Georgiadis, M. C. (2015). An integrated stochastic multi-regional long-term energy planning model incorporating autonomous power systems and demand response. *Energy*, *82*, 865–888. doi:10.1016/j.energy.2015.01.097

Koopmans, T. C. (1949). Optimum utilization of the transportation system. *Econometrica, 17*, 136-146. http://www.jstor.org/stable/1907301

Kopsidas, K., & Abogaleela, M. (2019, May). Utilizing Demand Response to Improve Network Reliability and Ageing Resilience. *IEEE Transactions on Power Systems*, *34*(3), 2216–2227. doi:10.1109/TPWRS.2018.2883612

Korhonen, P. J., & Luptacik, M. (2004). Eco-efficiency analysis of power plants: An extension of data envelopment analysis. *European Journal of Operational Research, 154*(2), 437–446. doi:10.1016/S0377-2217(03)00180-2

Kumar, P. S. (2010). *A comparative study on transportation problem in fuzzy environment* [M.Phil thesis]. Jamal Mohamed College, Tiruchirappalli, India.

Kumar, P. S. (2017b). *Algorithmic approach for solving allocation problems under intuitionistic fuzzy environment* [PhD thesis]. Jamal Mohamed College, Tiruchirappalli, India.

Kumar, P. S. (2021). Finding the solution of balanced and unbalanced intuitionistic fuzzy transportation problems by using different methods with some software packages. In Handbook of Research on Applied AI for International Business and Marketing Applications (pp. 278-320). doi:10.4018/978-1-7998-5077-9.ch015

Kumar, P. S., & Hussain, R. J. (2014b). A method for finding an optimal solution of an assignment problem under mixed intuitionistic fuzzy environment. In *Proceedings in International Conference on Mathematical Sciences (ICMS-2014)* (pp. 417-421). Academic Press.

Kumar, P. S., & Hussain, R. J. (2016a). Computationally simple approach for solving fully intuitionistic fuzzy real life transportation problems. *International Journal of System Assurance Engineering and Management, 7*(S1), 90-101. doi:10.1007/s13198-014-0334-2

Kumar, P.S. (2017a). PSK method for solving type-1 and type-3 fuzzy transportation problems. *Fuzzy Systems,* 367–392. doi:10.4018/978-1-5225-1908-9.ch017

Kumari, N. (2020). Zero accomplishment method for finding an optimal more-for-less solution of transportation problem with mixed constraints. *International Journal of Mathematics Trends and Technology*, *66*(8), 144–149.

Kumar, P. S. (2016a). PSK method for solving type-1 and type-3 fuzzy transportation problems. *International Journal of Fuzzy System Applications*, *5*(4), 121–146. doi:10.4018/IJFSA.2016100106

Kumar, P. S. (2016b). A simple method for solving type-2 and type-4 fuzzy transportation problems. *International Journal of Fuzzy Logic and Intelligent Systems*, *16*(4), 225–237. doi:10.5391/IJFIS.2016.16.4.225

Kumar, P. S. (2018a). PSK method for solving intuitionistic fuzzy solid transportation problems. *International Journal of Fuzzy System Applications*, *7*(4), 62–99. doi:10.4018/IJFSA.2018100104

Kumar, P. S. (2018b). A note on 'a new approach for solving intuitionistic fuzzy transportation problem of type-2'. *International Journal of Logistics Systems and Management*, *29*(1), 102–129. doi:10.1504/IJLSM.2018.088586

Kumar, P. S. (2018c). Linear programming approach for solving balanced and unbalanced intuitionistic fuzzy transportation problems. *International Journal of Operations Research and Information Systems*, *9*(2), 73–100. doi:10.4018/IJORIS.2018040104

Kumar, P. S. (2018d). Search for an optimal solution to vague traffic problems using the PSK method. In *Handbook of Research on Investigations in Artificial Life Research and Development* (pp. 219–257). IGI Global. doi:10.4018/978-1-5225-5396-0.ch011

Kumar, P. S. (2018e). A simple and efficient algorithm for solving type-1 intuitionistic fuzzy solid transportation problems. *International Journal of Operations Research and Information Systems*, *9*(3), 90–122. doi:10.4018/IJORIS.2018070105

Kumar, P. S. (2019a). PSK method for solving mixed and type-4 intuitionistic fuzzy solid transportation problems. *International Journal of Operations Research and Information Systems*, *10*(2), 20–53. doi:10.4018/ijoris.2019040102

Kumar, P. S. (2019b). Intuitionistic fuzzy solid assignment problems: A software-based approach. *International Journal of System Assurance Engineering and Management, 10*(4), 661–675. doi:10.1007/s13198-019-00794-w

Kumar, P. S. (2020a). Intuitionistic fuzzy zero point method for solving type-2 intuitionistic fuzzy transportation problem. *International Journal of Operational Research, 37*(3), 418–451. doi:10.1504/IJOR.2020.105446

Kumar, P. S. (2020b). The PSK method for solving fully intuitionistic fuzzy assignment problems with some software tools. In *Theoretical and Applied Mathematics in International Business* (pp. 149–202). IGI Global. doi:10.4018/978-1-5225-8458-2.ch009

Kumar, P. S. (2020c). Developing a new approach to solve solid assignment problems under intuitionistic fuzzy environment. *International Journal of Fuzzy System Applications, 9*(1), 1–34. doi:10.4018/IJFSA.2020010101

Kumar, P. S. (2020d). Algorithms for solving the optimization problems using fuzzy and intuitionistic fuzzy set. *International Journal of System Assurance Engineering and Management, 11*(1), 189–222. doi:10.1007/s13198-019-00941-3

Kumar, P. S. (2022). Computationally simple and efficient method for solving real-life mixed intuitionistic fuzzy 3D assignment problems. *International Journal of Software Science and Computational Intelligence*. Advance online publication. doi:10.4018/ijssci.291715

Kumar, P. S. (2023a). The PSK method: A new and efficient approach to solving fuzzy transportation problems. In *Transport and Logistics Planning and Optimization* (pp. 149–197). IGI Global. doi:10.4018/978-1-6684-8474-6.ch007

Kumar, P. S. (2023b). The theory and applications of the software-based PSK method for solving intuitionistic fuzzy solid transportation problems. In M. Habib (Ed.), *Perspectives and Considerations on the Evolution of Smart Systems* (pp. 137–186). IGI Global. doi:10.4018/978-1-6684-7684-0.ch007

Kumar, P. S. (2024a). AI-Driven Decision Support System for Intuitionistic Fuzzy Assignment Problems. In T. Nguyen & N. Vo (Eds.), *Using Traditional Design Methods to Enhance AI-Driven Decision Making* (pp. 352–398). IGI Global. doi:10.4018/979-8-3693-0639-0.ch016

Kumar, P. S. (2024b). Theory and applications of the software-based PSK method for solving intuitionistic fuzzy solid assignment problems. In *Applications of New Technology in Operations and Supply Chain Management*. IGI Global.

Kumar, P. S., & Hussain, R. J. (2014a). A systematic approach for solving mixed intuitionistic fuzzy transportation problems. *International Journal of Pure and Applied Mathematics, 92*(2), 181–190. doi:10.12732/IJPAM.V92I2.4

Kumar, P. S., & Hussain, R. J. (2014c). New algorithm for solving mixed intuitionistic fuzzy assignment problem. *Elixir Appl. Math, 73*, 25971–25977.

Kumar, P. S., & Hussain, R. J. (2015). A method for solving unbalanced intuitionistic fuzzy transportation problems. *Notes on Intuitionistic Fuzzy Sets, 21*(3), 54–65.

Kumar, P. S., & Hussain, R. J. (2016b). A simple method for solving fully intuitionistic fuzzy real life assignment problem. *International Journal of Operations Research and Information Systems, 7*(2), 39–61. doi:10.4018/IJORIS.2016040103

Kumar, P. S., & Hussain, R. J. (2016c). An algorithm for solving unbalanced intuitionistic fuzzy assignment problem using triangular intuitionistic fuzzy number. *The Journal of Fuzzy Mathematics, 24*(2), 289–302.

Lange, G.-M., Wodon, Q., & Carey, K. (2018). *The Changing Wealth of Nations 2018: Building a Sustainable Future*. World Bank. doi:10.1596/978-1-4648-1046-6

Le Berre, D., & Parrain, A. (2010). The sat4j library, release 2.2, system description. *Journal on Satisfiability, Boolean Modeling and Computation, 7*(2-3), 59–64. doi:10.3233/SAT190075

Lee, Y. W., Tay, K. G., & Choy, Y. Y. (2018). Forecasting electricity consumption using time series model. *International Journal of Engineering & Technology, 7*(4.30), 218.

Leszkiewicz, A., Hormann, T., & Krafft, M. (2022). Smart Business and the Social Value of AI. In Smart Industry – Better Management (Advanced Series in Management, Vol. 28). Emerald Publishing Limited. doi:10.1108/S1877-636120220000028004

Li, S., Wen, D., Feng, B., Li, F., Yue, D., Zhang, Q., Wang, J., & Feng, Z. (2023). Numerical optimization of geothermal energy extraction from deep karst reservoir in North China. *Renewable Energy, 202*, 1071-1085. doi:10.1016/j.renene.2022.12.016

Li, D. F., Nan, J. X., & Zhang, M. J. (2010). A ranking method of triangular intuitionistic fuzzy numbers and application to decision making. *International Journal of Computational Intelligence Systems*, *3*(5), 522–530. doi:10.1080/18756891.2010.9727719

Lieder, M., & Rashid, A. (2016). Towards circular economy implementation: A comprehensive review in context of manufacturing industry. *Journal of Cleaner Production*, *115*, 36–51. doi:10.1016/j.jclepro.2015.12.042

Li, J.-h., Dai, G.-x., & Li, H.-h. (2009). Mutation analysis for testing finite state machines. *Proceedings of the 2009 Second International Symposium on Electronic Commerce and Security*, 620–624. 10.1109/ISECS.2009.158

Li, K., & Zhang, T. (2018). Forecasting electricity consumption using an improved grey prediction model. *Information (Basel)*, *9*(8), 204. doi:10.3390/info9080204

Li, M., Patiño-Echeverri, D., & Zhang, J. J. (2019). Policies to promote energy efficiency and air emissions reductions in China's electric power generation sector during the 11th and 12th five-year plan periods: Achievements, remaining challenges, and opportunities. *Energy Policy*, *125*, 429–444. doi:10.1016/j.enpol.2018.10.008

Lin, Y., Chen, Y.-W., & Yang, J.-T. (2023). Optimized thermal management of a battery energy-storage system (BESS) inspired by air-cooling inefficiency factor of data centers. *International Journal of Heat and Mass Transfer, 200*. https://www.sciencedirect.com/science/article/pii/S0017931022008572 doi:10.1016/j.ijheatmasstransfer.2022.123388

Lindberg, O., Lingfors, D., Arnqvist, J., van der Meer, D., & Munkhammar, J. (2023). Day-ahead probabilistic forecasting at a co-located wind and solar power park in Sweden: Trading and forecast verification. *Advances in Applied Energy, 9*. https://www.sciencedirect.com/science/article/pii/S2666792422000385 doi:10.1016/j.adapen.2022.100120

Lin, Z., Wang, P., Ren, S., & Zhao, D. (2023). Comprehensive Impact Assessment of Carbon Neutral Pathways and Air pollution Control Policies in Shaanxi Province of China. *Resources, Conservation & Recycling Advances*, *200143*. Advance online publication. doi:10.1016/j.rcradv.2023.200143

Liu, J., Liu, H., Yao, X.-L., & Liu, Y. (2016). Evaluating the sustainability impact of consolidation policy in China's coal mining industry: A data envelopment analysis. *Journal of Cleaner Production*, *112*, 2969–2976. doi:10.1016/j.jclepro.2015.08.011

Liu, X., Huang, L., & Li, Q. (2019). Electricity consumption forecasting in China using seasonal hybrid ARIMA-ETS model. *Energies*, *12*(6), 1084.

Liu, X., Wang, P., & Loh, P. C. (2010). A hybrid AC/DC micro-grid. *2010 Conference Proceedings IPEC*, 746–751. 10.1109/IPECON.2010.5697024

Liu, Y., Lei, S., & Hou, Y. (2019, September). Restoration of Power Distribution Systems With Multiple Data Centers as Critical Loads. *IEEE Transactions on Smart Grid*, *10*(5), 5294–5307. doi:10.1109/TSG.2018.2880255

Li, W. (2010). Real-Time Simulation of a Wind Turbine Generator Coupled With a Battery Supercapacitor Energy Storage System. *IEEE Transactions on Industrial Electronics*, *57*(4), 1137–1145. doi:10.1109/TIE.2009.2037103

Longo, L., Colantoni, A., Castellucci, S., Carlini, M., Vecchione, L., Savuto, E., Pallozzi, V., Di Carlo, A., Bocci, E., Moneti, M., Cocchi, S., & Boubaker, K. (2015). DEA (data envelopment analysis)-assisted supporting measures for ground coupled heat pumps implementing in Italy: A case study. *Energy*, *90*, 1967–1972. doi:10.1016/j.energy.2015.07.024

Lozano, S., Villa, G., & Brännlund, R. (2009). Centralised reallocation of emission permits using DEA. *European Journal of Operational Research*, *193*(3), 752–760. doi:10.1016/j.ejor.2007.07.029

Luo, J., Yang, Y., & Chen, Y. (2012). Optimizing the drilled well patterns for CBM recovery via numerical simulations and data envelopment analysis. *International Journal of Mining Science and Technology*, *22*(4), 503–507. doi:10.1016/j.ijmst.2012.01.011

Ma, H., Ge, M., Wu, G., Du, B., & Liu, Y. (2021). Formulas of the optimized yaw angles for cooperative control of wind farms with aligned turbines to maximize the power production. *Applied Energy*, *303*. doi:10.1016/j.apenergy.2021.117691

Madhav, A. V. S., & Tyagi, A. K. (2023). Explainable Artificial Intelligence (XAI): Connecting Artificial Decision-Making and Human Trust in Autonomous Vehicles. In *Proceedings of Third International Conference on Computing, Communications, and Cyber-Security. Lecture Notes in Networks and Systems* (vol. 421). Springer. 10.1007/978-981-19-1142-2_10

Madhav, A. V. S., & Tyagi, A. K. (2022). The World with Future Technologies (Post-COVID-19): Open Issues, Challenges, and the Road Ahead. In A. K. Tyagi, A. Abraham, & A. Kaklauskas (Eds.), *Intelligent Interactive Multimedia Systems for e-Healthcare Applications*. Springer. doi:10.1007/978-981-16-6542-4_22

Mahapatra, G. S., & Roy, T. K. (2009). Reliability evaluations using triangular intuitionistic fuzzy numbers, arithmetic operations. *International Scholarly and Scientific Research & Innovation*, *3*(2), 422–429.

Mahapatra, G. S., & Roy, T. K. (2013). Intuitionistic fuzzy number and its arithmetic operation with application on system failure. *Journal of Uncertain Systems*, *7*(2), 92–107.

Mahmoudi, F., & Nasseri, S. H. (2019). A new approach to solve fully fuzzy linear programming problem. *Journal of Applied Research on Industrial Engineering*, *6*(2), 139–149. doi:10.22105/JARIE.2019.183391.1090

Mahmoud, M. A., Tang, Y. C., Maseleno, A., Lim, F.-C., Kasim, H., & Yong, C. (2020). *Towards the Development of a Smart Energy Grid.* Springer, LLC. doi:10.1007/978-3-030-33582-3_63

Ma, J., Silva, V., Belhomme, R., Kirschen, D. S., & Ochoa, L. F. (2013, July). *Evaluating and planning flexibility in sustainable power systems. In 2013 IEEE power & energy society general meeting*. IEEE.

Malik, A. S. (2007). Impact on power planning due to demand-side management (DSM) in commercial and government sectors with rebound effect—A case study of central grid of Oman. *Energy*, *32*(11), 2157–2166. doi:10.1016/j.energy.2007.05.004

Malik, S. M., Ai, X., Sun, Y., Zhengqi, C., & Shupeng, Z. (2017). Voltage and frequency control strategies of hybrid AC/DC microgrid: A review. *IET Generation, Transmission & Distribution*, *11*(2), 303–313. doi:10.1049/iet-gtd.2016.0791

Malik, S., Bansal, R., & Tyagi, A. K. (Eds.). (2022). *Impact and Role of Digital Technologies in Adolescent Lives*. IGI Global. doi:10.4018/978-1-7998-8318-0

Malini, P. (2019). A new ranking technique on heptagonal fuzzy numbers to solve fuzzy transportation problem. *International Journal of Mathematics in Operational Research*, *15*(3), 364–371. doi:10.1504/IJMOR.2019.102078

Mandal, M. K., Kumar, A., & Burnwal, A. P. (2020). A fuzzy approach for with quadratic from of achievement function for transportation problem. *Journal of Information and Computational Science*, *10*(9), 316–326.

Mapanga, Kumar, Makondo, Kushboo, Kadebu, & Chanda. (2017). Design and implementation of an intrusion detection system using MLP-NN for MANET. *2017 IST-Africa Week Conference (IST-Africa),* 1-12. . doi:10.23919/ISTAFRICA.2017.8102374

Mateus, R., & Bragança, L. (2011). Sustainability assessment and rating of buildings: Developing the methodology SBToolPT-H. *Building and Environment*, *46*(10), 1962–1971. doi:10.1016/j.buildenv.2011.04.023

Mathur, N., & Srivastava, P. K. (2020). An inventive approach to optimize fuzzy transportation problem. *International Journal of Mathematical, Engineering and Management Sciences*, *5*(5), 985–994. doi:10.33889/IJMEMS.2020.5.5.075

Maurya, R., Prakash, S., & Singh, A. K. (2022). Challenges, Configuration, Control, and Scope of DC Microgrid Systems: A Review. *Journal of Electrical Engineering & Technology*. Advance online publication. doi:10.1007/s42835-022-01304-y

McGregor, J. (2010). *Testing a software product line. In Testing Techniques in Software Engineering* (Vol. 6153). Springer.

Meegahapola, L. G., Robinson, D., Agalgaonkar, A. P., Perera, S., & Ciufo, P. (2014). Microgrids of Commercial Buildings: Strategies to Manage Mode Transfer From Grid Connected to Islanded Mode. *IEEE Transactions on Sustainable Energy*, *5*(4), 1337–1347. doi:10.1109/TSTE.2014.2305657

Meena, M., Shubham, S., Paritosh, K., Pareek, N., & Vivekanand, V. (2021). Production of biofuels from biomass: Predicting the energy employing artificial intelligence modelling. *Bioresource Technology, 340*. doi:10.1016/j.biortech.2021.125642

Meinhardt, M., & Wimmer, D. (2001). Multistring-converter: The next step in evolution of string converter. *Conference: 9th European Conference on Power Electronics and Applications.*

Mekaroonreung, M., & Johnson, A. L. (2012). Estimating the shadow prices of SO2 and NOx for U.S. coal power plants: A convex nonparametric least squares approach. *Energy Economics*, *34*(3), 723–732. doi:10.1016/j.eneco.2012.01.002

Mendonca, M., Branco, M., & Cowan, D. (2009). *S.p.l.o.t.: software product lines online tools*. http://doi.acm.org/10.1145/1639950.1640002

Mendonca, M., Wasowski, A., & Czarnecki, K. (2009). *Sat-based analysis of feature models is easy. In Proceedings of the 13th International Software Product Line Conference, ser. SPLC '09*. Carnegie Mellon University. http: //dl.acm.org/citation.cfm?id=1753235.1753267

Midha, S., & Tripathi, K. (2019). Extended TLS security and Defensive Algorithm in OpenFlow SDN. *2019 9th International Conference on Cloud Computing, Data Science & Engineering (Confluence),* 141-146. 10.1109/CONFLUENCE.2019.8776607

Midha, S., & Tripathi, K. (2020). Remotely Triggered Blackhole Routing in SDN for Handling DoS. In *Proceedings of International Conference on IoT Inclusive Life (ICIIL 2019), NITTTR Chandigarh, India. Lecture Notes in Networks and Systems* (vol. 116). Springer. 10.1007/978-981-15-3020-3_1

Midha, S., Kaur, G., & Tripathi, K. (2017). Cloud deep down — SWOT analysis. *2017 2nd International Conference on Telecommunication and Networks (TEL-NET),* 1-5. 10.1109/TEL-NET.2017.8343560

Midha, S., & Tripathi, K. (2021). Extended Security in Heterogeneous Distributed SDN Architecture. In G. Hura, A. Singh, & L. Siong Hoe (Eds.), *Advances in Communication and Computational Technology. Lecture Notes in Electrical Engineering* (Vol. 668). Springer. doi:10.1007/978-981-15-5341-7_75

Midha, S., Tripathi, K., & Sharma, M. K. (2021, April). Practical Implications of Using Dockers on Virtualized SDN. *Webology.*, *18*(1), 312–330. doi:10.14704/WEB/V18SI01/WEB18062

Miller, D. D., & Brown, E. W. (2018). Artificial intelligence in medical practice: The question to the answer? *The American Journal of Medicine*, *131*(2), 129–133. doi:10.1016/j.amjmed.2017.10.035 PMID:29126825

Mishra, M., Chandak, S., & Rout, P. K. (2019). Taxonomy of islanding detection techniques for distributed generation in microgrid. *Renewable Energy Focus*, *31*, 9–30. doi:10.1016/j.ref.2019.09.001

Mishra, M., Panda, D., & Hota, S. (2020). Unbalanced transportation problems with multiple fuzzy goals. *Materials Today: Proceedings*. Advance online publication. doi:10.1016/j.matpr.2020.09.362

Mishra, S., & Tyagi, A. K. (2022). The Role of Machine Learning Techniques in Internet of Things-Based Cloud Applications. In S. Pal, D. De, & R. Buyya (Eds.), *Artificial Intelligence-based Internet of Things Systems. Internet of Things (Technology, Communications and Computing)*. Springer. doi:10.1007/978-3-030-87059-1_4

Mitchell, H. B. (2004). Ranking intuitionistic fuzzy numbers. *International Journal of Uncertainty, Fuzziness and Knowledge-based Systems*, *12*(3), 377–386.

Mladjao, M., & Al Anfaf, M. (2016). *Contribution à la modélisation et à l'optimisation de systèmes énergétiques multi-sources et multi charges* [Thèse de doctorat]. l'Université de Lorraine.

Modu, B., Abdullah, M. P., Sanusi, M. A., & Hamza, M. F. (2023). DC-based microgrid: Topologies, control schemes, and implementations. *Alexandria Engineering Journal*, *70*, 61–92. doi:10.1016/j.aej.2023.02.021

Mohamad, H., Mokhlis, H., & Ping, H. W. (2011). A review on islanding operation and control for distribution network connected with small hydro power plant. *Renewable & Sustainable Energy Reviews*, *15*(8), 3952–3962. doi:10.1016/j.rser.2011.06.010

Mohammad, Shams, Niaz, Won, Lee, & Liu. (2022). Risk-based multistage stochastic mixed-integer optimization for biofuel supply chain management under multiple uncertainties. *Renewable Energy, 200*, 694-705. doi:10.1016/j.renene.2022.10.003

Mohammadi, J., Badrkhani Ajaei, F., & Stevens, G. (2019). Grounding the DC Microgrid. *IEEE Transactions on Industry Applications*, *55*(5), 4490–4499. doi:10.1109/TIA.2019.2928278

Mohammed, B. F. (2023). (Tc) Technique for finding optimal solution to transportation problem. *Mathematical Theory and Modeling*, *13*(1), 30–38.

Mohideen, S. I., & Kumar, P. S. (2010a). A comparative study on transportation problem in fuzzy environment. *International Journal of Mathematics Research*, *2*(1), 151–158.

Mohideen, S. I., & Kumar, P. S. (2010b). A comparative study on transportation problem in fuzzy environment. *International Conference on Emerging Trends in Mathematics and Computer Applications (ICETMCA2010)*, 226-229.

Morocco-Energy. (2019). Retrieved September 25, 2022, from International Trade Administration | Trade.gov website: https://www.trade.gov/country-commercial-guides/morocco-energy#:~:text=Per%20the%20state%2Downed%20power,%2C%20others%20(2%20percent)

Motie, S., Keynia, F., Ranjbar, M. R., & Maleki, A. (2016). Generation expansion planning by considering energy-efficiency programs in a competitive environment. *International Journal of Electrical Power & Energy Systems*, *80*, 109–118. doi:10.1016/j.ijepes.2015.11.107

Motter, D., & de Melo Vieira, J. C. (2018). The setting map methodology for adjusting the DG anti-islanding protection considering multiple events. *IEEE Transactions on Power Delivery*, *33*(6), 2755–2764. doi:10.1109/TPWRD.2018.2816400

Mouelhi, T., Fleurey, F., & Baudry, B. (2008). A generic metamodel for security policies mutation. *Software Testing Verification and Validation Workshop, ICSTW '08. IEEE International Conference on*, 278–286.

Muljadi, E., Singh, M., & Gevorgian, V. (2013). PSCAD modules representing PV generator (No. NREL/TP-5500-58189). National Renewable Energy Lab.

Munawer, M. E. (2018). Human health and environmental impacts of coal combustion and post-combustion wastes. *Journal of Sustainable Mining*, *17*(2), 87–96. doi:10.1016/j.jsm.2017.12.007

Munksgaard, J., Pade, L.-L., & Fristrup, P. (2005). Efficiency gains in Danish district heating. Is there anything to learn from benchmarking? *Energy Policy*, *33*(15), 1986–1997. doi:10.1016/j.enpol.2004.03.019

Muselli, M., Notton, G., & Louche, A. (1999). Design of hybrid-photovoltaic power generator, with optimization of energy management. Solar Energy, 65(3), 143–157. doi:10.1016/S0038-092X(98)00139-X

Musleh Al-Sartawi, A. M., Hussainey, K., & Razzaque, A. (2022). The role of artificial intelligence in sustainable finance. *Journal of Sustainable Finance & Investment*, 1–6. doi:10.1080/20430795.2022.2057405

Nabavi-Pelesaraei, A., Rafiee, S., Mohtasebi, S. S., Hosseinzadeh-Bandbafha, H., & Chau, K. W. (2018). Integration of artificial intelligence methods and life cycle assessment to predict energy output and environmental impacts of paddy production. *The Science of the Total Environment*, *631*, 1279–1294. doi:10.1016/j.scitotenv.2018.03.088 PMID:29727952

Nair, M. M., & Tyagi, A. K. (2021). Privacy: History, Statistics, Policy, Laws, Preservation and Threat Analysis. Journal of Information Assurance & Security, 16(1), 24-34.

Nale, R., Biswal, M., & Kishor, N. (2022). A passive communication based islanding detection technique for AC microgrid. *International Journal of Electrical Power & Energy Systems*, *137*, 107657. doi:10.1016/j.ijepes.2021.107657

Nasraoui, K., Lakhoua, M. N., & El Amraoui, L. (2016). Analysis and Modeling of a Wind Power System based on SysML. *Proceedings of the International Conference on Recent Advances in Electrical Systems*.

Nawaz, A., Zhou, M., Wu, J., & Long, C. (2022). A comprehensive review on energy management, demand response, and coordination schemes utilization in multi-microgrids network. *Applied Energy*, *323*, 119596. doi:10.1016/j.apenergy.2022.119596

Nehi, H. M. (2010). A new ranking method for intuitionistic fuzzy numbers. *International Journal of Fuzzy Systems*, *12*(1), 80–86.

Neumann, M. B., & Emanuel, K. (2018). The nature of future risks: Changing threats in the 21st century. *Global Environmental Change*, *53*, 1–9.

Neureiter, C., Uslar, M., Engel, D., & Lastro, G. (2016). A standards-based approach for domain specific modelling of Smart Grid system architectures. *11th System of Systems Engineering Conference*. 10.1109/SYSOSE.2016.7542888

Ngastiti, P. T. B., & Surarso, B. (2018). Zero point and zero suffix methods with robust ranking for solving fully fuzzy transportation problems. *Journal of Physics: Conference Series*, *1022*(1), 012005.

Nichiforov, C., Stamatescu, I., Făgărăşan, I., & Stamatescu, G. (2017, October). Energy consumption forecasting using ARIMA and neural network models. In *2017 5th International Symposium on Electrical and Electronics Engineering (ISEEE)* (pp. 1-4). IEEE. 10.1109/ISEEE.2017.8170657

Niet, I., van Est, R., & Veraart, F. (2021). Governing AI in electricity systems: Reflections on the EU artificial intelligence bill. *Frontiers in Artificial Intelligence*, *4*, 1–7. doi:10.3389/frai.2021.690237 PMID:34396090

Niksirat, M. (2022). A new approach to solve fully fuzzy multi-objective transportation problem. *Fuzzy Information and Engineering*, *14*(4), 456–467. doi:10.1080/16168658.2022.2152836

Nishant, R., Kennedy, M., & Corbett, J. (2020). Artificial intelligence for sustainability: Challenges, opportunities, and a research agenda. *International Journal of Information Management*, *53*, 102104. doi:10.1016/j.ijinfomgt.2020.102104

Nizami, S., Khan, N. A., & Mahmood, K. (2018). ARIMA-based electricity consumption forecasting: A case study of Pakistan. *Journal of Energy and Power Engineering*, *12*(5), 307–313.

Nti, E. K., Cobbina, S. J., Attafuah, E. E., Opoku, E., & Gyan, M. A. (2022). Environmental Sustainability Technologies in Biodiversity, Energy, Transportation and Water Management using Artificial Intelligence: A Systematic Review. *Sustainable Futures : An Applied Journal of Technology, Environment and Society*, *100068*. Advance online publication. doi:10.1016/j.sftr.2022.100068

OECD. (2018). Oslo Manual 2018: Guidelines for Collecting, Reporting and Using Data on Innovation, 4th Edition, The Measurement of Scientific, Technological and Innovation Activities. OECD Publishing. https://doi.org/ doi:10.1787/978

Offutt, J. (2011). A mutation carol: Past, present and future. *Information and Software Technology*, *53*(10), 1098–1107. doi:10.1016/j.infsof.2011.03.007

Ofori-Boadu, A., Owusu-Manu, D. G., Edwards, D., & Holt, G. (2012). Exploration of management practices for LEED projects: Lessons from successful green building contractors. *Structural Survey*, *30*(2), 145–162. doi:10.1108/02630801211228743

Okorie, O., Salonitis, K., Charnley, F., Moreno, M., Turner, C., & Tiwari, A. (2018). Digitisation and the circular economy: A review of current research and future trends. *Energies*, *11*(11), 3009. doi:10.3390/en11113009

Orji, U., Schantz, C., Leeb, S. B., Kirtley, J. L., Sievenpiper, B., Gerhard, K., & McCoy, T. (2017). Adaptive Zonal Protection for Ring Microgrids. *IEEE Transactions on Smart Grid*, *8*(4), 1843–1851. doi:10.1109/TSG.2015.2509018

Ouedraogo, K. E., Ekim, P. O., & Demirok, E. (2023). Feasibility of low-cost energy management system using embedded optimization for PV and battery storage assisted residential buildings. *Energy*, *271*. doi:10.1016/j.energy.2023.126922

Owens, B., Macken, C., Rohloff, A., & Roseberg, H. (2013). LEED v4 - Impact Category and Point Allocation Process. USGBC.

Ozkok, B. A. (2019). Finding fuzzy optimal and approximate fuzzy optimal solutionof fully fuzzy linear programming problems with trapezoidal fuzzy numbers. *Journal of Intelligent & Fuzzy Systems*, *36*(2), 1389–1400. doi:10.3233/JIFS-18016

Palmintier, B. S., & Webster, M. D. (2015). Impact of operational flexibility on electricity generation planning with renewable and carbon targets. *IEEE Transactions on Sustainable Energy*, *7*(2), 672–684. doi:10.1109/TSTE.2015.2498640

Pandey, U., & Singh, S. (2021). Environmental performance evaluation of European farms by assessing polluting factors in joint production. *Journal of Cleaner Production*, *328*, 129457. doi:10.1016/j.jclepro.2021.129457

Pandian, P. (2014). Realistic method for solving fully intuitionistic fuzzy transportation problems. [AMS]. *Applied Mathematical Sciences*, *8*(113), 5633–5639.

Pan, J., Jain, R., & Paul, S. (2014). A Survey of Energy Efficiency in Buildings and Microgrids using Networking Technologies. *IEEE Communications Surveys and Tutorials*, *16*(3), 1709–1731. doi:10.1109/SURV.2014.060914.00089

Papadakis, M., & Traon, Y. L. (2012). Using mutants to locate "unknown" faults. *Software Testing, Verification, and Validation, 2008 International Conference on,* 691–700. 10.1109/ICST.2012.159

Papadakis, M., & Malevris, N. (2012, September). Mutation based test case generation via a path selection strategy. *Inf. Softw. Technol., 54*(9), 915–932. Advance online publication. doi:10.1016/j.infsof.2012.02.004

Park, H., & Baldick, R. (2015). Stochastic generation capacity expansion planning reducing greenhouse gas emissions. *IEEE Transactions on Power Systems, 30*(2), 1026–1034. doi:10.1109/TPWRS.2014.2386872

Pathomsiri, S., Haghani, A., Dresner, M., & Windle, R. J. (2008). Impact of undesirable outputs on the productivity of US airports. *Transportation Research Part E, Logistics and Transportation Review, 44*(2), 235–259. doi:10.1016/j.tre.2007.07.002

Pedregal, D. J., & Trapero, J. R. (2010). Mid-term hourly electricity forecasting based on a multi-rate approach. *Energy Conversion and Management, 51*(1), 105–111. doi:10.1016/j.enconman.2009.08.028

Pérez-Lombard, L., Ortiz, J., González, R., & Maestre, I. R. (2009). A review of benchmarking, rating and labelling concepts within the framework of building energy certification schemes. *Energy and Building, 41*(3), 272–278. doi:10.1016/j.enbuild.2008.10.004

Perrouin, G., Sen, S., Klein, J., Baudry, B., & Traon, Y. L. (2010). *Automated and scalable t-wise test case generation strategies for software product lines. In ICST*. IEEE Computer Society.

Perucica, N., & Andjelkovic, K. (2022). Is the future of AI sustainable? A case study of the European Union. Transforming Government: People. *Transforming Government, 16*(3), 347–358. doi:10.1108/TG-06-2021-0106

Piao, S.-R., Li, J., & Ting, C.-J. (2019). Assessing regional environmental efficiency in China with distinguishing weak and strong disposability of undesirable outputs. *Journal of Cleaner Production, 227*, 748–759. doi:10.1016/j.jclepro.2019.04.207

Pickett, S. T., Cadenasso, M. L., & Grove, J. M. (Eds.). (2018). *Resilient cities: Meaning, models, and metaphor for integrating the ecological, socio-economic, and planning realms*. Springer.

Pina, A., Silva, C., & Ferrão, P. (2012). The impact of demand side management strategies in the penetration of renewable electricity. *Energy, 41*(1), 128–137. doi:10.1016/j.energy.2011.06.013

Pinto, S. J., & Panda, G. (2015). Wavelet technique based islanding detection and improved repetitive current control for reliable operation of grid-connected PV systems. *International Journal of Electrical Power & Energy Systems, 67*, 39–51. doi:10.1016/j.ijepes.2014.11.008

Planas, E., Andreu, J., Gárate, J. I., Martínez de Alegría, I., & Ibarra, E. (2015). AC and DC technology in microgrids: A review. *Renewable & Sustainable Energy Reviews, 43*, 726–749. doi:10.1016/j.rser.2014.11.067

Pohl, K., Bockle, G., & van der Linden, F. J. (2005). *Software Product Line Engineering: Foundations, Principles and Techniques*. Springer-Verlag. doi:10.1007/3-540-28901-1

Pollitt, H., & Mercure, J.-F. (2018). The role of money and the financial sector in energy-economy models used for assessing climate and energy policy. *Climate Policy, 18*(2), 184–197. doi:10.1080/14693062.2016.1277685

Pombo, D. V., Rincón, M. J., Bacher, P., Bindner, H. W., Spataru, S. V., & Sørensen, P. E. (2022). Assessing stacked physics-informed machine learning models for co-located wind–solar power forecasting. *Sustainable Energy, Grids and Networks, 32*. https://www.sciencedirect.com/science/article/pii/S2352467722001886 doi:10.1016/j.segan.2022.100943

Popeangă, J., & Lungu, I. (2014). Forecasting Final Energy Consumption using the Centered Moving Average Method and Time Series Analysis. *Database Systems Journal*, *5*(1), 42–50.

Porter, M. E. (1990). The Competitive Advantage of Nations. *Harvard Business Review*. Advance online publication. doi:10.1080/13602389400000006

Pourofoghi, F., Salokolaei, D. D., & Abraham, A. (2021, August). Designing transportation problem under grey linear programming based on sensitivity analysis. In *International Conference on Intelligent and Fuzzy Systems* (pp. 666-675). Springer. doi:10.1007/978-3-030-85626-7_78

Pramod, A. (2022). *Emerging Innovations in the Near Future Using Deep Learning Techniques, Book: Advanced Analytics and Deep Learning Models*. Wiley Scrivener. doi:10.1002/9781119792437.ch10

Priyadarshini Padhi, P., & Deepa, K. (2021). DC Microgrids: Architecture and Challenges. *IOP Conference Series. Materials Science and Engineering*, *1070*(1), 012098. doi:10.1088/1757-899X/1070/1/012098

Priyadharshini, P., & Anju, A. (2020). Type-2 duality trapezoidal fuzzy fractional transportation problem using fuzzy optimization techniques. In AIP Conference Proceedings (Vol. 2261, No. 1, p. 030058). AIP Publishing LLC. doi:10.1063/5.0016927

Pustina, L., Biral, F., & Serafini, J. (2022). A novel Economic Nonlinear Model Predictive Controller for power maximisation on wind turbines. *Renewable and Sustainable Energy Reviews, 170*. doi:10.1016/j.rser.2022.112964

Raab, R. L., & Lichty, R. W. (2002). Identifying Subareas That Comprise A Greater Metropolitan Area: The Criterion of County Relative Efficiency. *Journal of Regional Science*, *42*(3), 579–594. doi:10.1111/1467-9787.00273

Rajesh, K., Bhuvanesh, A., Kannan, S., & Thangaraj, C. (2016). Least cost generation expansion planning with solar power plant using differential evolution algorithm. *Renewable Energy*, *85*, 677–686. doi:10.1016/j.renene.2015.07.026

Rajesh, K., Karthikeyan, K., Kannan, S., & Thangaraj, C. (2016). Generation expansion planning based on solar plants with storage. *Renewable & Sustainable Energy Reviews*, *57*, 953–964. doi:10.1016/j.rser.2015.12.126

Raj, M. E. A., Sivaraman, G., & Vishnukumar, P. (2022). A novel kind of arithmetic operations on trapezoidal fuzzy numbers and its applications to optimize the transportation cost. *International Journal of Fuzzy Systems*, 1–8. doi:10.1007/s40815-022-01425-6

Ranchhodbhai, P.N., & Tripathi, K. (2019). Identifying and Improving the Malicious Behavior of Rushing and Blackhole Attacks using Proposed IDSAODV Protocol. *International Journal of Recent Technology and Engineering, 8*(3), 6554-6562.

Redclift, M. (2005). An Oxymoron Comes of Age. *Sustainable Development (Bradford)*, *13*(4), 212–227. doi:10.1002/sd.281

Reed, B. (2007). Shifting from 'sustainability' to regeneration. *Building Research and Information*, *35*(6), 674–680. doi:10.1080/09613210701475753

Rees, W., & Wackernagel, M. (2013). The shoe fits, but the footprint is larger than Earth: Urban ecological footprints show that cities can be unsustainable. *Environment and Urbanization*, *25*(2), 481–499.

Reinhard, S., Knox Lovell, C. A., & Thijssen, G. J. (2000). Environmental efficiency with multiple environmentally detrimental variables; estimated with SFA and DEA. *European Journal of Operational Research*, *121*(2), 287–303. doi:10.1016/S0377-2217(99)00218-0

Rennings, K., Ziegler, A., Ankele, K., & Hoffmann, E. (2006). The influence of different characteristics of the EU environmental management and auditing scheme on technical environmental innovations and economic performance. *Ecological Economics*, *57*(1), 45–59. doi:10.1016/j.ecolecon.2005.03.013

Reza Salehizadeh, M., Rahimi-Kian, A., & Oloomi-Buygi, M. (2015). Security-based multi-objective congestion management for emission reduction in power system. *International Journal of Electrical Power & Energy Systems*, *65*, 124–135. doi:10.1016/j.ijepes.2014.09.031

Rezkallah, M., Ibrahim, H., Chandra, A. F., Hamadi, A., & Hazzab, A. (2023). Control of Hybrid Off-Microgrid Configurations Based on Renewable and Non-renewable Energy Sources. doi:10.1007/978-981-19-7956-9_8

Robichaud, L. B., & Anantatmula, V. S. (2011). Greening Project Management Practices for Sustainable Construction. *Journal of Management Engineering*, *27*(1), 48–57. doi:10.1061/(ASCE)ME.1943-5479.0000030

Rocha, M. F. (2018). The future of cities in the 21st century: Implications of urban energy use and greenhouse gas emissions. *Energy Procedia*, *147*, 72–77.

Rochayata, K. S. B., Widodasih, R. W. K., & Gunardi, H. (2022). Optimization of transportation costs with the implementation of the Milk-Run method. *Enrichment: Journal of Management*, *12*(5), 3662–3669.

Rodriguez-Diaz, E., Savaghebi, M., Vasquez, J. C., & Guerrero, J. M. (2015). An overview of low voltage DC distribution systems for residential applications. *2015 IEEE 5th International Conference on Consumer Electronics - Berlin (ICCE-Berlin)*, 318–322. 10.1109/ICCE-Berlin.2015.7391268

Rosenzweig, C., Solecki, W., Romero-Lankao, P., Mehrotra, S., & Dhakal, S. (Eds.). (2018). *Climate change and cities: Second assessment report of the Urban Climate Change Research Network*. Cambridge University Press. doi:10.1017/9781316563878

Ruiz-Alvarez, S., Patiño, J., Marquez, A., & Espinosa, J. (2017). Optimal design for an electrical hybrid micro grid in Colombia under fuel price variation. *Int. J. Renew. Energy Res*, *7*(4), 1535–1545.

Sabzehgar, R. (2017). Overview of Technical Challenges, Available Technologies and Ongoing Developments of AC/DC Microgrids. In *Development and Integration of Microgrids*. InTech. doi:10.5772/intechopen.69400

Saeed, M. H., Fangzong, W., Kalwar, B. A., & Iqbal, S. (2021). A Review on Microgrids' Challenges & Perspectives. *IEEE Access : Practical Innovations, Open Solutions*, *9*, 166502–166517. doi:10.1109/ACCESS.2021.3135083

Saheb-Koussa, D. (2007). Production d'électricité sans interruption moyennant un système hybride (éolien – photovoltaïque – diesel). Revue des Energies Renouvelables, 121 – 128.

Sahoo, S. K., Sinha, A. K., & Kishore, N. K. (2018). Control Techniques in AC, DC, and Hybrid AC–DC Microgrid: A Review. *IEEE Journal of Emerging and Selected Topics in Power Electronics*, *6*(2), 738–759. doi:10.1109/JESTPE.2017.2786588

Sai, G. H., Tripathi, K., & Tyagi, A. K. (2023). Internet of Things-Based e-Health Care: Key Challenges and Recommended Solutions for Future. In *Proceedings of Third International Conference on Computing, Communications, and Cyber-Security. Lecture Notes in Networks and Systems* (vol. 421). Springer. 10.1007/978-981-19-1142-2_37

Sai, G. H., Tyagi, A. K., & Sreenath, N. (2023). Biometric Security in Internet of Things Based System against Identity Theft Attacks. *2023 International Conference on Computer Communication and Informatics (ICCCI)*, 1-7. 10.1109/ICCCI56745.2023.10128186

Salas, J., Patterson, G., & de Barros Vidal, F. (2022). A Systematic Mapping of Artificial Intelligence Solutions for Sustainability Challenges in Latin America and the Caribbean. *Revista IEEE América Latina*, *20*(11), 2312–2329. doi:10.1109/TLA.2022.9904756

Salazar-Ordóñez, M., Pérez-Hernández, P. P., & Martín-Lozano, J. M. (2013). Sugar beet for bioethanol production: An approach based on environmental agricultural outputs. *Energy Policy*, *55*, 662–668. doi:10.1016/j.enpol.2012.12.063

Saleh, S. A., Aljankawey, A. S., Meng, R., Meng, J., Chang, L., & Diduch, C. P. (2015). Apparent power-based anti-islanding protection for distributed cogeneration systems. *IEEE Transactions on Industry Applications*, *52*(1), 83–98. doi:10.1109/TIA.2015.2464307

Sam'an, M., Surarso, B., & Irwanto, B. (2020). Solving of fuzzy transportation problem using fuzzy analytical hierarchy process (AHP). In *The 2nd International Seminar on Science and Technology (ISSTEC 2019)* (pp. 10-15). Atlantis Press.

Sam'an, M., Dasril, Y., & Bujang, N. B., & Farikhin. (2022). Improved total difference method (ITDM): A new approach to solving transportation problem based on modifications of total difference method 1 and integration of total ratio cost matrix. *International Journal of Computing Science and Mathematics*, *16*(1), 13–23. doi:10.1504/IJCSM.2022.10051689

Sam'an, M., Hariyanto, S., & Surarso, B. (2018). Optimal solution of full fuzzy transportation problems using total integral ranking. *Journal of Physics: Conference Series*, *983*(1), 012075.

Samuel, A. E. (2012). Improved zero point method (IZPM) for the transportation problems. *Applied Mathematical Sciences*, *6*(109), 5421–5426.

Sanders, N. R., Boone, T., Ganeshan, R., & Wood, J. D. (2019). Sustainable supply chains in the age of AI and digitization: Research challenges and opportunities. *Journal of Business Logistics*, *40*(3), 229–240. doi:10.1111/jbl.12224

Sarin, A., Gupta, R., & Jituri, V. V. (2018). Solar Residential Rooftop Systems (SRRS) in South Delhi: A Strategic Study with Focus on Potential Consumers' Awareness. *International Journal of Renewable Energy Research*, *8*(2), 954–963.

Sarkis, J., & Cordeiro, J. J. (2012). Ecological modernization in the electrical utility industry: An application of a bads–goods DEA model of ecological and technical efficiency. *European Journal of Operational Research*, *219*(2), 386–395. doi:10.1016/j.ejor.2011.09.033

Sarkis, J., Zhu, Q., & Lai, K. H. (2011). An organizational theoretic review of green supply chain management literature. *International Journal of Production Economics*, *130*(1), 1–15. doi:10.1016/j.ijpe.2010.11.010

Sarma, D., Das, A., & Bera, U. K. (2020). Generalized type-2 intuitionistic fuzzy approaches for allocation and redistribution of resources in the disaster operation. *Studies in Computational Intelligence*, *863*, 327–341. doi:10.1007/978-3-030-34152-7_25

Sassen, S. (2019). *Cities in a world of cities: The globalization of urban governance*. Yale University Press.

Schaltegger, S., & Wagner, M. (2011). Sustainable entrepreneurship and sustainability innovation: Categories and interactions. *Business Strategy and the Environment*, *20*(4), 222–237. doi:10.1002/bse.682

Sciarretta, E., Mancini, R., & Greco, E. (2022). Artificial Intelligence for Healthcare and Social Services: Optimizing Resources and Promoting Sustainability. *Sustainability (Basel)*, *14*(24), 16464. doi:10.3390/su142416464

SE4ALL. (2016). *2015 SEforALL Global Tracking Framework, Sustainable Energy for All (SE4ALL)*. http://gtf.esmap.org/data/files/downloaddocuments/gtf-2105-full-report.pdf

Sebastien, H., Heike, A., Nishil, B., & Arpad, H. (2007). Leadership in Energy and Environmental Design (LEED) - A critical evaluation by LCA and recommendations for improvement. *International Journal of Life Cycle Assessment*, *12*(Special Issue 1), 25. doi:10.11436/mssj.15.250

Sedaghat, A., Mostafaeipour, A., Rezaei, M., Jahangiri, M., & Mehrabi, A. (2020). A new semi-empirical wind turbine capacity factor for maximizing annual electricity and hydrogen production. *International Journal of Hydrogen Energy, 45*(32), 15888-15903.

Seghiri. (2016). *Modélisation et simulation d'une architecture d'entreprise - application aux Smart Grids.* Université Paris-Saclay préparée à centrale Supelec.

Shabani, A., & Jamkhaneh, E. B. (2014). A new generalized intuitionistic fuzzy number. *Journal of Fuzzy Set Valued Analysis*, *24*, 1–10. doi:10.5899/2014/JFSVA-00199

Shaik, M., Shaik, A. G., & Yadav, S. K. (2022). Hilbert–Huang transform and decision tree based islanding and fault recognition in renewable energy penetrated distribution system. *Sustainable Energy. Grids and Networks*, *30*, 100606.

Sharma, M. K. (2020). Zero-point maximum allocation method for solving intuitionistic fuzzy transportation problem. *International Journal of Applied and Computational Mathematics*, *6*(4), 1–11. doi:10.1007/s40819-020-00867-6

Shaw, A. K., & Roy, T. K. (2012). Some arithmetic operations on triangular intuitionistic fuzzy number and its application on reliability evaluation. *International Journal of Fuzzy Mathematics and Systems*, *2*(4), 363–382.

Sheth, H. S. K. (2022). Deep Learning, Blockchain based Multi-layered Authentication and Security Architectures. *2022 International Conference on Applied Artificial Intelligence and Computing (ICAAIC),* 476-485. 10.1109/ICAAIC53929.2022.9793179

Shin, H., Chae, S. H., & Kim, E. H. (2020). Design of microgrid protection schemes using PSCAD/EMTDC and ETAP programs. *Energies*, *13*(21), 5784. doi:10.3390/en13215784

Shortall, O. K., & Barnes, A. P. (2013). Greenhouse gas emissions and the technical efficiency of dairy farmers. *Ecological Indicators*, *29*, 478–488. doi:10.1016/j.ecolind.2013.01.022

Shreyas Madhav, A. V., Ilavarasi, A. K., & Tyagi, A. K. (2022). The Heroes and Villains of the Mix Zone: The Preservation and Leaking of USer's Privacy in Future Vehicles. In V. Arunachalam & K. Sivasankaran (Eds.), *Microelectronic Devices, Circuits and Systems. ICMDCS 2022. Communications in Computer and Information Science* (Vol. 1743). Springer. doi:10.1007/978-3-031-23973-1_12

Shu, T., Song, D., & Joo, Y. H. (2022). Non-centralised coordinated optimisation for maximising offshore wind farm power via a sparse communication architecture. *Applied Energy, 324*. doi:10.1016/j.apenergy.2022.119705

Silva, J. G., Ferrari, R., & van Wingerden, J.-W. (2023). Wind farm control for wake-loss compensation, thrust balancing and load-limiting of turbines. *Renewable Energy, 203*, 421-433. doi:10.1016/j.renene.2022.11.113

Singh, V., Fozdar, M., Singh, A. K., & Singh, S. (2020). Analysis of anti-islanding protection methods integrated in distributed generation. *Intelligent Computing Techniques for Smart Energy Systems Proceedings of ICTSES, 2018*, 663–671.

Smith, J. C., Milligan, M. R., DeMeo, E. A., & Parsons, B. (2007). Utility wind integration and operating impact state of the art. *IEEE Transactions on Power Systems*, *22*(3), 900–908. doi:10.1109/TPWRS.2007.901598

Somisetti, K., Tripathi, K., & Verma, J. K. (2020). Design, Implementation, and Controlling of a Humanoid Robot. *2020 International Conference on Computational Performance Evaluation (ComPE),* 831-836. 10.1109/ComPE49325.2020.9200020

Song, M., Gao, C., Shahidehpour, M., Li, Z., Yang, J., & Yan, H. (2019, July). State Space Modeling and Control of Aggregated TCLs for Regulation Services in Power Grids. *IEEE Transactions on Smart Grid, 10*(4), 4095–4106. doi:10.1109/TSG.2018.2849321

Song, M., Zhang, J., & Wang, S. (2015). Review of the network environmental efficiencies of listed petroleum enterprises in China. *Renewable & Sustainable Energy Reviews*, *43*, 65–71. doi:10.1016/j.rser.2014.11.050

Song, Y., & Zegeye, A. (2017). Urbanization and the environment: An overview. In *The Urbanization and the Environment* (pp. 1–16). Springer.

Sonia, N. A., & Chidambaran, S. (2023). Time paradox in transportation problems. *International Journal of Operational Research*, *45*(4), 540–552. doi:10.1504/IJOR.2022.128397

Spiegel. (2018). *Neuzulassungen: Erstmals mehr als eine Million E-Autos (New Registrations: First time more than one Million Electric Vehicles)*. https://www.spiegel.de/auto/aktuell/emobilitaet-erstmals-ueber-eine-million-e-autoszugelassen-a-1197636.html

Srivastava, P. K., & Bisht, D. C. (2019). An efficient fuzzy minimum demand supply approach to solve fully fuzzy transportation problem. *Mathematics in Engineering, Science and Aerospace*, *10*(2), 253–269.

Stallard, T., Rothschild, R., & Aggidis, G. A. (2008). A comparative approach to the economic modelling of a large-scale wave power scheme. *European Journal of Operational Research*, *185*(2), 884–898. doi:10.1016/j.ejor.2007.01.021

Steffens, Oster, Lochau, & Fogdal. (2012). Industrial evaluation of pairwise spl testing with moso-polite. *VaMoS*, 55–62.

Stoyanov, L. (2011). *Etude de différentes structures de systèmes hybrides à sources d'énergie renouvelables*. Université Pascal Paoli.

Subasree, S., & Sakthivel, N. K. (2022). Combining the advantages of radiomic features based feature extraction and hyper parameters tuned RERNN using LOA for breast cancer classification. *Biomedical Signal Processing and Control*, *72*(A). doi:10.1016/j.bspc.2021.103354

Subburaj, D. V., Dash, R., Reddy, J., & CN, R. (2022). *International Congress on Renewable Energy (I CORE 2021)*. 10.13052/rp-9788770227537

Sueyoshi, T., & Goto, M. (2012a). Data envelopment analysis for environmental assessment: Comparison between public and private ownership in petroleum industry. *European Journal of Operational Research*, *216*(3), 668–678. doi:10.1016/j.ejor.2011.07.046

Sueyoshi, T., & Goto, M. (2012b). Weak and strong disposability vs. natural and managerial disposability in DEA environmental assessment: Comparison between Japanese electric power industry and manufacturing industries. *Energy Economics*, *34*(3), 686–699. doi:10.1016/j.eneco.2011.10.018

Sueyoshi, T., & Goto, M. (2014). Photovoltaic power stations in Germany and the United States: A comparative study by data envelopment analysis. *Energy Economics*, *42*, 271–288. doi:10.1016/j.eneco.2014.01.004

Sueyoshi, T., & Goto, M. (2015a). DEA environmental assessment in time horizon: Radial approach for Malmquist index measurement on petroleum companies. *Energy Economics*, *51*, 329–345. doi:10.1016/j.eneco.2015.07.010

Sueyoshi, T., & Goto, M. (2015b). Environmental assessment on coal-fired power plants in U.S. north-east region by DEA non-radial measurement. *Energy Economics*, *50*, 125–139. doi:10.1016/j.eneco.2015.04.016

Sueyoshi, T., Yuan, Y., & Goto, M. (2017). A literature study for DEA applied to energy and environment. *Energy Economics*, *62*, 104–124. doi:10.1016/j.eneco.2016.11.006

Suneela, S., & Chakraverty, S. (2019). New ranking function for fuzzy linear programming problem and system of linear equations. *Journal of Information and Optimization Sciences*, *40*(1), 141–156. doi:10.1080/02522667.2018.1450196

Sun, T., Zhang, T., Teng, Y., Chen, Z., & Fang, J. (2019). Monthly electricity consumption forecasting method based on X12 and STL decomposition model in an integrated energy system. *Mathematical Problems in Engineering*, *2019*, 1–16. doi:10.1155/2019/9012543

Talapur, G. G., Suryawanshi, H. M., Xu, L., & Shitole, A. B. (2018). A Reliable Microgrid With Seamless Transition Between Grid Connected and Islanded Mode for Residential Community With Enhanced Power Quality. *IEEE Transactions on Industry Applications*, *54*(5), 5246–5255. doi:10.1109/TIA.2018.2808482

Tanatvanit, S., Limmeechokchai, B., & Shrestha, R. M. (2004). CO2 mitigation and power generation implications of clean supply-side and demand-side technologies in Thailand. *Energy Policy*, *32*(1), 83–90. doi:10.1016/S0301-4215(02)00260-4

Tan, C., Tan, Y., & Zhang, Y. (2020). Hybrid ARIMA-LSSVM model for electricity consumption forecasting. *Energy Procedia*, *158*, 4155–4160.

Tan, Y., Shen, L., & Yao, H. (2011). Sustainable construction practice and contractors' competitiveness: A preliminary study. *Habitat International*, *35*(2), 225–230. doi:10.1016/j.habitatint.2010.09.008

Tao, L., Schwaegerl, C., Narayanan, S., & Zhang, J. H. (2011). From laboratory Microgrid to real markets — Challenges and opportunities. *8th International Conference on Power Electronics - ECCE Asia*, 264–271. 10.1109/ICPE.2011.5944600

Tarmudi, Z., & Rahman, N. A. (2020). Diverse ranking approach in MCDM based on trapezoidal intuitionistic fuzzy numbers. In *Proceedings of the Tenth International Conference on Soft Computing and Pattern Recognition (SoCPaR 2018).* Springer. doi:10.1007/978-3-030-17065-3_2

Tavakoli, M., Shokridehaki, F., Funsho Akorede, M., Marzband, M., Vechiu, I., & Pouresmaeil, E. (2018). CVaR-based energy management scheme for optimal resilience and operational cost in commercial building microgrids. *International Journal of Electrical Power & Energy Systems*, *100*, 1–9. doi:10.1016/j.ijepes.2018.02.022

Tekkanat, B., Yuksel, Y. E., & Ozturk, M. (2023). The evaluation of hydrogen production via a geothermal-based multigeneration system with 3E analysis and multi-objective optimization. *International Journal of Hydrogen Energy*, *48*(22), 8002-8021. doi:10.1016/j.ijhydene.2022.11.185

Thabor, M. Z. (1988). Small diesel power. *International Conference on Small Power Supplies, Tasmanie.*

Thanoon, F. H. (2022). Proposed method for solving transportation problems. *Wasit Journal of Pure Sciences*, *1*(3), 9-14. doi:10.31185/WJPS.62

Toloo, M., & Babaee, S. (2015). On variable reductions in data envelopment analysis with an illustrative application to a gas company. *Applied Mathematics and Computation*, *270*, 527–533. doi:10.1016/j.amc.2015.06.122

Tone, K. (2016). Advances in DEA Theory and Applications: With Examples in Forecasting Models. *Advances in DEA Theory and Applications: With Examples in Forecasting Models*. doi:10.1002/9781118946688

Tone, K., Chang, T.-S., & Wu, C.-H. (2020). Handling negative data in slacks-based measure data envelopment analysis models. *European Journal of Operational Research*, *282*(3), 926–935. doi:10.1016/j.ejor.2019.09.055

Tone, K., & Tsutsui, M. (2007). Decomposition of cost efficiency and its application to Japanese-US electric utility comparisons. *Socio-Economic Planning Sciences*, *41*(2), 91–106. doi:10.1016/j.seps.2005.10.007

Toole, T. M., Hallowell, M., & Chinowsky, P. (2013). A tool for enhancing innovation in construction organizations. *Engineering Project Organization Journal*, *3*(1), 32–50. doi:10.1080/21573727.2012.717531

Traneva, V., & Tranev, S. (2020). Intuitionistic fuzzy transportation problem by zero point method. *Proceedings of the 2020 Federated Conference on Computer Science and Information Systems, 21*, 349–358. doi:10.15439/2020F61

Traneva, V., & Tranev, S. (2021). An intuitionistic fuzzy zero suffix method for solving the transportation problem. In I. Dimov & S. Fidanova (Eds.), *Advances in High Performance Computing. HPC 2019. Studies in Computational Intelligence* (Vol. 902). Springer. doi:10.1007/978-3-030-55347-0_7

Trindade, F. C., do Nascimento, K. V., & Vieira, J. C. (2013). Investigation on voltage sags caused by DG anti-islanding protection. *IEEE Transactions on Power Delivery*, *28*(2), 972–980. doi:10.1109/TPWRD.2012.2237419

Tsolakis, N., Schumacher, R., Dora, M., & Kumar, M. (2022). Artificial intelligence and blockchain implementation in supply chains: A pathway to sustainability and data monetisation? *Annals of Operations Research*, 1–54. PMID:35755830

Tulumello, S., & Hernández-Molina, G. (2018). Smart cities as techno-social assemblages: A framework for understanding the rise of smart urbanism. *Frontiers in Sociology*, *3*, 30.

Tyagi & Aswathy. (2021). Autonomous Intelligent Vehicles (AIV): Research statements, open issues, challenges and road for future. *International Journal of Intelligent Networks, 2*, 83-102. doi:10.1016/j.ijin.2021.07.002

Tyagi & Sreenath. (2021). Cyber Physical Systems: Analyses, challenges and possible solutions. *Internet of Things and Cyber-Physical Systems, 1*, 22-33. doi:10.1016/j.iotcps.2021.12.002

Tyagi, A. K., & Bansal, R. (2023). A Step-To-Step Guide to Write a Quality Research Article. In Intelligent Systems Design and Applications. ISDA 2022. Lecture Notes in Networks and Systems (vol. 717). Springer. doi:10.1007/978-3-031-35510-3_36

Tyagi, A. K., Nair, M. M., Niladhuri, S., & Abraham, A. (2020). Security, Privacy Research issues in Various Computing Platforms: A Survey and the Road Ahead. Journal of Information Assurance & Security, 15(1), 1-16.

Tyagi. (2022a). *Handbook of Research on Technical, Privacy, and Security Challenges in a Modern World.* IGI Global. doi:10.4018/978-1-6684-5250-9

Tyagi. (2022b). *Using Multimedia Systems, Tools, and Technologies for Smart Healthcare Services.* IGI Global. doi:10.4018/978-1-6684-5741-2

Tyagi, A. K. (Ed.). (2021). *Multimedia and Sensory Input for Augmented, Mixed, and Virtual Reality.* IGI Global. doi:10.4018/978-1-7998-4703-8

Tyagi, A. K., Agarwal, D., & Sreenath, N. (2022). SecVT: Securing the Vehicles of Tomorrow using Blockchain Technology. *2022 International Conference on Computer Communication and Informatics (ICCCI),* 1-6. 10.1109/ICCCI54379.2022.9740965

Tyagi, A. K., Chandrasekaran, S., & Sreenath, N. (2022). Blockchain Technology:– A New Technology for Creating Distributed and Trusted Computing Environment. *2022 International Conference on Applied Artificial Intelligence and Computing (ICAAIC),* 1348-1354. 10.1109/ICAAIC53929.2022.9792702

Tyagi, A. K., & Sreenath, N. (2023a). Security, Privacy, and Trust Issues in Intelligent Transportation System. In *Intelligent Transportation Systems: Theory and Practice. Disruptive Technologies and Digital Transformations for Society 5.0.* Springer. doi:10.1007/978-981-19-7622-3_8

Tyagi, A. K., & Sreenath, N. (2023b). Artificial Intelligence—Internet of Things-Based Intelligent Transportation System. In *Intelligent Transportation Systems: Theory and Practice. Disruptive Technologies and Digital Transformations for Society 5.0.* Springer. doi:10.1007/978-981-19-7622-3_10

Tyagi, A. K., & Sreenath, N. (2023c). Future Intelligent Vehicles: Open Issues, Critical Challenges, and Research Opportunities. In *Intelligent Transportation Systems: Theory and Practice. Disruptive Technologies and Digital Transformations for Society 5.0*. Springer. doi:10.1007/978-981-19-7622-3_15

Tyagi, A. K., & Sreenath, N. (2023d). Intelligent Transportation System: Past, Present, and Future. In *Intelligent Transportation Systems: Theory and Practice. Disruptive Technologies and Digital Transformations for Society 5.0*. Springer. doi:10.1007/978-981-19-7622-3_2

Tyagi, A., Kukreja, S., Nair, M. M., & Tyagi, A. K. (2022). Machine Learning: Past, Present and Future. *NeuroQuantology : An Interdisciplinary Journal of Neuroscience and Quantum Physics*, *20*(8). Advance online publication. doi:10.14704/nq.2022.20.8.NQ44468

Tyagi, Aswathy, Aghila, & Sreenath. (2021, October). AARIN: Affordable, Accurate, Reliable and INnovative Mechanism to Protect a Medical Cyber-Physical System using Blockchain Technology. *IJIN*, *2*, 175–183.

Tyagi, & Chahal. (2020). *Artificial Intelligence and Machine Learning Algorithms. In Challenges and Applications for Implementing Machine Learning in Computer Vision*. IGI Global. doi:10.4018/978-1-7998-0182-5.ch008

Unamuno, E., & Barrena, J. A. (2015a). Hybrid ac/dc microgrids—Part I: Review and classification of topologies. *Renewable & Sustainable Energy Reviews*, *52*, 1251–1259. doi:10.1016/j.rser.2015.07.194

Unamuno, E., & Barrena, J. A. (2015b). Hybrid ac/dc microgrids—Part II: Review and classification of control strategies. *Renewable & Sustainable Energy Reviews*, *52*, 1123–1134. doi:10.1016/j.rser.2015.07.186

United Nations. (1998). *Kyoto Protocol to the United Nations Framework Convention on Climate Chance. Technical report*. United Nations.

USGBC. (2018). *Leed and the International Green Construction Code*. https://www.usgbc.org/about/programs/green-codes

Van Wynsberghe, A. (2021). Sustainable AI: AI for sustainability and the sustainability of AI. *AI and Ethics*, *1*(3), 213–218. doi:10.1007/s43681-021-00043-6

Varghese, A., & Kuriakose, S. (2012). Centroid of an intuitionistic fuzzy number. *Notes on Intuitionistic Fuzzy Sets*, *18*(1), 19–24.

Varsha, R., Nair, S. M., Tyagi, A. K., & Aswathy, S. U. (2021). The Future with Advanced Analytics: A Sequential Analysis of the Disruptive Technology's Scope. In Hybrid Intelligent Systems. HIS 2020. Advances in Intelligent Systems and Computing (vol. 1375). Springer. doi:10.1007/978-3-030-73050-5_56

Vegunta, S. C., Higginson, M. J., Kenarangui, Y. E., Li, G. T., Zabel, D. W., Tasdighi, M., & Shadman, A. (2021). AC Microgrid Protection System Design Challenges—A Practical Experience. *Energies*, *14*(7), 2016. doi:10.3390/en14072016

Vishnuram, G., Tripathi, K., & Kumar Tyagi, A. (2022). Ethical Hacking: Importance, Controversies and Scope in the Future. *2022 International Conference on Computer Communication and Informatics (ICCCI),* 1-6. 10.1109/ICCCI54379.2022.9740860

Visser, L., AlSkaif, T., Hu, J., Louwen, A., & van Sark, W. (2023). On the value of expert knowledge in estimation and forecasting of solar photovoltaic power generation. *Solar Energy, 251*, 86-105. doi:10.1016/j.solener.2023.01.019

Vyas, S., Kumar, R., & Kavasseri, R. (2017). Data analytics and computational methods for anti-islanding of renewable energy based distributed generators in power grids. *Renewable & Sustainable Energy Reviews*, *69*, 493–502. doi:10.1016/j.rser.2016.11.116

Wang, J., Zhao, Z., Liu, G., & Xu, H. (2022). A robust optimization approach of well placement for doublet in heterogeneous geothermal reservoirs using random forest technique and genetic algorithm. *Energy, 254*(C). doi:10.1016/j.energy.2022.124427

Wang, F., Lei, Z., Xu, X., & Shu, X. (2017). Topology Deduction and Analysis of Voltage Balancers for DC Microgrid. *IEEE Journal of Emerging and Selected Topics in Power Electronics*, *5*(2), 672–680. doi:10.1109/JESTPE.2016.2638959

Wang, G., & Peng, J. (2019). Fuzzy optimal solution of fuzzy number linear programming problems. *International Journal of Fuzzy Systems*, *21*(3), 865–881. doi:10.1007/s40815-018-0594-0

Wang, H., Wang, M., Cheng, Q., Lv, S., & Ji, X. (2022). Modeling simulation and inverter control strategy research of microgrid in grid-connected and island mode. *Energy Reports*, *8*, 206–218. doi:10.1016/j.egyr.2022.09.117

Wang, W., Yang, D., Hong, T., & Kleissl, J. (2022). *An archived dataset from the ECMWF Ensemble Prediction System for probabilistic solar power forecasting* (Vol. 248). Solar Energy. doi:10.1016/j.solener.2022.10.062

Wertani, H., & Lakhoua, M. N. (2018). *Overview of Smart Grids Architecture and Design*. SmartNets2018, Hammamet, Tunisia. 10.1109/SMARTNETS.2018.8707404

Wertani, H., Ben Salem, J., & Lakhoua, M. N. (2020a, July). Analysis and supervision of a smart grid system with a systemic tool. *The Electricity Journal*, *33*(6), 106784. doi:10.1016/j.tej.2020.106784

Wertani, H., Ben Salem, J., & Lakhoua, M. N. (2020b, May). Application of Bond Graph of Renewable Energy Sources in Smart Grids. *Journal of Electrical and Electronics Engineering (Oradea)*, *13*.

Winardi, B., Sinuraya, E. W., & Nugroho, A. (2021). Design and Implementation of Rooftop off Grid Solar Panel for Household Electricity Loads. *2021 8th International Conference on Information Technology, Computer and Electrical Engineering (ICITACEE)*, 258–263. doi:10.1109/ICITACEE53184.2021.9617551

World Bank. (2008). *Environmental sustainability: An evaluation of World Bank group support. Technical report*. World Bank.

World Energy Outlook 2022. (n.d.). www.iea.org/t&c/

Woroghi, I. (2015). Modeling a Smart Grid using Objects Interaction. *Proceedings of the IEEE power and energy society general meeting*.

Xiao, H. F., Fang, Z., Xu, D., Venkatesh, B., & Singh, B. (2017). Anti-islanding protection relay for medium voltage feeder with multiple distributed generators. *IEEE Transactions on Industrial Electronics*, *64*(10), 7874–7885. doi:10.1109/TIE.2017.2694394

Xia, Y., Wei, W., Yu, M., Wang, X., & Peng, Y. (2018). Power Management for a Hybrid AC/DC Microgrid With Multiple Subgrids. *IEEE Transactions on Power Electronics*, *33*(4), 3520–3533. doi:10.1109/TPEL.2017.2705133

Xue, Z., Yao, S., Ma, H., Zhang, C., Zhang, K., & Chen, Z. (2023). Thermo-economic optimization of an enhanced geothermal system (EGS) based on machine learning and differential evolution algorithms. *Fuel, 340*. doi:10.1016/j.fuel.2023.127569

Xu, L., Guerrero, J., Lashab, A., Wei, B., Bazmohammadi, N., Vasquez, J., & Abusorrah, A. (2022). A Review of DC Shipboard Microgrids—Part I: Power Architectures, Energy Storage, and Power Converters. *IEEE Transactions on Power Electronics*, *37*(5), 5155–5172. doi:10.1109/TPEL.2021.3128417

Xu, M., Melnik, R. V., & Borup, U. (2004). Modeling anti-islanding protection devices for photovoltaic systems. *Renewable Energy*, *29*(15), 2195–2216. doi:10.1016/j.renene.2004.04.005

Xu, Z., Yang, P., Zheng, C., Zhang, Y., Peng, J., & Zeng, Z. (2018). Analysis on the organization and Development of multi-microgrids. *Renewable & Sustainable Energy Reviews*, *81*, 2204–2216. doi:10.1016/j.rser.2017.06.032

Yadav, N., & Tummuru, N. R. (2023). Application of Filter Capacitor Dynamics Based Short-Circuit Fault Detection Method in Ring-Type DC Microgrid. *IEEE Transactions on Industrial Electronics*, *70*(5), 5286–5295. doi:10.1109/TIE.2022.3183363

Yadav, V. K., Chauhan, Y. K., Padhy, N. P., & Gupta, H. O. (2013). A novel power sector restructuring model based on Data Envelopment Analysis (DEA). *International Journal of Electrical Power & Energy Systems*, *44*(1), 629–637. doi:10.1016/j.ijepes.2012.08.013

Yamashita, D. Y., Vechiu, I., & Gaubert, J.-P. (2020). A review of hierarchical control for building microgrids. *Renewable & Sustainable Energy Reviews*, *118*, 109523. doi:10.1016/j.rser.2019.109523

Yang, H., & Pollitt, M. (2009). Incorporating both undesirable outputs and uncontrollable variables into DEA: The performance of Chinese coal-fired power plants. *European Journal of Operational Research*, *197*(3), 1095–1105. doi:10.1016/j.ejor.2007.12.052

Yang, H., & Pollitt, M. (2010). The necessity of distinguishing weak and strong disposability among undesirable outputs in DEA: Environmental performance of Chinese coal-fired power plants. *Energy Policy*, *38*(8), 4440–4444. doi:10.1016/j.enpol.2010.03.075

Yao, J., Hedengren, J. D., Gao, T., & Powell, K. M. (2023). A two-level optimization framework for battery energy storage systems to enhance economics and minimize long-term capacity fading. *Journal of Energy Storage, 63*. doi:10.1016/j.est.2023.106943

Yee, F. M., Shaharudin, M. R., Ma, G., Zailani, S. H. M., & Kanapathy, K. (2021). Green purchasing capabilities and practices towards Firm's triple bottom line in Malaysia. *Journal of Cleaner Production*, *307*, 127268. doi:10.1016/j.jclepro.2021.127268

Ye, X., Han, S., & Lin, A. (2010). A note on the connection between the Primal-Dual and the A* algorithm. *International Journal of Operations Research and Information Systems*, *1*(1), 73–85. doi:10.4018/JORIS.2010101305

Ye, X., Han, S., & Lin, A. (2012). A note on the connection between the Primal-Dual and the A* algorithm. In J. Wang (Ed.), *Innovations in Information Systems for Business Functionality and Operations Management* (pp. 170–181). IGI Global. doi:10.4018/978-1-4666-0933-4.ch010

Yildiz Çankaya, S., & Sezen, B. (2019). Effects of green supply chain management practices on sustainability performance. *Journal of Manufacturing Technology Management*, *30*(1), 98–121. doi:10.1108/JMTM-03-2018-0099

Yilmaz, C., & Sen, O. (2022). Thermoeconomic analysis and artificial neural network based genetic algorithm optimization of geothermal and solar energy assisted hydrogen and power generation. *International Journal of Hydrogen Energy, 47*(37), 16424-16439. doi:10.1016/j.ijhydene.2022.03.140

Yoo, S., & Harman, M. (2012, March). Regression testing minimization, selection and prioritization: A survey. *Softw. Test. Verif. Reliab., 22*(2), 67–120. Advance online publication. doi:10.1002/stv.430

Yu, Y., Niu, T., Wang, J., & Jiang, H. (2023). Intermittent solar power hybrid forecasting system based on pattern recognition and feature extraction. *Energy Conversion and Management, 277*. https://www.sciencedirect.com/science/article/pii/S0196890422013577 doi:10.1016/j.enconman.2022.116579

Yusuf, Y. Y., Gunasekaran, A., Musa, A., El-Berishy, N. M., Abubakar, T., & Ambursa, H. M. (2013). The UK oil and gas supply chains: An empirical analysis of adoption of sustainable measures and performance outcomes. *International Journal of Production Economics*, *146*(2), 501–514. doi:10.1016/j.ijpe.2012.09.021

Zachar, M., & Daoutidis, P. (2018). Energy management and load shaping for commercial microgrids coupled with flexible building environment control. *Journal of Energy Storage*, *16*, 61–75. doi:10.1016/j.est.2017.12.017

Zadeh, L. A. (1965). Fuzzy sets. *Information and Control*, *8*(3), 338–353. doi:10.1016/s0019-9958(65)90241-x

Zeeshan, A, M., Subash, A., & C S, A. (2022). Model employing the VAM-MODI method for risk management in the transportation of commodities. *International Journal of Research and Review*, *9*(11), 96–99. doi:10.52403/IJRR.20221116

Zhang, L., & Zoli, E. (2014). *Leaning Against the Wind: Macroprudential Policy in Asia* (IMF Working Papers). International Monetary Fund (IMF).

Zhang, Z., Wen, K., & Sun, W. (2023). Optimization and sustainability analysis of a hybrid diesel-solar-battery energy storage structure for zero energy buildings at various reliability conditions. *Sustainable Energy Technologies and Assessments, 55*. doi:10.1016/j.seta.2022.102913

Zhang, F., Meng, C., Yang, Y., Sun, C., Ji, C., Chen, Y., Wei, W., Qiu, H., & Yang, G. (2015). Advantages and challenges of DC microgrid for commercial building a case study from Xiamen university DC microgrid. *2015 IEEE First International Conference on DC Microgrids (ICDCM)*, 355–358. 10.1109/ICDCM.2015.7152068

Zhang, H.-Y., Ji, Q., & Fan, Y. (2013). An evaluation framework for oil import security based on the supply chain with a case study focused on China. *Energy Economics*, *38*, 87–95. doi:10.1016/j.eneco.2013.03.014

Zhang, Y., Zhang, X., & Wang, Z. (2019). Short-term power load forecasting using ARIMA and LSTM hybrid model. In *Proceedings of the 2019 IEEE International Conference on Smart Grid and Clean Energy Technologies (ICSGCE)* (pp. 122-126). IEEE.

Zhou, P., Ang, B. W., & Poh, K. L. (2008). A survey of data envelopment analysis in energy and environmental studies. *European Journal of Operational Research*, *189*(1), 1–18. doi:10.1016/j.ejor.2007.04.042

Zhu, K., Wang, Y., Lu, X., & Vieira, J. C. M. (2017). A new DG anti-islanding protection technique based on harmonic impedance measurement. *International Transactions on Electrical Energy Systems*, *27*(2), e2251. doi:10.1002/etep.2251

Zhu, Z., Wang, K., & Zhang, B. (2014). Applying a network data envelopment analysis model to quantify the eco-efficiency of products: A case study of pesticides. *Journal of Cleaner Production*, *69*, 67–73. doi:10.1016/j.jclepro.2014.01.064

Ziolkowska, J. R. (2013). Evaluating sustainability of biofuels feedstocks: A multi-objective framework for supporting decision making. *Biomass and Bioenergy, 59*, 425-440. doi:10.1016/j.biombioe.2013.09.008

Zografidou, E., Petridis, K., Arabatzis, G., & Dey, P. K. (2016). Optimal design of the renewable energy map of Greece using weighted goal-programming and data envelopment analysis. *Computers & Operations Research*, *66*, 313–326. doi:10.1016/j.cor.2015.03.012

Zolfaghari, M., Gharehpetian, G. B., Shafie-khah, M., & Catalão, J. P. S. (2022). Comprehensive review on the strategies for controlling the interconnection of AC and DC microgrids. *International Journal of Electrical Power & Energy Systems*, *136*, 107742. doi:10.1016/j.ijepes.2021.107742

About the Contributors

Joshua Thomas received his PhD. degree from University Sains Malaysia (USM), School of Computer Sciences, in 2015. He worked as research assistant at the Artificial Intelligence Lab in University Sains Malaysia. His research interests include scheduling algorithms, machine learning algorithms, data analytics, deep learning, and visual analytics, and Chemoinformatics. Dr. J. Joshua has authored several publications in leading international conferences and journals. Some of his research work were published in conferences including, IEEE, ICONIP, IVIC, IV, COMPSE, ICO. He has funded external, internal, short term research grants and industry collaborative projects. He has been invited as plenary speaker at IAIM2019, delivered, conduct Workshop's (IVIC19) at International conferences. He is an Associate Editor for the journal of Energy Optimization and Engineering (IJEOE), and invited as guest editor for JVLC-Elsevier, IJDSA-Springer, IJCC-IGI-Global, IJIRR-Inderscience.

Roman Rodriguez-Aguilar is a professor in the Panamerican University. His research is on large-scale mathematical optimization, evolutionary computation, statistical modeling, mathematical finance, competition and market regulation. He received his Ph.D. at School of Economics at National Polytechnic Institute, Mexico. A master's degree in Engineering from the School of Engineering at the National University of Mexico (UNAM), a master's degree in Administration and Public Policy in the School of Government and Public Policy at Monterrey Institute of Technology and Higher Education. A postgraduate in applied statistics at the Research Institute in Applied Mathematics and Systems of the UNAM and his degree in Economics at the UNAM. Prior to joining to Anahuac University, he has worked as a specialist in economics, statistics, simulation, finance and optimization, occupying different management positions in various public entities such as the Ministry of Energy, Ministry of Finance, and Ministry of Health. He has coauthored research articles in science citation index journals, conference proceedings, presentations and book chapters.

Pandian Vasant (H index: 34) holds PhD in Computational Intelligence (UNEM, Costa Rica), MSc (UMS, Malaysia, Engineering Mathematics) and BSc (Hons) in Mathematics (MU, Malaysia). His research interests include Soft Computing, Hybrid Optimization, Innovative Computing and Applications. He has co-authored research articles in journals, conference proceedings, presentation, special issues guest editor, book chapters (293 publications indexed in Research-Gate). In the year 2009 and 2015, Professor Pandian Vasant was awarded top reviewer and outstanding reviewer for the journal of Applied Soft Computing (Elsevier). He has 30 years of working experience at the universities. Currently he is an Editor-in-Chief of International Journal of Energy Optimization & Engineering, and Member of AMS

(USA), NAVY Research Group (TUO, Czech Republic) and Senior Research Associate at MERLIN Research Center (TDTU, Vietnam).

* * *

Karima Belmabrouk received her engineering degree in 1998 and her magister degree in 2006 and her Phd degree in 2017 from the University of Science and Technology of Oran, USTO-MB. Since December 2006, she is an associate professor and researcher in the computer science department of the USTO-MB. Her research interests are based on agent-based modeling and simulation, Web services composition, ontology and semantic Web as well as planning in artificial intelligence. Currently, she is working on cybersecurity and supply chain optimization.

Latifa Dekhici is an associate professor in computer sciences and a vice dean of external relations in mathematics and computer science faculty in the university of sciences and technology of Oran. She is also a researcher in LDREI, Smart Grid Develepement laboratory in the Higher School of Electrical Engineering and Energetic of Oran. Her research interests include power dispatch, power system design, healthcare scheduling and green optimization as well as optimization methods especially metaheuristics and event discrete simulation. DEKHICI has been a member of Inside Project (Inclusion of Students With impairements in Distance Education) in 2023 and a Responsable of Digitalisation Strategy Office in her university. She supervised Phd thesis about software engineering,dependability and risk managment and households utilities optimal consumption. She partipated in many sustainability; digitalization; accessibilty and inclusivity activities.

Jorge Gomes is a researcher at ADVANCE, ISEG, School of Economics & Management of the Universidade de Lisboa. He holds a PhD in Management from ISEG and a Masters in Management Sciences from ISCTE-IUL, He also have a post-graduation in Project Management from INDEG/ISCTE, and a degree in Geographic Engineering from the Faculty of Sciences of the Universidade de Lisboa. During the past 30 years, he has worked as an engineer, project manager, quality auditor and consultant. Teaches Management at ULHT, Lisboa. His research interests include Benefits Management, Project Management, Project Success, Maturity Models, IS/IT Investments, IS/IT in Healthcare and IS/IT Management.

Khaled Guerraiche received the engineering diploma in Electrical engineering from University of Sciences and the Technology, Oran, Algeria, in June 2004, and the M.E. and Ph.D. degrees in electrical engineering from University of Sciences and the Technology, Oran, Algeria (USTO), in 2008 and 2016, respectively. He is currently working as an Associate Professor in the Electrical Engineering Department, Higher School of Electrical Engineering and Energetic of Oran, Oran, Algeria. His research interests include operations, planning and economics of electric energy systems, systems dependability and fault tolerance as well as optimization theory and its applications in power electrical engineering.

P. Senthil Kumar is an Associate Professor (Mathematics) at the Amity School of Engineering and Technology of Amity University, Bengaluru, Karnataka, India. He received BSc, MSc, and MPhil degrees from Jamal Mohamed College (JMC), Tiruchirappalli, in 2006, 2008, and 2010, respectively. Furthermore, he received a B.Ed. degree from the Jamal Mohamed College of Teacher Education in 2009. Additionally, he completed PGDCA in 2011 at Bharathidasan University and PGDAOR in 2012 at Annamalai

University, Tamil Nadu, India. He completed his PhD at JMC in 2017. He has published many research papers in peer-reviewed journals. He has also presented research papers at various conferences. He has introduced the PSK (P. Senthil Kumar) method and the PSK theorem for solving transportation problems (TPs) in uncertain environments. And also, he has proved the supporting theorem, which states that the solution obtained by the PSK method for solving fuzzy TPs is always optimal. Its relevant articles have been published in Springer and IGI Global Journals, which are indexed in Scopus and WoS. His areas of research interest include operations research and fuzzy and intuitionistic fuzzy optimization.

Rinki Maurya received Diploma in Instrumentation and Control Engineering from IERT Allahabad in 2011, a B.Tech degree in Electronics & Communication Engineering from UPTU Lucknow in 2014, M.Tech in Control and Instrumentation Engineering from RTU Kota, Rajasthan, India in 2017 and She is currently pursuing a Ph.D. degree in the Department of Electronics and Communication Engineering, IIIT Allahabad, India. She had published many research works in a reputed journal and conferences. Her research areas are Control systems, PID, FOPID, Reinforcement learning, Fuzzy, etc.

K. S. Sastry Musti is presently with Namibia University of Science and Technology, Windhoek, Namibia. He has than 30 years of professional experience. He obtained his BTech in Electrical Engineering from JNTU, Kakinada (A.P.) in 1990, MTech and PhD from NIT, Warangal in the years 1996 and 2002 respectively. Some of his earlier employments include University of the West Indies, St Augustine, Trinidad, ESIGELEC, Rouen, France and Sophia University, Tokyo, Japan. His research interests include Power and Energy Systems, Information Systems and Engineering Education.

Surya Prakash received his master and doctorate degree from the Indian Institute of Technology Bombay, India. Presently he is associated as an Assistant Professor in the Department of Electronics & Communication Engineering at IIIT Allahabad, India. His research interests are Artificial Intelligence, Renewable Energy, etc.

Dharmbir Prasad was born in 1986 at Nalanda, Bihar, India. He received his B.Tech. degree in electrical engineering from Hooghly Engineering & Technology College (under West Bengal University of Technology), Hooghly, India and M.Tech. in Power System from Dr. B.C. Roy Engineering College (under West Bengal University of Technology), Durgapur, India, respectively. Currently, he is working in the capacity of assistant professor in the department of electrical engineering, Asansol Engineering College, Asansol, West Bengal, India. He has completed Ph.D. degree from Indian School of Mines, Dhanbad, Jharkhand, India. His research interest includes economic operation of power system.

Mário José Batista Romão is an Associate Professor of Information Systems at ISEG – University of Lisbon. He is Director of the Masters program in Computer Science and Management. He holds a PhD in Management Sciences by ISCTE-IUL and by Computer Integrated Manufacturing at Cranfield University (UK). He also holds a MsC in Telecommunications and Computer Science, at IST - Instituto Superior Técnico, University of Lisbon. He is Pos-Graduated in Project Management and holds the international certification Project Management Professional (PMP), by PMI – Project Management International. He has a degree in Electrotecnic Engineer by IST.

Priya Sharma received her B.Tech degree in ECE in 2014 and her M.Tech in 2016. Her areas of interest are MIcrowaves and Antennas, Wireless sensor Networks, and Wireless Communication.

Ashutosh Singh is currently associated with the Department of ECE, IIIT Allahabad as an Assistant Professor. His working areas are Electrical, Control, and Instrumentation engineering. In addition to this Dr. Singh is a project investigator of three projects, funded by external agencies/government. Dr. Singh has published three books and more than 75 research articles in reputed journals and conferences.

Sanjeet Singh joined IIM Lucknow in March 2020 as Professor in the Decision Sciences Area. Prior to joining IIML, he was working as Professor since 2016 in the Operations Management Group at IIM Calcutta. He has worked at IIM Calcutta for 13 years at different academic positions (Assistant, Associate, and Full Professor). Previously, he has also worked in DRDO as a Scientist from 2002 to 2006, and Lecturer at the University of Delhi from 2000 to 2002. He was also visiting faculty at the University of Waterloo, Canada, in 2012. He obtained his B.Sc. (Hons.) in Mathematics, M.Sc., and Ph.D. in Operational Research from the University of Delhi. He has been the Gold Medalist both in B.Sc. (Hons.) and M.Sc. His primary research interests include Data Envelopment Analysis, Mathematical Programming, Sports Modelling, Performance Evaluation of Education Sector, Distribution and Vehicle Routing, Bilevel Programming, Environmental Efficiency, and Fuzzy Optimization. His research has been continuously published in internationally reputed journals such as European Journal of Operational Research, Annals of Operations Research, Journal of the Operational Research Society, International Journal of Production Research, Applied Mathematics and Computation, Evaluation and; Program Planning, RAIRO-Operations Research, to name a few.

Saransh Tiwari is a PhD research scholar in the Decision Sciences Area at Indian Institute of Management (IIM) Lucknow, India. He is the recipient of University Gold Medal from Babasaheb Bhimrao Ambedkar University (A Central University), Lucknow for his outstanding performance during graduation. His research interests include Data envelopment analysis, Environmental performance evaluation, and Transportation policy evaluation. His research has been continuously published in internationally reputed journals as Managerial and Decision Economics and Applied Functional Materials. He is also the reviewer of the Journal of Transportation in Developing Economies.

Shrikant Tiwari (Senior Member, IEEE) was born in Karuwa Village, Maihar, Madhya Pradesh, India in 15th Aug. 1983. He received his Ph.D. in the Department of Computer Science & Engineering (CSE) from the Indian Institute of Technology (Banaras Hindu University), Varanasi (India) in 2012 and M. Tech. in Computer Science and Technology from the University of Mysore (India) in 2009. Currently, he is working as an Associate Professor in the School of Computing Science and Engineering (SCSE), Galgotias University, Greater Noida, Gautam Budha Nagar, Uttar Pradesh-203201 (India). He has authored and co-author more than 50 national and international journal publications, book chapters, and conference articles. He has five patents filed to his credit. His research interests include machine learning, deep learning, computer vision, medical image analysis, pattern recognition, and biometrics. Dr. Tiwari is a FIETE and member of ACM, IET, CSI, ISTE, IAENG, SCIEI. He is also a guest editorial board member and a reviewer for many international journals of repute.

Amit Kumar Tyagi is working as Assistant Professor, at National Institute of Fashion Technology, 110016, New Delhi, India. Previously he has worked as Assistant Professor (Senior Grade 2), and Senior Researcher at Vellore Institute of Technology (VIT), Chennai Campus, 600127, Chennai, Tamilandu, India for the period of 2019-2022. He received his Ph.D. Degree (Full-Time) in 2018 from Pondicherry Central University, 605014, Puducherry, India. About his academic experience, he joined the Lord Krishna College of Engineering, Ghaziabad (LKCE) for the periods of 2009-2010, and 2012-2013. He was an Assistant Professor and Head- Research, Lingaya's Vidyapeeth (formerly known as Lingaya's University), Faridabad, Haryana, India for the period of 2018-2019. His supervision experience includes more than 10 Masters' dissertations and one PhD thesis. He has contributed to several projects such as "AARIN" and "P3- Block" to address some of the open issues related to the privacy breaches in Vehicular Applications (such as Parking) and Medical Cyber Physical Systems (MCPS). He has published over 100 papers in refereed high impact journals, conferences and books.

Index

A

AC Microgrid 197, 225, 229-230, 250
Air Pollution 204, 210, 213, 225, 228, 232-233, 244-246, 248
Artificial Intelligence (AI) 199, 210, 217

B

Balanced IFTP 1
Biomass 74, 77, 105, 117, 255, 265-266
Building 74, 77, 82, 129-133, 135-138, 142-146, 201-202, 209, 219, 244, 250-251, 255-256

C

Carbon 61, 63, 71-72, 74, 77, 81, 93, 104, 111-112, 124, 127, 202-203, 219, 239, 241-243, 248, 257, 259, 279
Circular Economy 56, 59, 62, 69, 201, 204, 212, 215-216
Climate Change 56-57, 67, 72, 84, 108, 110, 112, 129, 145, 199-206, 208, 210, 218, 221-222, 245
Collaboration 66, 129-131, 135-136, 140, 143, 145, 202, 206-208, 216-218, 291

D

Data Analytics 62, 198-199, 209-214, 217, 256
Data Envelopment Analysis 68, 83-84, 109-116
DC Microgrid 196, 225, 229, 231-232, 246-251

E

Economic Performance 56, 60, 66, 146
Energy Efficiency 60, 71-72, 75, 78, 80-81, 113, 118, 126-127, 148, 150-151, 201, 207, 209, 215, 219, 225, 231, 233, 239, 241-243, 249, 253, 256-259, 280, 293
Energy Markets 147
Environmental efficiency 83-84, 105, 109-111, 114-115
Environmental Performance 56, 59-60, 66, 85, 95, 105-110, 113-114, 116, 131

F

Feature Models 267-269, 271, 276-277
Fossil 71, 73-74, 76, 78-80, 93, 105, 124, 202, 213, 226, 228, 232, 234-235, 238, 244, 253, 279
Functional Analysis 279-280, 291-292, 294, 297-298, 302
Fuzzy 1, 3-7, 9, 14-16, 23, 27, 30-39, 68, 108-110, 148-149, 247, 255

G

Generation Expansion Planning 117-118, 125-127
Grasshopper Optimization Algorithm 120
Green Building Certificates 129-131
Green Project Champion 129, 133, 141-143
Green Supply Chain Management 70
Greenhouse Gas 72, 74, 77, 115, 117, 127, 200, 202-203, 208, 213, 221-222, 235-236, 245, 253, 279
Grid 124, 127, 149, 168-174, 176-177, 179-180, 186, 188-189, 195-197, 209, 226-232, 234-239, 241-244, 246-250, 253, 255-257, 259, 261, 279-280, 282-283, 285-287, 291, 299-300, 303-304

H

Hydro 117, 126, 197, 219

I

IBFS 4, 31
IFN 1, 3, 6-7, 9, 30
IFNs 1, 3-5, 9, 12, 14, 27-30
IFTP 4-6, 14-16, 18-19, 27-29
Industry 2, 63, 65, 69, 71-72, 74, 77, 105, 109-115,

197, 202, 207, 211, 217, 221, 248-249
Innovation 36, 81, 103, 124, 126, 129-132, 134-137, 139-140, 142-146, 200, 207, 213, 217-220, 253
Internet of Things (IoT) 199, 209, 217
Intuitionistic Fuzzy 1, 3-7, 9, 14-16, 23, 27, 30-39
Islanding 170-177, 188-189, 196-198

L

LEED Certification 132-133, 136, 138
LINGO 1, 14, 20, 24, 28, 30, 44, 46-47, 50-53
Load Forecast 118-119, 147

M

Machine Learning 57-59, 63, 147, 149-150, 163-164, 168, 210, 217, 220, 222-223, 253, 265-266
Malmquist productivity index 101, 103
MATLAB 1, 6, 14, 20, 24, 28, 30, 42-44, 49-50, 173, 291
Microgrid 118, 170, 172-173, 175, 180, 196-198, 225-226, 228-251, 256, 258, 262, 264
Mutation 267-278

O

Optimization 2, 4-5, 8, 27-28, 30-31, 33, 35, 38-39, 43, 50, 62, 107-110, 118, 120, 125, 151, 166, 210, 212-213, 246, 254-259, 261-266, 302, 304

P

Performance evaluation 83, 85, 95, 105-109, 114, 223
Pollution 57, 60-61, 63, 67, 83-84, 110, 130, 199-201, 203-206, 208, 210-211, 213, 218, 225, 228, 232-233, 244-246, 248, 262-263
Power 61, 71, 73-74, 77-78, 80-81, 93, 96, 104, 106-120, 123-128, 147-151, 154, 156, 161-162, 167-180, 183, 185-189, 195-198, 209, 212, 219, 225-243, 246-266, 275, 280-284, 286, 288, 290-291, 294, 299-304
Power Quality 172, 225, 231-232, 246, 249, 301
Power System 78, 115, 117-119, 124-126, 147-149, 170, 173-174, 177, 229, 257, 299-300, 302, 304
Protection 59, 62, 79, 130, 170-174, 176-177, 180, 189, 196-198, 201, 204, 207, 210-211, 217, 220, 232, 247-250, 252-253, 263
PSK Method 1, 4, 15, 18, 34-36
PSK Theorem 15

R

Renewable 59, 62, 71-82, 105, 110-111, 115-118, 124-125, 127-128, 148, 168, 170, 172-175, 196-198, 200, 203-204, 206-207, 213, 216, 219, 225, 228, 230-233, 235-236, 238-239, 241-242, 244-250, 252-259, 263-265, 279, 286, 288, 290-291, 293, 295, 301-304
Renewable Energies 62, 279, 293, 301-302
Renewable Energy 59, 71-75, 77-82, 105, 111, 116-118, 124-125, 127-128, 168, 170, 172-173, 175, 197-198, 200, 203-204, 206-207, 213, 216, 219, 225, 228, 230-233, 235-236, 239, 241-249, 252-259, 263-265, 286, 288, 291, 303-304
Renewable Energy Resources 73, 117-118, 170, 243-244, 246
Resource Depletion 67, 199, 201, 203-204, 215, 218
RGui 1, 6, 14, 20, 24, 28, 30, 41-42, 48-49
RStudio 1, 6, 14, 20, 24, 28, 30, 40-41, 47-48

S

SADT 279-280, 294-296, 302
SA-RT 279-280, 296-299, 301-302
Similarity 267, 269, 274-277
Smart Environment 61, 199, 201-202, 209-210, 212, 214-216, 218, 220
Smart Grids 213, 228, 279, 304
Social Performance 58, 60, 66, 217
Software 1, 5, 14-15, 24, 26, 28-30, 35, 111, 170, 196, 209, 256, 267-270, 275-278, 291, 302
Software Product Lines 267, 276-278
Solar 71-75, 78, 80-81, 105, 117-118, 124-125, 127-128, 148, 170-173, 175-176, 178, 196, 219, 226, 238, 244, 250, 252-258, 264-266, 284-285, 291-292, 303-304
Solar Power 125, 127, 170, 173, 175-176, 252-254, 258, 264-266
Supply Chain Management 36, 56, 59, 61, 70, 113, 216, 255, 265
Sustainability 56-64, 66-71, 83-84, 108, 111, 113-114, 116, 131-134, 136, 141, 144-146, 151, 199-202, 206-207, 209-212, 215-220, 235-236, 247, 252-

253, 255, 263, 266
Sustainability-Oriented Innovation 134, 143, 145
Sustainable Development 60, 66, 68, 84, 108, 110, 129-131, 133-134, 136, 140-142, 144, 146, 200, 206-208, 211
Sustainable Environment 64, 199-201, 220
System Disturbance 170
System Modeling 279, 291, 303

T

Testing 162-163, 267-268, 270-271, 275-278
TIFN 1, 3, 6-9, 11-12, 20-21, 23, 27, 29-30
Time-Series Analysis 147
TP 1-4, 6, 12, 14, 18, 20-23, 26-27, 29-30
Transportation Problem 1, 3-5, 22, 31-39

U

UBIFTPs 3, 5, 13, 28-30
Uncertainty 1, 3-5, 29, 31, 37, 110, 118-119, 124-126, 135, 154, 264
Undesirable outputs 85, 93-98, 100, 105-109, 112-114, 116

W

Weak disposability 83, 94-95, 97, 100-101, 106-108, 111
Wind 73, 75, 78, 81-82, 105, 110, 113, 118, 124-126, 128, 148-149, 151, 172, 196, 219, 228, 230, 238, 244, 252, 254, 256-258, 264-265, 280-284, 286, 289-290, 292, 303-304

www.ingramcontent.com/pod-product-compliance
Lightning Source LLC
LaVergne TN
LVHW080635011026
849160LV00014B/1224
* 9 7 8 1 6 6 8 4 9 1 3 1 7 *